Brain and Music

Brain and Music

First Edition

Stefan Koelsch

WILEY-BLACKWELL

A John Wiley & Sons, Ltd., Publication

This edition first published 2013
© 2013 by John Wiley & Sons, Ltd

Wiley-Blackwell is an imprint of John Wiley & Sons, formed by the merger of Wiley's global Scientific, Technical and Medical business with Blackwell Publishing.

Registered Office
John Wiley & Sons Ltd, The Atrium, Southern Gate, Chichester, West Sussex, PO19 8SQ, UK

Editorial Offices
9600 Garsington Road, Oxford, OX4 2DQ, UK
The Atrium, Southern Gate, Chichester, West Sussex, PO19 8SQ, UK
111 River Street, Hoboken, NJ 07030-5774, USA

For details of our global editorial offices, for customer services and for information about how to apply for permission to reuse the copyright material in this book please see our website at www.wiley.com/wiley-blackwell.

Library of Congress Cataloguing-in-Publication Data applied for.

ISBN hardback: 9780470683408
ISBN paperback: 9780470683392
ISBN ePDF: 9781118374023
ISBN ePub: 9781119943112
ISBN mobi: 9781119943129

A catalogue record for this book is available from the British Library.

Set in 10.5/13pt Galliard by Aptara Inc., New Delhi, India

First   2013

Contents

Preface   ix

Part I   Introductory Chapters   1

1   Ear and Hearing   3
    1.1   The ear   3
    1.2   Auditory brainstem and thalamus   6
    1.3   Place and time information   8
    1.4   Beats, roughness, consonance and dissonance   9
    1.5   Acoustical equivalency of timbre and phoneme   11
    1.6   Auditory cortex   12

2   Music-theoretical Background   17
    2.1   How major keys are related   17
    2.2   The basic in-key functions in major   20
    2.3   Chord inversions and Neapolitan sixth chords   21
    2.4   Secondary dominants and double dominants   21

3   Perception of Pitch and Harmony   23
    3.1   Context-dependent representation of pitch   23
    3.2   The representation of key-relatedness   26
    3.3   The developing and changing sense of key   29
    3.4   The representation of chord-functions   30
    3.5   Hierarchy of harmonic stability   31
    3.6   Musical expectancies   35
    3.7   Chord sequence paradigms   36

4 From Electric Brain Activity to ERPs and ERFs   40
   4.1 Electro-encephalography (EEG)   43
        4.1.1 The 10–20 system   43
        4.1.2 Referencing   45
   4.2 Obtaining event-related brain potentials (ERPs)   45
   4.3 Magnetoencephalography (MEG)   48
        4.3.1 Forward solution and inverse problem   49
        4.3.2 Comparison between MEG and EEG   49

5 ERP Components   51
   5.1 Auditory P1, N1, P2   51
   5.2 Frequency-following response (FFR)   53
   5.3 Mismatch negativity   54
        5.3.1 MMN in neonates   57
        5.3.2 MMN and music   57
   5.4 N2b and P300   59
   5.5 ERP-correlates of language processing   59
        5.5.1 Semantic processes: N400   60
        5.5.2 Syntactic processes: (E)LAN and P600   63
        5.5.3 Prosodic processes: Closure Positive Shift   67

6 A Brief Historical Account of ERP Studies of Music Processing   70
   6.1 The beginnings: Studies with melodic stimuli   70
   6.2 Studies with chords   74
   6.3 MMN studies   75
   6.4 Processing of musical meaning   76
   6.5 Processing of musical phrase boundaries   77
   6.6 Music and action   77

7 Functional Neuroimaging Methods: fMRI and PET   79
   7.1 Analysis of fMRI data   81
   7.2 Sparse temporal sampling in fMRI   84
   7.3 Interleaved silent steady state fMRI   85
   7.4 'Activation' vs. 'activity change'   85

Part II Towards a New Theory of Music Psychology   87

8 Music Perception: A Generative Framework   89

9 Musical Syntax   98
   9.1 What is musical syntax?   98
   9.2 Cognitive processes   102

9.3		The early right anterior negativity (ERAN)	109
	9.3.1	The problem of confounding acoustics and possible solutions	113
	9.3.2	Effects of task-relevance	120
	9.3.3	Polyphonic stimuli	121
	9.3.4	Latency of the ERAN	127
	9.3.5	Melodies	127
	9.3.6	Lateralization of the ERAN	129
9.4		Neuroanatomical correlates	131
9.5		Processing of acoustic vs. music-syntactic irregularities	133
9.6		Interactions between music- and language-syntactic processing	138
	9.6.1	The Syntactic Equivalence Hypothesis	145
9.7		Attention and automaticity	147
9.8		Effects of musical training	149
9.9		Development	151
10		**Musical Semantics**	**156**
10.1		What is musical semantics?	156
10.2		Extra-musical meaning	158
	10.2.1	Iconic musical meaning	158
	10.2.2	Indexical musical meaning	159
		Excursion: Decoding of intentions during musinc listening	161
	10.2.3	Symbolic musical meaning	162
10.3		Extra-musical meaning and the N400	163
10.4		Intra-musical meaning	170
		Excursion: Posterior temporal cortex and processing of meaning	166
	10.4.1	Intra-musical meaning and the N5	171
10.5		Musicogenic meaning	177
	10.5.1	Physical	177
	10.5.2	Emotional	179
	10.5.3	Personal	180
10.6		Musical semantics	181
	10.6.1	Neural correlates	181
	10.6.2	Propositional semantics	182
	10.6.3	Communication vs. expression	182
	10.6.4	Meaning emerging from large-scale relations	183
	10.6.5	Further theoretical accounts	184
11		**Music and Action**	**186**
11.1		Perception–action mediation	186
11.2		ERP correlates of music production	189

12	Emotion	203
12.1	What are 'musical emotions'?	204
12.2	Emotional responses to music – underlying mechanisms	207
12.3	From social contact to spirituality – The Seven Cs	208
12.4	Emotional responses to music – underlying principles	212
12.5	Musical expectancies and emotional responses	216
12.5.1	The tension-arch	218
12.6	Limbic and paralimbic correlates of music-evoked emotions	219
12.6.1	Major–minor and happy–sad music	225
12.6.2	Music-evoked dopaminergic neural activity	226
12.6.3	Music and the hippocampus	227
12.6.4	Parahippocampal gyrus	231
12.6.5	A network comprising hippocampus, parahippocampal gyrus, and temporal poles	232
12.6.6	Effects of music on insular and anterior cingulate cortex activity	232
12.7	Electrophysiological effects of music-evoked emotions	233
12.8	Time course of emotion	234
12.9	Salutary effects of music making	235
13	Concluding Remarks and Summary	241
13.1	Music and language	241
13.2	The music-language continuum	244
13.3	Summary of the theory	249
13.4	Summary of open questions	258
References		267
Index		303

Preface

Music is part of human nature. Every human culture that we know about has music, suggesting that, throughout human history, people have played and enjoyed music. The oldest musical instruments discovered so far are around 30 000 to 40 000 years old (flutes made of vulture bones, found in the cave *Hohle Fels* in Geissenklösterle near Ulm in Southern Germany).[1] However, it is highly likely that already the first individuals belonging to the species homo sapiens (about 100 000 to 200 000 years ago) made musical instruments such as drums and flutes, and that they made music cooperatively together in groups. It is believed by some that music-making promoted and supported social functions such as communication, cooperation and social cohesion,[2] and that the human musical abilities played a key phylogenetic role in the evolution of language.[3] However, the adaptive function of music for human evolution remains controversial (and speculative). Nevertheless, with regard to human ontogenesis, we now know that newborns (who do not yet understand the syntax and semantics of words and sentences) are equipped with musical abilities, such as the ability to detect changes of musical rhythms, pitch intervals, and tonal keys.[4] By virtue of these abilities, newborn infants are able to decode acoustic features of voices and prosodic features of languages.[5] Thus, infants' first steps into language are based on prosodic information (that is, on the musical aspects of speech). Moreover, musical communication in

---

[1] Conard *et al.* (2009)
[2] Cross & Morley (2008), Koelsch *et al.* (2010a)
[3] Wallin *et al.* (2000)
[4] Winkler *et al.* (2009b) Stefanics *et al.* (2007), Perani *et al.* (2010)
[5] Moon *et al.* (1993)

early childhood (such as parental singing) appears to play an important role in the emotional, cognitive, and social development of children.[6]

Listening to music, and music making, engages a large array of psychological processes, including perception and multimodal integration, attention, learning and memory, syntactic processing and processing of meaning information, action, emotion, and social cognition. This richness makes music an ideal tool to investigate human psychology and the workings of the human brain: Music psychology inherently covers, and connects, the different disciplines of psychology (such as perception, attention, memory, language, action, emotion, etc.), and is special in that it can combine these different disciplines in coherent, integrative frameworks of both theory and research. This makes music psychology *the* fundamental discipline of psychology.

The neuroscience of music is music psychology's tool to understanding the human brain. During the last few years, neuroscientists have increasingly used this tool, which has led to significant contributions to social, cognitive, and affective neuroscience. The aim of this book is to inform readers about the current state of knowledge in several fields of the neuroscience of music, and to synthesize this knowledge, along with the concepts and principles developed in this book, into a new theory of music psychology.

The first part of this book consists of seven introductory chapters. Their main contents are identical to those of a 'first edition' of this book (the publication of my PhD thesis), but I have updated the chapters with regard to scientific developments in the different areas. These chapters introduce the ear and hearing, a few music-theoretical concepts, perception of pitch and harmony, neurophysiological mechanisms underlying the generation of electric brain potentials, components of the event-related brain potential (ERP), the history of electrophysiological studies investigating music processing, and functional neuroimaging techniques. The purpose of these introductory chapters is to provide individuals from different disciplines with essential knowledge about the neuroscientific, music-theoretical, and music-psychological concepts required to understand the second part of the book (so that individuals without background knowledge in either of these areas can nevertheless understand the second part). I confined the scope of these chapters to those contents that are relevant to understanding the second part, rather than providing exhaustive accounts of each area. Scholars already familiar with those areas can easily begin right away with the second part.

The second part begins with a chapter on a model of music perception (Chapter 8). This model serves as a theoretical basis for processes and concepts developed in the subsequent chapters, and thus as a basis for the construction of the theory of music psychology introduced in this book. The chapter is followed by a chapter on music-syntactic processing (Chapter 9). In that chapter, I first tease apart different cognitive operations underlying music-syntactic processing.

---

[6]   Trehub (2003)

In particular, I advocate differentiating between: (a) processes that do not require (long-term) knowledge, (b) processes that are based on long-term knowledge and involve processing of local, but not long-distance, dependencies, and (c) processing of hierarchically organized structures (including long-distance dependencies). Then, I provide a detailed account on studies investigating music-syntactic processing using the *early right anterior negativity* (ERAN). One conclusion of these studies is the *Syntactic Equivalence Hypothesis* which states that there exist cognitive operations (and neural populations mediating these operations) that are required for music-syntactic, language-syntactic, action-syntactic, as well as mathematical-syntactic processing, and that are neither involved in the processing of acoustic deviance, nor in the processing of semantic information.

Chapter 10 deals with music-semantic processing. Here I attempt to tease apart the different ways in which music can either communicate meaning, or evoke processes that have meaning for the listener. In particular, I differentiate between *extra-musical* meaning (emerging from iconic, indexical, and symbolic sign quality), *intra-musical* meaning (emerging from structural relations between musical elements), and *musicogenic* meaning (emerging from music-related physical activity, emotional responses, and personality-related responses). One conclusion is that processing of extra-musical meaning is reflected electrically in the N400 component of the ERP, and processing of intra-musical meaning in the N5 component. With regard to musicogenic meaning, a further conclusion is that music can evoke sensations which, *before* they are 'reconfigured' into words, bear greater inter-individual correspondence than the words that an individual uses to describe these sensations. In this sense, music has the advantage of defining a sensation without this definition being biased by the use of words. I refer to this musicogenic meaning quality as *a priori musical meaning*.

Chapter 11 deals with neural correlates of music and action. The first part of that chapter reviews studies investigating premotor processes evoked by listening to music. The second part reviews studies investigating action with ERPs. These studies investigated piano playing in expert pianists, with a particular focus on (a) ERP correlates of errors that the pianists made during playing, and (b) processing of false feedback (while playing a correct note). Particularly with regard to its second part, this chapter is relatively short, due to the fact that only few neuroscientific studies are yet available in this area. However, I regard the topic of music and action as so important for the neuroscience of music, that I felt that something was missing without this chapter.

Chapter 12 is a chapter on music-evoked emotions and their neural correlates. It first provides theoretical considerations about principles underlying the evocation of emotion with music. These principles are not confined to music, but can be extrapolated to emotion psychology in general. I also elaborate on several social functions that are engaged when making music in a group. One proposition is that music is special in that it can activate all of these social functions at the same time. Engaging in these functions fulfils human needs, and can, therefore, evoke strong emotions. Then, a detailed overview of functional neuroimaging studies investigating emotion with music is provided. These studies show that music-evoked emotions can modulate activity in virtually all so-called limbic/paralimbic brain structures. This indicates, in my view, that music-evoked emotions touch the core of evolutionarily adaptive neuroaffective mechanisms, and reflects that music satisfies basic human needs. I also argue that experiences of fun and reward have different neural correlates than experiences of joy, happiness, and love. With regard to the latter emotions, I endorse the hypothesis that they are generated in the hippocampus (and that, on a more general level, the hippocampus generates tender positive emotions related to social attachments). In the final section of that chapter, I present a framework on salutary effects of music making. Due to the scarcity of studies, that framework is thought of as a basis for further research in this area.

In the final chapter, I first provide a concluding account on 'music' and 'language'. I argue that there is no design feature that distinctly separates music and language, and that even those design features that are more prominent in either language or music also have a transitional zone into the respective other domain. Therefore, the use of the words 'music' and 'language' seems adequate for our everyday language, but for scientific use I suggest the term *music-language-continuum*.

Then, the different processes and concepts developed in the preceding chapters are summarized, and synthesized into a theory of music perception. Thus, readers with very limited time can skip to page 201 and read only Section 13.3, for these few pages contain the essence of the book. In the final section, the research questions raised in the previous chapters are summarized. That summary is meant as a catalogue of research questions that I find most important with regard to the topics dealt with in the second part of this book. This catalogue is also meant to provide interested students and scientists who are new to the field with possible starting points for research.

The theory developed in this book is based on the model of music perception described in Chapter 8; that model describes seven stages, or dimensions, of music perception. The principles underlying these dimensions are regarded here as so fundamental for music psychology (and psychology in general), that processes and concepts of other domains (such as music perception, syntactic processing, musical meaning, action, emotion, etc.) were developed and conceptualized in such a way that they correspond to the dimensions of music perception.

This led to a theory that integrates different domains (such as music, language, action, emotion, etc.) in a common framework, implying numerous shared processes and similarities, rather than treating 'language', 'music', 'action', and 'emotion' as isolated domains.[7] That is, different to what is nowadays common in psychology and neuroscience, namely doing research in a particular domain without much regard to other domains, the music-psychological approach taken in

---

[7] See also Siebel *et al.* (1990).

this book aims at bringing different domains together, and integrating them both theoretically and empirically into a coherent theory. In this regard, notably, this book is about understanding human psychology and the human brain (it is *not* about understanding music, although knowledge about how music is processed in the brain can open new perspectives for the experience of music). In my view, we do not need neuroscience to explain, or understand music (every child can understand music, and Bach obviously managed to write his music without any brain scanner). However, I do believe that we need music to understand the brain, and that our understanding of the human brain will remain incomplete unless we have a thorough knowledge about how the brain processes music.

Many of my friends and colleagues contributed to this book through valuable discussions, helpful comments, and numerous corrections (in alphabetical order): Matthias Bertsch, Rebecca Chambers-Lepping, Ian Cross, Philipp Engel, Thomas Fritz, Thomas Gunter, Thomas Hillecke, Sebastian Jentschke, Carol Lynne Krumhansl, Moritz Lehne, Eun-Jeong Lee, Giacomo Novembre, Burkhard Maess, Clemens Maidhof, Karsten Müller, Jaak Panksepp, Uli Reich, Tony Robertson, Martin Rohrmeier, María Herrojo Ruiz, Daniela Sammler, Klaus Scherer, Walter Alfred Siebel, Stavros Skouras, and Kurt Steinmetzger. Aleksandra Gulka contributed by obtaining the reprint permissions of figures. It is with great joy that I see this book now finalized and in its entirety, and I hope that many readers will enjoy reading this book.

Leipzig, 1st October, 2011

Part I

Introductory Chapters

1
Ear and Hearing

1.1   The ear

The human ear has striking abilities of detecting and differentiating sounds. It is
sensitive to a wide range of frequencies as well as intensities and has an extremely
high temporal resolution (for detailed descriptions see, e.g., Geisler, 1998; Moore,
2008; Pickles, 2008; Plack, 2005; Cook, 2001). The ear consists of three parts:
The outer, the middle, and the inner ear. The outer ear acts as a receiver and
filters sound waves on their way to the ear drum (tympanic membrane) via the
ear canal (meatus), amplifying some sounds and attenuating others (depending
on the frequency and direction of these sounds). Sound waves (i.e., alternating
compression and rarefaction of air) cause the tympanic membrane to vibrate, and
these vibrations are subsequently amplified by the middle ear. The middle ear
is composed of three linked bones: The malleus, incus, and stapes. These tiny
bones help transmit the vibrations on to the oval window of the cochlea, a small
membrane-covered opening in the bony wall of the inner ear and the interface
between the air-filled middle-ear and the fluid-filled inner ear (Figure 1.1).

   The cochlea has three fluid-filled compartments, the scala tympani, the scala
media, and the scala vestibuli (which is continuous with the scala tympani at the
helicotrema). Scala media and scala tympani are separated by the basilar membrane
(BM). The organ of Corti rests on the BM and contains the auditory sensory
receptors that are responsible for transducing the sound stimulus into electrical
signals. The vibration of the stapes results in varying pressures on the fluid in
the scala vestibuli, causing oscillating movements of scala vestibuli, scala media

Brain and Music, First Edition. Stefan Koelsch.
© 2013 John Wiley & Sons, Ltd. Published 2013 by John Wiley & Sons, Ltd.

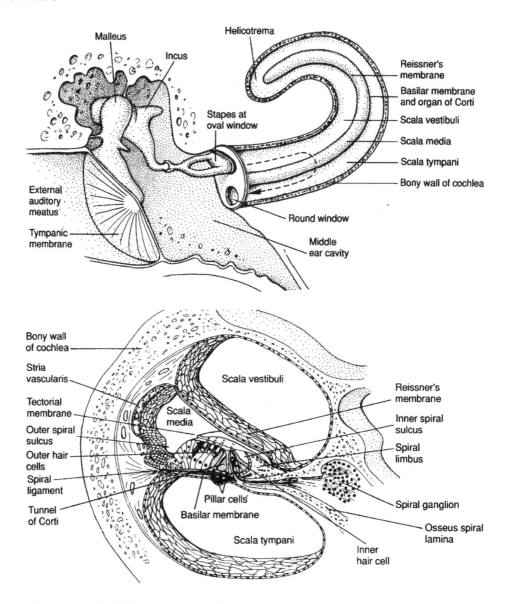

Figure 1.1   Top: The major parts of the human ear. In the Figure, the cochlea has been uncoiled for illustration purposes. Bottom: Anatomy of the cochlea (both figures from Kandel *et al.*, 2000).

(including BM) and scala tympani (for detailed descriptions see, e.g., Geisler, 1998; Pickles, 2008).

The organ of Corti contains the sensory receptor cells of the inner ear, the hair cells (bottom of Figure 1.1). There are two types of hair cells, inner hair cells and outer hair cells. On the apical surface of each hair cell is a bundle of around 100 stereocilia (mechanosensing organelles which respond to fluid motion or fluid pressure changes). Above the hair cells is the tectorial membrane that attaches to the longest stereocilia of the outer hair cells. The sound-induced movement of the scalae fluid (see above) causes a relative shearing between the tectorial membrane and BM, resulting in a deflection of the stereocilia of both inner and outer hair cells. The deflection of the stereocilia is the adequate stimulus of a hair cell, which then depolarizes (or hyperpolarizes, due to the direction of deflection) by opening an inward current (for detailed information see Steel & Kros, 2001).

The inner hair cells then release glutamate (Nouvian *et al.*, 2006)[1] at their basal ends where the hair cells are connected to the peripheral branches of axons of neurons whose bodies lie in the spiral ganglion. The central axons of these neurons constitute the auditory nerve. The release of glutamate by the hair cells excites the sensory neurons and this in turn initiates action potentials in the cell's central axon in the auditory nerve. Oscillatory changes in the potential of a hair cell thus result in oscillatory release of transmitter and oscillatory firing in the auditory nerve (for details see, e.g., Pickles, 2008; Geisler, 1998). The duration of an acoustic stimulus is encoded by the duration of activation of an auditory nerve fibre.

Different frequencies of sounds are selectively responded to in different regions of the cochlea. Each sound initiates a travelling wave along the length of the cochlea. The mechanical properties of the basilar membrane vary along the length of the cochlea; the BM is stiff and thin at the basal end (and vibrates more to high frequency sounds, similar to the high e-string on a guitar, which resonates at a sound frequency of ∼330 Hz), whereas at the apex the BM is thicker and less stiff (and resonates at sounds with lower frequencies, similar to the low e-string on a guitar, which resonates at a sound frequency of ∼82 Hz). Different frequencies of sound produce different travelling waves with peak amplitudes at different points along the BM. Higher frequencies result in peak amplitudes of oscillations of the BM that are located nearer to the base of the cochlea, lower frequencies result in oscillatory peaks near the apex of the cochlea (for more details see, e.g., Pickles, 2008; Geisler, 1998).

The outer hair cells specifically sharpen the peak of the travelling wave at the frequency-characteristic place on the BM (e.g., Fettiplace & Hackney, 2006). Interestingly, outer hair cells achieve the changes in tuning of the local region in the organ of Corti by increasing or decreasing the length of their cell bodies (thereby affecting the mechanical properties of the organ of Corti; Fettiplace & Hackney, 2006). This change in length is an example of the active processes occurring within the organ of Corti while processing sensory information. Moreover, the outer hair cells are innervated by efferent nerve fibres from the central nervous system, and it appears that the changes in length are at least partly influenced by top-down processes (such processes may even originate from neocortical areas of the brain).

---

[1]   The postsynaptic receptors at the afferent synapse to the inner hair cells have been identified as AMPA receptors, and glutamate transporters have been found in nearby supporting cells that dispose of excess glutamate.

Therefore, the dynamics of the cochlea (determining the processing of acoustic information) appears to be strongly influenced by the brain. The dynamic activity of the outer hair cells is a necessary condition for a high frequency-selectivity (which, in turn, is a prerequisite for both music and speech perception).

Corresponding to the tuning of the BM, the frequency-characteristic excitation of inner hair cells gives rise to action potentials in different auditory nerve fibres. Therefore, an auditory nerve fibre is most sensitive to a particular frequency of sound, its so-called *characteristic frequency*. Nevertheless, an individual auditory nerve fibre (which is innervated by several inner hair cells) still responds to a range of frequencies, because a substantial portion of the BM moves in response to a single frequency. The sound pressure level (SPL, for explanation and medical relevance see, e.g., Moore, 2008) is then encoded (1) by the action potential rate of afferent nerve fibres, and (2) by the number of neighbouring afferent nerve fibres that release action potentials (because the number of neurons that release action potentials increases as the intensity of an auditory stimulus increases). The brain decodes the spatio-temporal pattern consisting of the individual firing rates of all activated auditory nerve fibres (each with its characteristic frequency) into information about intensity and frequency of a stimulus (decoding of frequency information is dealt with in more detail further below).

1.2   Auditory brainstem and thalamus

The cochlear nerve enters the central nervous system in the brain stem (cranial nerve VIII).[2] Within the brain stem, information originating from the hair cells is propagated via both contra- and ipsilateral connections between the nuclei of the central auditory pathway (for a detailed description see Nieuwenhuys *et al.*, 2008). For example, some of the secondary auditory fibres that originate from the ventral cochlear nucleus project to the ipsilateral superior olivary nucleus and to the medial superior olivary nucleus of both sides (both superior olivary nuclei project to the inferior colliculus). Other secondary auditory fibres project to the contralateral nucleus of the trapezoid body (that sends fibres to the ipsilateral superior olivary nucleus; see Figure 1.2). The pattern of contra- and ipsilateral connections is important for the interpretation of interaural differences in phase and intensity for the localization of sounds.

The inferior colliculus (IC) is connected with the medial geniculate body of the thalamus. The cells in the medial geniculate body send most of their axons via the radiatio acustica to the ipsilateral primary auditory cortex (for a detailed description see Nieuwenhuys *et al.*, 2008). However, neurons in the medial division of the medial geniculate body (mMGB) also directly project to the lateral amygdala (LeDoux, 2000); specifically those mMGB neurons receive ascending inputs from

---

[2]   This can easily be remembered, because both 'ear' and 'eight' start with an 'e'.

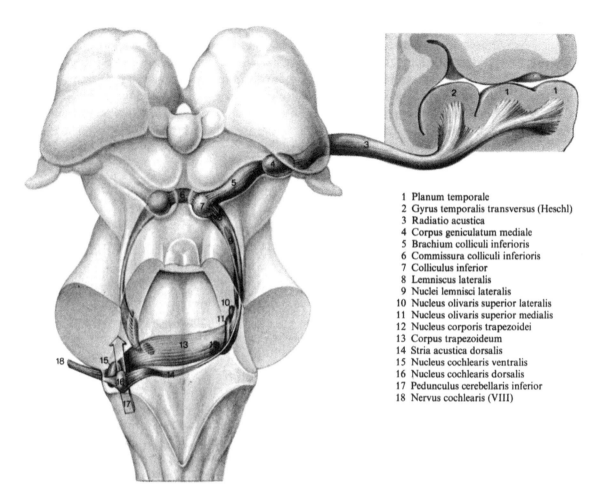

Figure 1.2   Dorsal view of nerve, nuclei, and tracts of the auditory system (from Nieuwenhuys *et al.*, 2008).

1 Planum temporale
2 Gyrus temporalis transversus (Heschl)
3 Radiatio acustica
4 Corpus geniculatum mediale
5 Brachium colliculi inferioris
6 Commissura colliculi inferioris
7 Colliculus inferior
8 Lemniscus lateralis
9 Nuclei lemnisci lateralis
10 Nucleus olivaris superior lateralis
11 Nucleus olivaris superior medialis
12 Nucleus corporis trapezoidei
13 Corpus trapezoideum
14 Stria acustica dorsalis
15 Nucleus cochlearis ventralis
16 Nucleus cochlearis dorsalis
17 Pedunculus cerebellaris inferior
18 Nervus cochlearis (VIII)

the inferior colliculus and are likely to be, at least in part, acoustic relay neurons (LeDoux *et al.*, 1990). The MGB, and presumably the IC as well, are involved in conditioned fear responses to acoustic stimuli. Moreover, already the IC plays a role in the expression of acoustic-motor as well as acoustic-limbic integration (Garcia-Cairasco, 2002), and chemical stimulation of the IC can evoke defence behaviour (Brandão *et al.*, 1988). It is for these reasons that the IC and the MGB are not simply acoustic relay stations, but that these structures are involved in the detection of auditory signals of danger.

What is often neglected in descriptions of the auditory pathway is the important fact that auditory brainstem neurons also project to neurons of the reticular formation. For example, intracellular recording and tracing experiments have shown that giant reticulospinal neurons in the caudal pontine reticular formation (PnC) can be driven at short latencies by acoustic stimuli, most presumably due to multiple and direct input from the ventral (and dorsal) cochlear nucleus (perhaps even from interstitial neurons of the VIII nerve root) and nuclei in the superior olivary complex (e.g., lateral superior olive, ventral periolivary areas; Koch *et al.*, 1992). These reticular neurons are involved in the generation of motor reflexes (by virtue of projections to spinal motoneurons), and it is conceivable that the projections from the auditory brainstem to neurons of the reticular formation contribute to the vitalizing effects of music, as well as to the (human) drive to move to music (perhaps in interaction with brainstem neurons sensitive for isochronous stimulation).[3]

1.3   Place and time information

The tonotopic excitation of the basilar membrane (BM),[4] is maintained as tonotopic structure (also referred to as *tonotopy*) in the auditory nerve, auditory brainstem, thalamus, and the auditory cortex. This tonotopy is an important source of information about the frequencies of tones. However, another important source is the temporal patterning of the action potentials generated by auditory nerve neurons. Up to frequencies of about 4–5 kHz, auditory nerve neurons produce action potentials that occur approximately in phase with the corresponding oscillation of the BM (although the auditory nerve neurons do not necessarily produce an action potential on every cycle of the corresponding BM oscillation). Therefore, up to about 4–5 kHz, the time intervals between action potentials of auditory neurons are approximately integer ratios of the period of a BM oscillation, and the timing of nerve activity codes the frequency of BM oscillation (and thus of the frequency of a tone, or partial of a tone, which elicits this BM oscillation). The brain uses both place information (i.e., information about which part/s, of the BM was/were oscillating) and time information (i.e., information about the frequency/ies of the BM oscillation/s). Note, however, (a) that time information is hardly available at frequencies above about 5 kHz, (b) that place information appears to be not accurate enough to decode differences in frequencies in the range of a few percent (e.g., between a tone of 5000 and 5050 Hz), and (c) that place information alone cannot explain the phenomenon of the pitch perception of tones with *missing fundamentals*[5] (for details about the *place theory* and *temporal theory* see, e.g., Moore, 2008).

---

[3]   A recent study by Zentner & Eerola (2010) suggests that this drive is already present in infants.

[4]   Recall that higher frequencies result in peak amplitudes closer to the base of the cochlea, and lower frequencies in peaks near the apex of the cochlea.

[5]   What is the perceived pitch of a tone consisting, for example, of the frequencies 200 Hz, 300 Hz, 400 Hz and 500 Hz? The answer is 100 Hz (not 200 Hz!), because all partials are integer multiples of a *missing fundamental* frequency of 100 Hz. Therefore, the perceived pitch of a complex tone consisting of the frequencies 400 Hz, 500, 600 Hz, and 700 Hz is also 100 Hz. That is, if a tone has enough overtones, then the fundamental frequency could be filtered out, and the pitch percept would remain unchanged (what would change, however, is the timbre of the sound).

The phenomenon of the perception of a 'missing fundamental' is an occurrence of *residue pitch*,[6] also referred to as *periodicity pitch*, *virtual pitch*, or *low pitch*. The value of a residue pitch equals the periodicity (i.e., the timing) of the waveform resulting from the superposition of sinusoids. Importantly, dichotically presented stimuli also elicit residue perception, arguing for the notion that temporal coding of sounds beyond the cochlea is important for pitch perception. Such temporal coding has been reported for neurons of the inferior colliculus (e.g., Langner *et al.*, 2002) and the auditory cortex (see below);[7] even neurons in the (dorsal) cochlear nucleus (DCN) are able to represent the periodicity of iterated rippled noise, supporting the notion that already the DCN is involved in the temporal representation of both envelope periodicity and pitch (Neuert *et al.*, 2005). However, note that two (or more) frequencies that can be separated (or 'resolved') by the BM, also generate (intermodulation) distortions on the BM with different frequencies, the one most easily audible having a frequency of f2 - f1 (usually referred to as difference *combination tone*). Usually both mechanisms (BM distortions generating combination tones, and temporal coding) contribute to the perception of residue pitch, although combination tones and residue pitch can also be separated (Schouten *et al.*, 1962).

1.4   Beats, roughness, consonance and dissonance

If two sinusoidal tones (or two partials of two tones with similar frequency) cannot be separated (or 'resolved') by the BM, that is, if two frequencies pass through the same *equivalent rectangular bandwidth* (ERB; for details see Moore, 2008; Patterson & Moore, 1986),[8] then the two frequencies are added together (or 'merged') by the BM. This results in an oscillation of the BM with a frequency equal to the mean frequency of the two components, and an additional *beat*[9] (see also von Helmholtz, 1870). Such beats are regular amplitude fluctuations occurring due to the changing phase relationship between the two initial sinusoids, which results in the phenomenon that the sinusoids alternately reinforce and cancel out each other. The frequency of the beat is equal to the frequency difference between the two initial sinusoids. For example, two sinusoidal tones with frequencies of 1000 and 1004 Hz add up to (and are then perceived as) a tone of 1002 Hz, with four beats occurring each second (similar to when turning a volume knob up and

---

[6]   The German term for 'residue pitch' is 'Residualton'.

[7]   As a note of caution, (McAlpine *et al.*, 2000) showed that some neural responses representing periodicity information at the level of the inferior colliculus may simply be due to cochlear (intermodulation) distortions.

[8]   Others have used the term *critical band* (Zwicker, 1961; Zwicker & Terhardt, 1980) or *auditory filter* (for a historical account and details see Moore, 2008).

[9]   The German word for beats with relatively low frequency (roughly below about 15–20 Hz) is *Schwebung*.

down four times in one second). When the beats have higher frequencies (above ~20 Hz), these beats are perceived as *roughness* (Plomp & Steeneken, 1968; Terhardt, 1974; 1978), and are a sensory basis for the so-called *sensory dissonance* (Terhardt, 1976, 1984; Tramo *et al.*, 2001). Western listeners tend to judge two sinusoidal tones as consonant as soon as their frequency separation exceeds about one ERB (Plomp & Levelt, 1965), which is typically between 11% and 17% of the centre frequency.

Ernst Terhardt (1976; 1984) distinguished two components of musical consonance/dissonance, namely *sensory consonance/dissonance* and *harmony*.[10] According to Terhardt, sensory consonance/dissonance represents the graded absence/presence of annoying factors (such as beats and roughness). Others (Tramo *et al.*, 2001) argued that consonance is also a positive phenomenon (not just a negative phenomenon that depends on the absence of roughness), one reason being that residue pitches produced by the auditory system contribute to the percept of consonance.[11] Tramo *et al.* (2001) argue that, in the case of consonant intervals, the most common interspike interval (ISI) distributions of auditory nerve fibres correspond (a) to the F0 frequencies of the tones, as well as (b) to the frequency (or frequencies) of the residue pitch(es). Moreover (c), all or most of the partials can be resolved. By contrast, for dissonant intervals, the most common ISIs in the distribution do not correspond (a) to either of the F0s, nor (b) to harmonically related residue pitch(es). Moreover (c), many partials cannot be resolved.

Harmony, according to Terhardt, represents the fulfilment, or violation, of musical regularities that, given a particular musical style, govern the arrangement of subsequent or simultaneously sounding tones ('tonal affinity, compatibility, and fundamental-note relation', Terhardt, 1984 p. 276).[12] The degree to which *harmony* is perceived as un/pleasant is markedly shaped by cultural experience, due to its relation to music- (and thus presumably also culture-) specific principles.

Sensory dissonance (i.e., the 'vertical dimension of harmony'; Tramo *et al.*, 2001) is universally perceived as less pleasant than consonance, but the degree to which sensory consonance/dissonance is perceived as pleasant/unpleasant is also significantly shaped by cultural experience. This notion has recently received support by a study carried out in Cameroon with individuals of the Mafa people who had presumably never listened to Western music before participating in the experiment (Fritz *et al.*, 2009). The Mafa showed a significant preference for original Western music over continuously dissonant versions of the same pieces.

---

[10]   (Tramo *et al.*, 2001) use the terms *vertical* and *horizontal* dimensions of harmony instead. They restrict use of the terms *consonance* and *dissonance* to the vertical dimension of harmony.

[11]   Note that merely a critical band account of consonance as the absence of roughness cannot explain why in experiments with pure tones the interval of a tritone is perceived as less consonant (or more dissonant) than the fourth or the fifth, although both pitches can clearly be resolved by the BM (for a review see Tramo *et al.*, 2001).

[12]   (Tramo *et al.*, 2001) use the term 'horizontal dimension of harmony' instead.

Notably, the difference in normalized pleasantness ratings between original music and the continuously dissonant versions was moderate, and far smaller than those made by a control group of Western listeners. That is, both Western and Mafa listeners preferred more consonant over continuously dissonant music, but whereas this preference was very strong in Western listeners, it was rather moderate in the Mafas. This indicates that the preference for mainly consonant music over continuously dissonant music is shaped by cultural factors.[13]

Beating sensations can not only occur monaurally (i.e., when different frequencies enter the same ear), but also binaurally (i.e., when each ear receives different frequencies, for example one frequency entering one ear, and another frequency entering the other ear). Binaural beats presumably emerge mainly from neural processes in the auditory brainstem (Kuwada *et al.*, 1979; McAlpine *et al.*, 2000), which are due to the continuously changing interaural phase that results from the superposition of two sinusoids, possibly related to sound localization.[14] Perceptually, binaural beats are somewhat similar to monaural beats, but not as distinct as monaural beats. Moreover, in contrast to monaural beats (which can be observed over the entire audible frequency range) binaural beats are heard most distinctly for frequencies between 300 and 600 Hz (and they become progressively more difficult to hear at higher frequencies; for details see Moore, 2008).

1.5 Acoustical equivalency of timbre and phoneme

With regard to a comparison between music and speech, it is worth mentioning that, in terms of acoustics, there is no difference between a phoneme and the timbre of a musical sound (and it is only a matter of convention if phoneticians use terms such as 'vowel quality' or 'vowel colour', instead of 'timbre').[15] Both are characterized by the two physical correlates of timbre: *Spectrum envelope* (i.e.,

---

[13] Interestingly, the cultural influence on preference of consonance/dissonance works both ways: Individuals who listen a lot to music with high degree of dissonance begin to prefer higher degrees of dissonance in music. This is reminiscent, for example, of the un/pleasantness caused by capsaicin (the alkaloid that makes paprika and chili taste hot); capsaicin is universally perceived as less pleasant than sugar (Rozin & Schiller, 1980), but individuals develop strong, culture-specific preferences for strong spices. In fact, adults across the world daily ingest substances that are innately rejected, such as bitter substances, or substances irritating the oral mucosa (e.g., coffee, beer, spirits, tobacco, and chili pepper; Rozin & Schiller, 1980).

[14] Contrary to what sellers of so-called *i-dosing* audio-files promise, it is almost certain that binaural beats themselves cannot evoke brain states that are even remotely comparable with those induced by drugs such as heroin, marijuana, etc. There is also lack of scientific evidence indicating that binaural beats have any systematic effect on relaxation, anxiety-reduction, etc.

[15] When two sounds are perceived as having the same pitch, loudness, duration, and location of origin, and 'a difference can still be heard between the two sounds, that difference is called timbre' (e.g., Moore, 2008). For example: Imagine that a clarinet, a saxophone, and a piano successively play a middle C at the same location, with the same loudness and the same duration. Each of these instruments has a unique sound quality. This difference is called timbre, tone colour, or simply sound

differences in the relative amplitudes of the individual harmonics) and *amplitude envelope* (also referred to as amplitude contour or energy contour of the sound wave, i.e., the way that the loudness of a sound changes, particularly with regard to the attack and the decay of a sound).[16] Aperiodic sounds can also differ in spectrum envelope (see, e.g., the difference between /ʃ/ and /s/), and timbre differences related to amplitude envelope play a role in speech, e.g., in the shape of the attack for /b/ vs. /w/ and /ʃ/ vs. /tʃ/.

1.6 Auditory cortex

The *primary auditory cortex* corresponds to the transverse gyrus of Heschl (or gyrus temporalis transversus) which is part of the superior temporal gyrus (STG). Most researchers agree that the primary auditory cortex (corresponding to Brodmann's area 41) consists of three sub-areas, referred to as AI (or A1), R, and RT by some authors (e.g., Kaas & Hackett, 2000; Petkov *et al.*, 2006; see also Figure 1.3), or Te1.0, Te1.1, and Te1.2 by others (Morosan *et al.*, 2001, 2005). The primary auditory cortex (or 'auditory core region') is surrounded by auditory belt and parabelt regions that constitute the *auditory association cortex* (Kaas & Hackett, 2000; Petkov *et al.*, 2006).[17,18]

Figure 1.3 shows these regions and their connectivity according to the nomenclature introduced by Kaas & Hackett (2000).[19] Note that, unlike what is shown in Figure 1.3, Nieuwenhuys *et al.* (2008) stated that the parabelt region also covers parts of the temporal operculum, that is, part of the medial (and not only the lateral) surface of the STG (p. 613). Nieuwenhuys *et al.* (2008) also noted that the precise borders of the posterior parabelt region (which grades in the left hemisphere into Wernicke's area) are not known, but that 'it is generally assumed that it includes the posterior portions of the planum temporale and superior

---

quality. There are also many examples of timbre differences in speech. For example, two vowels spoken with the same loudness and same pitch differ from one another in timbre.

[16]   E.g., sudden or slow attack or decay, such as in the sounds of plucked vs. bowed stringed instruments. Additional features include microtemporal variations such as jitter (microvariations in the F0 frequency) and shimmer (microvariations in the glottal pulse amplitude), which are also characteristic for both 'phonemes' and 'timbres.'

[17]   In terms of Brodmann's nomenclature, the auditory core region appears to correspond to Brodmann's area (BA) 41, the lateral auditory belt region to BA 42, medial belt region to BA 52, and auditory parabelt region to much of BA 22 (Hackett & Kaas, 2004, although parts of BA 22 may also constitute the auditory belt region).

[18]   Galaburda & Sanides (1980) reported that (in humans) regions of caudo-dorsal parakoniocortex (PaA c/d) extended from the posterior temporal plane (caudomedial end of the Sylvian fissure), *around* the retroinsular region and then dorsally onto the medial aspect of the parietal operculum. Thus, according to Galaburda & Sanides (1980), auditory cortex can also be found in the parietal operculum.

[19]   Others (e.g., Morosan *et al.*, 2001, 2005) refer to these regions as areas Te2.1, Te2.2, Te3, and Te4.

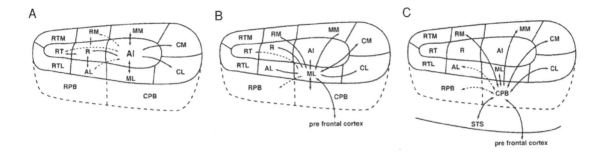

Figure 1.3 Subdivisions and connectivity of the auditory cortex. **(A)** The auditory core region (also referred to as primary auditory cortex) is comprised of the auditory area I (AI), a rostral area (R), and a rostrotemporal area (RT). Area AI, as well as the other two core areas, has dense reciprocal connections with adjacent areas of the core and belt (left panel, solid lines with arrows). Connections with nonadjacent areas are less dense (left panel, dashed lines with arrows). The core has few, if any, connections with the parabelt or more distant cortex. **(B)** shows auditory cortical connections of the middle lateral auditory belt area (ML). Area ML, as well as other belt areas, have dense connections with adjacent areas of the core, belt, and parabelt (middle panel, solid lines with arrows). Connections with nonadjacent areas tend to be less dense (middle panel, dashed lines with arrows). The belt areas also have topographically organized connections with functionally distinct areas in the prefrontal cortex. **(C)** Laterally adjacent to the auditory belt is a rostral (RPB) and a caudal parabelt area (CPB). Both these parabelt areas have dense connections with adjacent areas of the belt and RM in the medial belt (illustrated for CPB by the solid lines with arrows). Connections to other auditory areas tend to be less dense (dashed lines with arrows). The parabelt areas have few, if any, connections with the core areas. The parabelt also has connections with the polysensory areas in the superior temporal sulcus (STS) and with functionally distinct areas in prefrontal cortex. Further abbreviations: CL, caudolateral area; CM, caudomedial area; ML, middle lateral area; RM, rostromedial area; AL, anterolateral area; RTL, lateral rostrotemporal area; RTM, medial rostrotemporal area. Reprinted with permission from Kaas & Hackett (2000).

temporal gyrus, and the most basal parts of the angular and supramarginal gyri' (p. 613–614).

All of the core areas, and most of the belt areas, show a tonotopic structure, which is clearest in AI. The tonotopic structure of R seems weaker than that of AI, but stronger than that of RT. The majority of belt areas appear to show a tonotopic structure comparable to that of R and RT (Petkov *et al.*, 2006, reported that, in the macaque monkey, RTM and CL have only a weak, and RTL and RM no clear tonotopic structure).

The primary auditory cortex (PAC) is thought to be involved in several auditory processes. (1) The *analysis of acoustic features* (such as frequency, intensity, and timbral features). Compared to the brainstem, the auditory cortex is capable of performing such analysis with considerably higher resolution (perhaps with the exception of the localization of sound sources). Tramo *et al.* (2002) reported that a patient with bilateral lesions of the PAC (a) had normal detection thresholds for sounds (i.e., the patient could say whether there was a tone or not), but (b) had elevated thresholds for determining whether two tones have the same pitch or not (i.e., the patient had difficulties detecting minute frequency differences between two subsequent tones). (2) *Auditory sensory memory* (also referred to as 'echoic memory'). The auditory sensory memory is a short-term buffer that stores auditory information for a few instances (up to several seconds). (3) *Extraction of inter-sound relationships.* The study by Tramo *et al.* (2002) also reported that the patient with PAC lesions had markedly increased thresholds for determining the pitch direction (i.e., the patient had great difficulties in saying whether the second tone was higher or lower in pitch than the first tone, *even though* he could tell that both tones differed (see also Johnsrude *et al.*, 2000; Zatorre, 2001, for similar results obtained from patients with right PAC lesions). (4) *Stream segregation*, including discrimination and organization of sounds as well as of sound patterns (see also Fishman *et al.*, 2001). (5) *Automatic change detection.* Auditory sensory memory representations also serve the detection of changes in regularities inherent in the acoustic input. Such detection is thought to be reflected electrically as the mismatch negativity (MMN; see Chapter 5), and several studies indicate that the PAC is involved in the generation of the MMN (for an MEG-study localizing the MMN generators in the PAC see Maess *et al.*, 2007). (6) *Multisensory integration* (Hackett & Kaas, 2004), particularly integration of auditory and visual information. (7) The *transformation of acoustic features into auditory percepts*, that is, transformation of acoustic features such as frequency, intensity etc. into auditory percepts such as pitch height, pitch chroma, and loudness.[20] It appears that patients with (right) PAC lesions have lost the ability to perceive residue pitch (Zatorre, 1988), consistent with animal studies showing that bilateral lesions of the auditory cortex (in the cat) impair the discrimination of changes in the pitch of a missing fundamental (but not changes in frequency alone; Whitfield, 1980). Moreover, neurons in the anterolateral region of the PAC show responses to a missing fundamental frequency (Bendor & Wang, 2005, data were obtained from marmoset monkeys), and magnetoencephalographic data suggest that response properties in the PAC depend on whether or not a missing fundamental of a complex tone is perceived (Patel & Balaban, 2001, data were obtained from humans). In that study (Patel & Balaban, 2001) phase changes of the auditory steady-state response (aSSR) were related to the pitch percept of a sound.[21]

As mentioned above, combination tones emerge already in the cochlea (generated by the nonlinear mechanics of the basilar membrane), and the periodicity of complex tones is coded in the spike pattern of auditory brainstem neurons.[22] That is, different mechanisms contribute to the perception of residue pitch on at least

---

[20]   For example, a sound with the frequencies 200 Hz, 300 Hz, and 400 Hz is transformed into the pitch percept of 100 Hz.

[21]   The aSSR is an ongoing oscillatory brain signal resulting from continuous amplitude modulation (AM) of an acoustic stimulus; for example, in the study by Patel & Balaban (2001), complex tones were amplitude-modulated at a rate of 41.5 Hz. The aSSR presumably originates from the PAC (e.g., Ross *et al.*, 2000).

[22]   Responses in the PAC related to the perception of missing fundamental frequencies in the studies by Bendor & Wang (2005) and Patel & Balaban (2001) are presumably partly due to the periodicity information about the missing fundamental frequency coded in the spike pattern of collicular neurons.

three different levels: (1) On the basilar membrane (BM), (2) in the brainstem (due to temporal coding that leads to a periodicity of the neuronal spike pattern), and (3) in the auditory cortex.[23] However, the studies by Zatorre (2001) and Whitfield (1980) suggest that the auditory cortex plays a more prominent role for the transformation of acoustic features into auditory percepts than the brainstem (or the basilar membrane).

It is also worth noting that neurons in AI are responsive to both sinusoidal ('pure') tones and complex tones, as well as to noise stimuli, whereas areas outside AI become increasingly unresponsive to pure tones, and respond more strongly (or exclusively) to complex tones and noises. Therefore, it seems most plausible that accurate acoustic feature analysis, sound discrimination and pattern organization, as well as transformation of acoustic features into percepts are the results of close interactions between auditory core and belt areas. In addition, the auditory association cortex fulfils a large array of functions (many of which have just begun to be investigated systematically with neuroscientific methods) such as auditory scene analysis and stream segregation (De Sanctis *et al.*, 2008; Gutschalk *et al.*, 2007; Snyder & Alain, 2007), auditory memory (Näätänen *et al.*, 2010; Schonwiesner *et al.*, 2007), phoneme perception (Obleser & Eisner, 2009), voice perception (Belin *et al.*, 2004), speaker identification (von Kriegstein *et al.*, 2005), perception of the size of a speaker or an instrument (von Kriegstein *et al.*, 2007), audio-motor transformation (Warren *et al.*, 2005; Rauschecker & Scott, 2009), syntax processing (Friederici, 2009), or storage and activation of lexical representations (Lau *et al.*, 2008).

With regard to functional differences between the left and the right PAC, as well as neighbouring auditory association cortex, several studies indicate that the left auditory cortex (AC) has a higher resolution of temporal information than the right AC, and that the right AC has a higher spectral resolution than the left AC (Zatorre *et al.*, 2002; Hyde *et al.*, 2008). Furthermore, with regard to pitch perception, Warren *et al.* (2003) report that changes in pitch height as well as changes in pitch chroma (see p. 20 for description of the term 'pitch chroma') activate PAC, but that chroma changes involve auditory belt areas anterior of the PAC (covering parts of the planum polare) more strongly than changes in pitch height. Conversely, changes in pitch height activated auditory belt areas posterior of the PAC (covering parts of the planum temporale) more strongly than changes in pitch chroma.

With regard to the perception of the pitches of melodies, it appears that the analysis of the contour of a melody (which is part of the auditory Gestalt formation)[24] particularly relies on the right superior temporal gyrus (posterior rather

---

[23]   But note also that combination tones and residue pitch can be separated (Schouten *et al.*, 1962).

[24]   The formation of auditory Gestalten follows so-called Gestalt-principles, such as the principle of similarity, of proximity, or of continuation. For example, (1) the single tones of a chord are perceived as one auditory Gestalt (a chord) because they are played at the same time (principle of contiguity); (2) when a melody is played in a high register which is accompanied by chords in a low register, the

than anterior STG), whereas the use of more detailed interval information appears to involve both posterior and anterior areas of the supratemporal cortex bilaterally (Peretz & Zatorre, 2005; Liegeois-Chauvel *et al.*, 1998; Patterson *et al.*, 2002). The planum temporale especially has been implicated in the processing of pitch intervals and sound sequences (Patterson *et al.*, 2002; Zatorre *et al.*, 1994; Koelsch *et al.*, 2009), consistent with the notion that this region is a crucial structure for auditory scene analysis and stream segregation. An introduction to subjective measures of pitch perception is provided in Chapter 3.

---

tones of the melody are perceived as one Gestalt, and the tones of the chords as another, even if they have the same onsets (principle of proximity); (3) if the same melody is played in a low register by a cello, and the chords are played in a low register on a piano, then the cello tones are perceived as one Gestalt and the chords as another (principle of similarity); (4) if two cellos play two melodies, and both melodies cross, then one melody will be perceived as ascending and the other as descending (principle of continuity).

2

Music-theoretical Background

2.1 How major keys are related

In music theory, the distance between two tones is called an *interval*. When the relation between the fundamental frequencies of two tones is 1:2, the interval is called an octave (e.g., *c′* and *c″*, Figure 2.1). The higher tone of two tones building an octave is perceived twice as high than the lower one.

In the tempered intonation, the octave-range is divided into twelve equally-spaced *semitone*-steps.[1] The division of the octave-range into twelve semitone steps leads to a set of twelve different tones. These tones build the *chromatic scale* and are the basic elements of western tonal music (Figure 2.2).

The interval of two tones which are one semitone step apart from each other is a minor second. Two semitone steps build a major second (i.e., a whole tone),

Figure 2.1 Octave interval built by the tones *c′* (left) and *c″* (right).

---

[1] Non-tempered scale-systems which distinguish for instance between the tones *f sharp* and *g flat* are more complicated and will be neglected here. For detailed descriptions see e.g., Eggebrecht (1967); Apel (1970); Eggebrecht (1972); Dahlhaus & Eggebrecht (1979); Dahlhaus (1980).

Brain and Music, First Edition. Stefan Koelsch.
© 2013 John Wiley & Sons, Ltd. Published 2013 by John Wiley & Sons, Ltd.

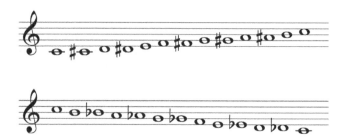

Figure 2.2 Ascending (top row), and descending (bottom row) chromatic scale: The octave ($c'-c''$) is divided into 12 semitone steps. In the tempered intonation, where e.g., *c sharp* is the same tone as *d flat*, such a division of an octave leads to twelve different tones (c'' is the octave of c' and thus not counted).

three semitone steps a minor third, four a major third, etc. (Figure 2.3). By combining steps of 1, 2, or 3 semitones within an octave-range, several scales can be constituted. These scales usually comprise seven tone-steps (the pentatonic scale comprises five). During the last centuries, four scales have become most prominent in western tonal music: A major scale, and three minor scales (harmonic minor, melodic minor, and natural minor, i.e., minor without raised sixth and seventh degrees).[2] The major scale consists of two tetrachords, each tetrachord with a degree progression of $1-1-\frac{1}{2}$ (i.e., whole tone step – whole-tone step – semitone step). Both tetrachords are separated by a whole-tone step (Figure 2.4). Because a major scale comprises seven tone steps (e.g., in *C* major: $C - d - e - f - g - a - b - c$), and an octave can be divided into twelve semitone steps, there are always five tones of the chromatic scale which do not belong to a given major scale.[3]

A tonal key exactly determines the tones which belong to this key. For example, the *C* major key determines exclusively the tones of the *C* major scale as belonging

Figure 2.3 Examples of intervals, from left: Minor second, major second, minor third, major third, perfect fourth, perfect fifth.

---

[2]   Especially in jazz music, as well as in some folk music styles, 'old' scales like doric, phrygic, lydic etc. are often employed. Several folk music styles also use scales such as the *Hungarian minor gypsy scale* (harmonic minor with raised fourth degree, e.g., a – b – c – d sharp – e – f – g sharp – a), or the *major gypsy scale* (major with flattened second and sixth degree, e.g., c – d flat – e – f – g – a flat – b – c, also referred to as *double harmonic*, *Arabic*, or *Byzantine scale*).

[3]   For example: C major does not comprise the tones *c sharp* (or *d flat*), *d sharp* (or *e flat*), *f sharp* (or *g flat*), *g sharp* (or *a flat*), and *a sharp* (or *b flat*).

Figure 2.4 Example of two tetrachords building the *C* major scale. The two tetrachords are separated by one whole-tone step.

to *C* major (no further tones belong to *C* major). Two different major keys never consist of exactly the same tones, though they may have tones in common. Importantly, each initial major key has exactly two neighbouring major keys which consist of the same tones *except one* with regard to the initial key. For example, *C* major shares six tones with *G* major ($c - d - e - g - a - b$). The missing (seventh) tone of *C* major is *f*, the missing tone of *G* major is *f sharp* (being one semitone step distant from *f*). The other major key which has six tones in common with *C* major is *F* major.[4]

Because both *G* major and *F* major share six tones with *C* major (more than any other major key), they are from a music-theoretical point of view the most closely related major keys of *C* major. In *C* major, *G* major is called the dominant key, *F* major the subdominant key. Vice versa, *C* major is the subdominant key of *G* major, and the dominant key of *F* major. In *F* major, the tone *c* is the fifth scale-tone above *f*. The interval between *f* and *c* is called a (perfect) *fifth*. Analogously: The fifth scale-tone in *C* major is *g*, the interval between *c* and *g* is also a fifth.[5]

The keys which are most closely related to an initial key have each for their part a further most closely related key. Continuing the example: *G* major has (besides the subdominant key *C* major) also a dominant key, which is *D* major. *D* major can be distinguished from *G* major by the tone *c* which belongs to *G* major, but not to *D* major. Note that *d* is the fifth scale-tone of *G* major, the interval $d - g$ is (again) a fifth. The example illustrates that every major key has two most closely related major neighbours (each having six tones in common with the initial key), two second-closest neighbours (each having five tones in common with the initial key), two third-closest neighbours, etc. Notably, the first scale-tone of an initial key (e.g., *c* in *C* major) is one fifth distant from the first scale-tone of both dominant and subdominant keys (which are the most closely related keys), two fifths distant from the second-closest keys, etc. The fifths-relatedness of major keys can nicely be described using the *circle of fifths* (Figure 2.5).[6]

---

[4]   These six tones are: $c - d - e - f - g - a$; the missing (seventh) tone is *b* in *C* major (*b flat* in *F* major, respectively).

[5]   Notably, the relationship between keys can not only be described in terms of fifths, but also in terms of tetrachords: The second tetrachord of *F* major is the first tetrachord of *C* major, and the second tetrachord of *C* major is the first of *G* major.

[6]   For another description than the circle of fifths see (Schönberg 1969;).

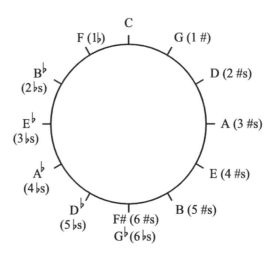

Figure 2.5 The circle of fifths (for major keys only).

2.2 The basic in-key functions in major

The tones of a scale are also referred to as *scale degrees*. The tonic tone of a scale, e.g., in *C* major the tone *c*, is called the first degree, the second tone (in *C* major the tone *d*) is called second degree, etc. An *in-key chord* is a triad in root position built on a scale degree with in-key tones (see Figure 2.6 for illustration). The triads built on the first, fourth, and fifth degree of a major scale are major chords, the triads built on the second, third, and sixth degree are minor chords. The triad on the seventh degree is a diminished chord. The in-key triad of the first degree is called tonic chord (or just *tonic*). The tonic chord is the best representative of its key: (a) the root tone of the tonic chord is also the root tone of the key, (b) the tonic chord contains the fifth of the key (which is the third overtone of the root tone), and (c) the tonic chord contains the third of the key, which determines the tonal genus (major or minor). The in-key triad of the fourth degree is called *subdominant*, of the fifth degree *dominant*. The minor triad on the sixth degree is the *submediant*, on the second degree *supertonic*, and on the third degree *mediant*. Tonic, subdominant, dominant, mediant, etc. are called *chord functions* (Figure 2.6). Chord functions can be denoted by roman numerals of the degrees

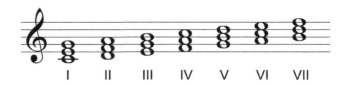

Figure 2.6 Chords built on the degrees of the *C* major scale, the degrees are indicated by roman numerals (chord functions from left: Tonic, supertonic, mediant, subdominant, dominant, submediant).

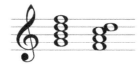

Figure 2.7   Dominant seventh chord (left) and subdominant with sixte ajoutée (right) in *C* major.

on which they are built (Figure 2.6), as well as by letters (e.g., T for major tonic, S for major subdominant).

The triad on the fifth scale degree with an added minor seventh is called a *dominant seventh chord* (in *C* major: *g – b – d – f*, the interval *g – f* is a minor seventh, Figure 2.7). The seventh is the *characteristic dissonance* of the dominant, and usually acts as an upper leading-tone to the third of the subsequent (tonic) chord. The characteristic dissonance of the subdominant is the *sixte ajoutée*, a major sixth added to a major triad (usually subdominant, Figure 2.7). According to Jean Philippe Rameau (1722) the three chords: Tonic, dominant seventh chord, and subdominant with sixte ajoutée build the harmonic centre ('Centre harmonique') of a tonal key.

2.3   Chord inversions and Neapolitan sixth chords

Each chord in root position (that is, with the root tone in the base; see Figure 2.8) can be inverted into a *sixth chord* by setting the base tone, e.g., into the top voice, so that the third becomes the base tone of the new chord. A repetition of this procedure leads to a *six – four chord* with the fifth of the chord in the base (Figure 2.8).

A minor subdominant with a minor sixth instead of a fifth is called a *Neapolitan sixth chord*. For example: In *c* minor the minor sixth of the subdominant is *d flat*. Thus, when the fifth of the subdominant is replaced by a minor sixth, the subdominant consists of the tones *f – a flat – d flat*. The Neapolitan sixth chord can also be interpreted as the mediant of a minor subdominant inverted into a sixth chord. That is, e.g., in *c* minor: The subdominant is *f – a flat – c*, the mediant of the minor subdominant is *d flat – f – a flat*, which is, once inverted, the sixth chord *f – a flat – d flat* (Figure 2.9).

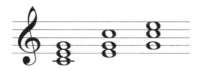

Figure 2.8   *C* major triad in root position (left), as sixth chord (middle), and as six – four chord (right).

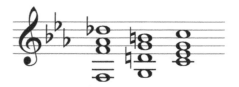

Figure 2.9   Neapolitan sixth chord in *c* minor (left), followed by the dominant (middle) and the tonic chord (right).

2.4   Secondary dominants and double dominants

As described before, each in-key chord has a harmonic function within a tonal key. However, chords may be paraphrased in a way that they temporarily take over a function within another tonal key. In case that a chord takes over the function as a tonic, this might be indicated by a preceding dominant seventh chord of that temporary tonic. In other words, an in-key chord may be preceded by a dominant seventh chord, so that the in-key chord temporarily functions as a tonic. Such a dominant seventh chord, which alters the function of the subsequent chord, is called a *secondary dominant*. A secondary dominant which is the dominant to the dominant is also denoted as *double dominant*. In major, a double dominant is also referred to as *chromatic supertonic*.

For example: The dominant of *C* major (*g − b − d*) may be preceded by the dominant seventh chord of *G* major (*d − f sharp − a − c*). The tone *f sharp* does not belong to *C* major, but to *G* major. Moreover, a seventh (*d − c*) is the characteristic dissonance of the dominant. Thus, the *G* major triad (*g − b − d*), which was formerly functioning as dominant of *C* major, now functions as a tonic (possibly only temporarily). This function change was introduced by the dominant seventh chord of *G* major: *d − f sharp − a − c* (Figure 2.10).

Figure 2.10   Example of a secondary dominant (in *C* major). From left: Tonic, secondary dominant to the dominant, dominant, dominant seventh chord, and tonic.

3

Perception of Pitch and Harmony

3.1 Context-dependent representation of pitch

The perception of pitch is fundamental for the perception of music. Pitch is a *morphometric medium* (e.g., Attneave & Olson, 1971; Shepard, 1999), that is, pitch is a medium capable of bearing forms (or *Gestalt*). For example, pitch patterns like melodies or harmonies can be moved up and down in pitch, and still be recognized as being the same pattern. What is relevant for psychological relations between pitches is the ratio of their physical frequencies, not their arithmetic differences (for a detailed description see, e.g., Krumhansl, 1979).[1]

Importantly, the perceived distance between two pitches is not only dependent on the (physical) frequencies of both pitches, but is influenced by numerous factors. For example, two sinusoidal tones separated by an octave are perceived as 'somewhat identical,' although they are physically different in terms of their frequencies, whereas two pitches separated by a semitone are perceived as quite different, although they are physically similar in terms of their frequencies. Pitch perception is thus not simply a linear phenomenon (and is not adequately described by a linear pitch scale). The left panel of Figure 3.1 shows a configuration in which the pitches are placed on a spiral (or *helix*), with the octaves lying on a vertical line. The vertical position on the pitch helix represents the *pitch height*, the

---

[1] See also Fechner (1873). Other than logarithmic representations have also been described with special experimental paradigms; see e.g., Stevens *et al.* (1937); Ellis (1965). For summaries see Shepard (1982a, 1999).

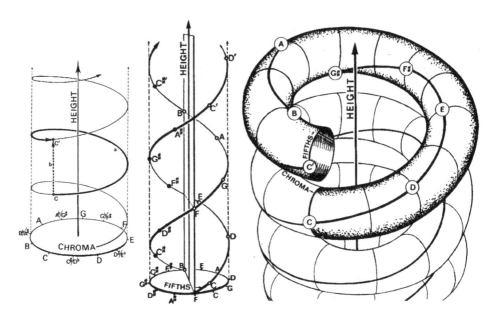

Figure 3.1   Left: Helical configuration of tones accounting for the increased similarity between tones separated by an octave. Pitch height is the vertical dimension, the chroma circle is the projection onto the horizontal plane. Middle: Double helix wound around a cylinder, illustrating the representation of fifth-relationship within the double-helical structure. Tones with the same pitch class are located on the same vertical axis (see dashed lines), with the neighbouring tones being one fifth distant (e.g., G and A are located most closely to D). Right: Five-dimensional configuration capturing pitch height, chroma, and the relationships of perfect fifths and octaves. Figures reprinted with permission from Shepard (1982b).

position within an octave around the cylinder defined by the helix represents the *pitch chroma*. The middle and right panels of Figure 3.1 show configurations that also take the musical importance of the (perfect) fifth into account (in addition to the importance of the octave). These configurations represent pitch height, pitch chroma, and the relationships of octaves and fifths using a double-helix wound around a helical cylinder (see also Shepard, 1982a; Deutsch, 1982).

Note that these configurations describe pitch perception in a non-musical context. In a series of experiments, Carol L. Krumhansl (1979, 1990) found that the presence of a musical context (namely, the establishment of a tonal key) modulates the psychological representation of musical pitch. In one of those experiments (Krumhansl, 1979), musically trained participants were in each trial presented with a pair of tones after a tonal key had been established (by presenting a tonic chord or a major scale). The participants were asked to judge how similar the first tone of the tone-pair was to the second tone in the tonal system suggested by the preceding musical context. A similarity matrix containing similarity-ratings of the tone-pairs revealed that the tones of the tonic chord were judged as more similar

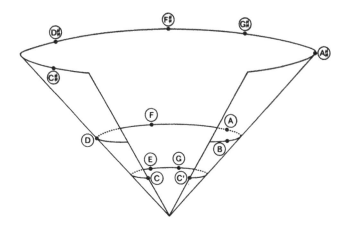

Figure 3.2 MDS of the averaged similarity matrix, three-dimensional solution (reprinted with permission from Krumhansl, 1979).

compared to the diatonic scale tones, and the diatonic scale tones were judged as more similar compared to the non-diatonic tones.[2]

All data were scaled using a non-metric multidimensional scaling method (MDS). The best MDS solution was obtained by a cone-shaped three-dimensional configuration (Figure 3.2). The cone had a radius equal to its height, with the major triad components falling on a circular cross-section of half the radius of the circular cross-section containing the other (diatonic) scale tones, and one-quarter the radius of the circular cross-section containing the non-diatonic tones. Results were taken to indicate that, *within a musical context*, the tones of the tonic chord were perceived as more similar (or proximal) to each other compared to the other diatonic tones, and the other diatonic tones as more similar to each other compared to the non-diatonic scale tones. Interestingly, with regard to musical meaning, Krumhansl (1979) stated also that 'in an explicitly musical context, musical listeners perceive a complex pattern of interrelationships among the individual tones,' and that 'tones acquire meaning through their relationships to other tones' (1979, p. 358, p. 370; intra-musical meaning will be dealt with in detail in Chapter 10).

The degree of relationship between tones correlated with the function of tones with regard to an established tonal key. Tones of the tonic chord were perceived as most closely related (and thus as structurally most stable), followed by the other diatonic tones, and the non-diatonic tones. This finding shows that, within a tonal context, tones are psychologically represented within a tonal hierarchy.

---

[2] That is, within C major, tone-pairs consisting of tonic-chord tones only (like $c - e$, $c - g$ or $e - g$) obtained the highest similarity ratings, whereas lower ratings were given to tone-pairs consisting of the remaining diatonic scale tones (like $d - f$ or $a - b$), and only poor similarity ratings to tone-pairs consisting of at least one non-diatonic tone (like $c\ sharp - f\ sharp$ or $c - f\ sharp$).

Another interesting finding of Krumhansl's experiment (1979) was that the similarity ratings also depended on the order in which the two tones were presented. That is, an asymmetry was found in the ratings of tone-pairs: When a diatonic tone was followed by a tone of the tonic-chord, tones were rated as more similar than when these tones were presented in reverse order.[3] Krumhansl suggested that these asymmetries reflect a tendency to perceive tones as moving over time toward (rather than away from) the vertex of the conical configuration, that is towards the tonic. This explanation nicely describes a dynamic aspect of music in time and will become important when discussing neural processes of integration of out-of-key notes into a musical context (Chapters 9 and 10).

3.2   The representation of key-relatedness

The experiment by Krumhansl (1979) investigated the perception of tones within the context of a (single) key. The perceived stability of tones is closely linked to the perception of the tonal stability of chords (because a chord is the simultaneous sounding of tones; Krumhansl & Kessler, 1982b). With regard to the relatedness of keys, Gottfried Weber (1817) was one of the first to create a schematic diagram of all major and minor keys (for a historical account see Purwins *et al.*, 2007), by linking relative and parallel relations (with thirds on the horizontal plane and fifths on the vertical). A strip of keys cut out from Weber's chart of tone centres is shown in the top panel of Figure 3.3. Purwins *et al.* (2007) pointed out that this strip can be transformed into a toroidal model of key relations by curling the strip in a way that identical keys are overlaid. For example, *g* minor occurs as the relative minor of *B* major, and as the parallel minor of *G* major. When the strip is curled as shown in the middle panel of Figure 3.3, these keys can be overlaid. Once the redundant strips overlap, the curls can be compacted into a single tube (see bottom panel of Figure 3.3), and the enharmonic equivalents at both ends of the tube can be wrapped horizontally to produce a three-dimensional torus.

This torus roughly represents major–minor tonal key space, and the shape of the torus is consistent with behavioural data obtained by Krumhansl & Kessler (1982b) in a study on the perception of relations between different keys, as well as on the perception of chords as tonal functions. In a first experiment, musically trained subjects rated how well, 'in a musical sense,' a probe tone went with a preceding musical element (for a similar study see Krumhansl & Shepard, 1979).[4]

---

[3]   And, analogously, when a non-diatonic tone was followed by a diatonic tone, as well as when a non-diatonic tone was followed by a tone of the tonic-chord. For example: In a *C* major context, the tones of the tone-pair *b″ − c‴* were judged to be more similar than the same tones presented in the reverse order (*c‴ − b″*), the tones of the pair *c sharp – d* as more similar compared to *d – c sharp*, and the tones *f sharp –g* more similar than *g – f sharp*.

[4]   A 'musical element' was either an ascending major or minor scale, or a major, minor, diminished or dominant seventh chord, or a three-chord cadence.

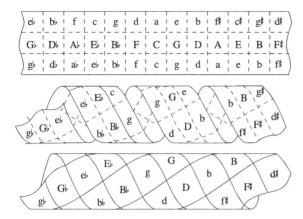

Figure 3.3 Toroidal model of key relations, derived from Gottfried Weber's chart of tone centres. For details see text. Figure adapted with permission from Purwins *et al.* (2007).

The presentation of a musical element was expected to evoke the representation of a tonal key. From the behavioural data, major and minor key profiles were obtained which indicate how strongly the tones of the chromatic scale are perceived to fit into a given key.[5] By shifting two key profiles to two different tonics (e.g., to C major and *a* minor), and then correlating the ratings of both profiles for each tone of the chromatic scale, measures of interkey distance were calculated.[6] The procedure of correlating key profiles was applied to all major–major, major–minor, and minor–minor key pairs, resulting in a correlation matrix of all major and minor keys. This correlation matrix was analysed using MDS. Interestingly, dimensions 1 and 2 of a four-dimensional solution yielded an arrangement of keys representing the circle of fifths (though of *either* major *or* minor keys; see below) which is suggested by music theory to describe interkey distances (Figure 3.4).[7]

However, another solution was suggested by Krumhansl & Kessler (1982b), in which all keys were arranged in a toroidal configuration. A flattened-out representation of this toroidal configuration is shown in Figure 3.5. In this configuration, the pattern of interkey distances becomes strikingly interpretable. All keys separated by fifths 'fall on a path wrapping three times around the torus before joining up with itself again; the major keys fall on such a path, and the minor keys on another, parallel path. These are lined up so that any major key is flanked by its

---

[5]   As to be expected from both music theory and the experiment from Krumhansl (1979), the tonic – tone received the highest ratings, the tones from the tonic – triad received higher ratings than the other diatonic scale tones, and the diatonic tones higher ratings compared to the non-diatonic tones.

[6]   For example: In the case of C major and *a* minor (*a* minor being the relative minor key of C major), the profiles gave a high correlation, in the case of C major and *F sharp* major (being only distantly related in the sense of the circle of fifths), the correlation was very low.

[7]   Notably, in music theory minor keys are arranged within the circle of fifths with respect to their relative major key. The placement obtained from the key-correlation matrix is different, and suggested by Krumhansl & Kessler as reflecting a 'compromise between the close tie of a major key to both its relative major and parallel minor keys' (1982b, p. 344).

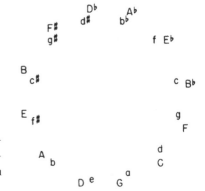

Figure 3.4 Dimensions 1 and 2 of the four-dimensional solution of the MDS- scaled key-correlation matrix (reprinted with permission from Krumhansl & Kessler, 1982b).

relative minor on one side and its parallel minor on the other' (Krumhansl & Kessler, 1982b, p.345), similar to Weber's chart of tone centres (Figure 3.3).

Compared to the circle of fifths, the torus has the advantage of depicting empirically obtained measures of psychologically represented interkey relations. Interestingly, analyses of the empirical data led to a depiction of *all* interkey relations. In the circle of fifths, merely relations of immediately surrounding single major or minor keys are represented.[8] The configuration shown in Figure 3.5 can

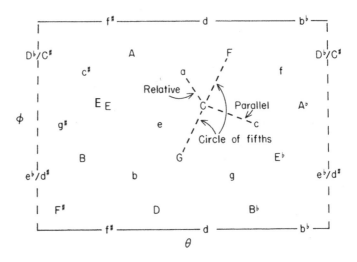

Figure 3.5 Flattened-out toroidal configuration of the multidimensionally scaled key correlation matrix (from Krumhansl & Kessler, 1982b). The opposite edges of the rectangle are identified. Examples of 'relative' and 'parallel' minor keys refer to *C* major.

---

[8] For example, the *C* major key is (psychologically as well as from a music–theoretical point of view) closer related to its minor relative (*a* minor) than to its minor parallel (*c* minor).

thus be taken as a spatial map of key regions and key distances. This spatial map will become important for approaching the issues of how chords relate to different tonal centres and how the sense of key develops and changes as listeners hear sequences of chords (see Chapters 9 and 10).

Key profiles strongly reminiscent of those obtained in the study by Krumhansl & Kessler (1982b) can also be obtained using (a) computational models that take acoustic similarity (including residue pitches) and auditory sensory memory information into account (Huron & Parncutt, 1993; Leman, 2000), as well as (b) computational models involving self-organizing neural networks (with layers corresponding to pitch classes, chords, and key; Tillmann *et al.*, 2000).[9] That is, results of the probe tone ratings were largely due to acoustic similarity and representations of the context tones in the auditory sensory memory (this issue will become relevant when discussing different cognitive operations underlying music–syntactic processing in Chapter 9). Krumhansl & Cuddy (2010) argued that the establishment of a representation of the major–minor tonal hierarchy is based on 'general perceptual predispositions' (p. 62), but that acculturation shapes such establishment (one argument being that such establishment has a developmental trajectory during childhood; see also Lamont & Cross, 1994).[10] Therefore, the ratings obtained by Krumhansl & Kessler (1982b) were presumably due to both (a) acoustic similarity and representations of tones in auditory sensory memory, as well as (b) cognitive, experience-shaped representations of major–minor tonal key-, chord-, and tone-relationships. That is, cultural experience is not a *necessary condition* for the establishment of representations of major–minor tonal hierarchies (including a sense of key, representations of distances between keys, and of relations between chords and keys), but a *sufficient condition* for the modulation of such representations (at least as long as a musical system is learned).

3.3   The developing and changing sense of key

Each chord has a tendency to be interpreted as the tonic of a key, the most simple solution from a functional point of view (as compared to the interpretation of a chord with regard to another chord functioning as tonic). Schenker (1956) stated that 'not only at the beginning of a composition but also in the midst of it, each (chord) manifests an irresistible urge to attain the value of the tonic for itself' (p. 256, also in Krumhansl & Kessler, 1982b). This *tonicization* is enhanced when a chord is preceded by its own dominant (especially by a dominant seventh chord), and during 'sections of greatest key ambiguity and instability' (Krumhansl & Kessler, 1982b).

---

[9]   For a review see Krumhansl & Cuddy (2010).

[10]   Also note that, so far, the tonal hierarchy has not been convincingly recovered using auditory modelling.

Krumhansl & Kessler (1982b) also investigated how a sense of key develops or changes during listening to a sequence of chords. Some sequences consisted of in-key chords only, and were employed to investigate a developing sense of key (a tonal key was established and supported with progressing in-key chords). During other chord sequences, the initial key changed to another key. Such a change from one key to another is called *modulation* (for another behavioural study employing modulating sequences as stimuli see Berent & Perfetti, 1993).[11]

It is important to note that during in-key chord sequences, participants perceived tones (which are elements of chords) more and more with regard to their stability within the established key (and less with regard to the last heard chord only). Moreover, the sense of key tended to be stronger than just the sense dependent on the relative functional proximity of chords. That is, the sense of key was stronger towards the end of a cadence, and that stronger sense of key included a perception of chords according to their stability within the established key. This increase in the strength of the sense of key reflects that a musical context build-up (music theoretically inherent in the cadences) was psychologically represented in listeners. It was also found that during modulating sequences, listeners shifted their key sense from the region of the first key toward the region of the second key (for further data and modelling approaches see also Janata, 2007; Toiviainen & Krumhansl, 2003). Notably, some residual effect of the first key was maintained throughout the entire sequence.[12]

3.4   The representation of chord-functions

The data obtained by Krumhansl & Kessler (1982b) were also used to investigate the perceived relations of chords as functions with regard to different abstract tonal centres. Chord-profiles were calculated for major, minor, diminished, and dominant seventh chords.[13] The profiles of major and minor chords correlated highly with a key profile when both chord profile and key profile were adjusted to the same reference tone, that is, when a chord was the tonic of a key (e.g., the profile of a C major chord correlated highly with the profile of the C major key). The correlations between a chord and all keys were analysed using an MDS-method, so that the psychological distance between a chord and all 24 keys could be

---

[11] For an approach using self-organizing maps (SOMs) to model the developing and changing sense of key see Toiviainen & Krumhansl (2003); for a review of different approaches modelling tonal key space see Janata (2007).

[12] For an fMRI study investigating the tracking of tonal space see Janata *et al.* (2002a). For an fMRI study using tonal modulations (changes in key) as stimuli see Koelsch *et al.* (2002a). For a study investigating electrophysiological correlates of the processing of tonal modulations see Koelsch *et al.* (2003b).

[13] Fit-ratings for probe tones were shifted to a reference chord, and then averaged across each of the following conditions: Probe tone preceded by a major chord, by a minor chord, by a diminished chord, and by a dominant-seventh chord.

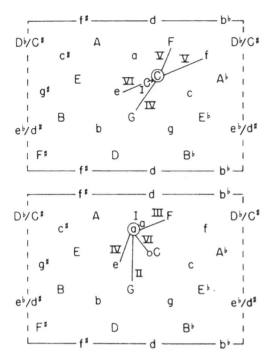

Figure 3.6 Placement of a C major, and an *a* minor chord in the toroidal configuration (from Krumhansl & Kessler, 1982b). Interestingly, the C major chord is not only located near the C major key, but also drawn slightly toward the F major key (and *f* minor key, respectively), in which it plays the functional role of the dominant. The greater distance between the C major chord and the G major key reflects that the harmonic function of a subdominant (of G major) is slightly weaker than that of a dominant. Analogously, the position of the *a* minor chord in the key distance map reflects its role as mediant (in F major), as submediant (in C major), as supertonic (in G major), and as subdominant (in *e* minor).

determined (see Krumhansl & Kessler, (1982b) p. 350 for further description).[14] Results are illustrated for the C major and *a* minor chords in Figure 3.6. The results show that the psychologically perceived position of a chord within the tonal system seems to be a compromise of the chords' function with regard to different keys. Interestingly, there seems to be a strong tendency for the *a* minor chord to the A major key, probably reflecting a general tendency for 'every passage in minor to be resolved in major' (Schenker, 1956), or–even more basic–the general tendency to move from dissonance to consonance.

3.5   Hierarchy of harmonic stability

Bharucha & Krumhansl (1983) investigated the mental representation of the organization of harmonic information (see Krumhansl *et al.*, 1982a & b, for similar studies). They summarized their results by stating six principles that describe hierarchies of harmonic stability which govern the perceived relationships between chords. Some of these principles were found to be dependent, and some to be independent of a tonal context. Similarity ratings were obtained from individuals

---

[14]   This procedure can be applied to $4 \times 12 = 48$ chords: Four chord types were investigated, and each chord can be adjusted to each chromatic scale tone.

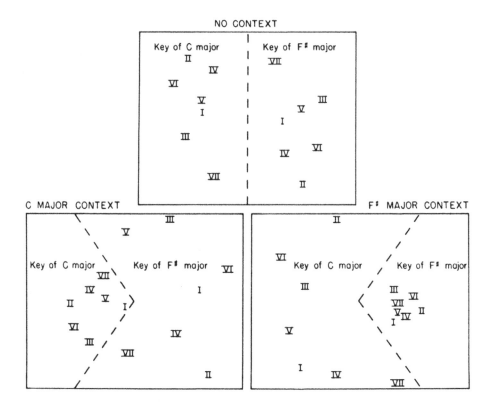

Figure 3.7 Two-dimensional scaling solutions. Top: No context condition, left: *C* major context condition, right: *F sharp* major context condition (reprinted with permission from Bharucha & Krumhansl, 1983).

with and without formal musical training. Participants were presented with test chords that either (a) followed a IV–V–I cadence in *C* major ('*C* major context'), or (b) a IV–V–I cadence in *F sharp* major ('*F sharp* major context'), or (c) were presented without any context. Then, they had to rate how well the second chord followed the first chord with regard to the previously heard cadence.[15]

Applying a multidimensional scaling method to the data, it was found that chords from the same key were judged to be more closely related to each other than chords that were not from the same key (independent of whether the test chords were preceded by a tonal context or not; see Figure 3.7). This context-independent effect was taken to reflect a principle that was termed *Key Membership*. This principle states that 'chords from the same key are perceived as more closely related than chords that are not from the same key' (Bharucha & Krumhansl, 1983 p.70).

---

[15]   This judgment was taken by Bharucha & Krumhansl (1983) as a similarity rating.

To formally state the harmonic principles, Bharucha & Krumhansl (1983) denoted the psychological distance between two directly succeeding chords C_1 and C_2 by $d(C_1, C_2)$, the set of the seven in-key chords by K, and the membership of a chord C in K by $C \in K$ ($C \notin K$ means that the chord C is not from K). The set of the three chords tonic, dominant, and subdominant was referred to as the *harmonic core*. The harmonic core was denoted by S. d_K refers to the psychological distance between chords when a key K was instantiated by a preceding context.[16] Formally, the principle of Key Membership was stated as:

$$d(C_1, C_2) < d(C_3, C_4), \text{ where } C_1, C_2 \in K,$$
$$\text{and there does not exist any } K' \text{ such that } C_3, C_4 \in K'.$$

It was also found that the I, V, and IV chords occupied central positions within each key. That is, the chords of the harmonic core clustered together, surrounded by the other four chords from the same key. This pattern was also found to be independent of harmonic context. Thus, independent of a tonal context, chords in the harmonic core were perceived as more closely related to each other than were the other chords from the key but not in the core (note, however, that chords in the core were major chords, and chords not in the core minor chords, or a diminished chord).[17] This effect was taken to reflect a principle termed *Intrakey Distance*, formally written as:

$$d(C_1, C_2) < d(C_3, C_4), \text{ where } C_1, C_2 \in S, C_3, C_4 \notin S, \text{ and}$$
$$C_1, C_2, C_3, C_4 \in K, C_3 \neq C_4.$$

Moreover, in the no-context-condition, chords were separated into two sets (corresponding to the two keys from which they were drawn). In the C major context condition, the chords from C major were perceived as more closely related, and the chords belonging to F *sharp* major were perceived as more dispersed than in the no context condition. By contrast, in the F *sharp* major context the chords from F *sharp* major clustered together more closely, and the C major chords were more separated from each other. Thus, two chords were perceived as most closely related when they were in the context key, moderately related when no tonal context was provided, and more distantly related when neither chord was in the context key. This effect was described by the principle of *Contextual Distance*

---

[16]   In a later formulation of the principles, Krumhansl (1990) generalized 'core membership' to 'stability,' which also applied to tones (Table 6.1 in Bharucha & Krumhansl, 1983).

[17]   Therefore, pitches of chords in the harmonic core fitted better in a common overtone series than pitches of chords outside the harmonic core, leading to the perception of greater similarity of chords within the harmonic core than of chords outside the harmonic core.

(also investigated in a study by Krumhansl *et al.*, 1982), formally stated as:

$$d_K(C_1, C_2) < d(C_1, C_2) < d_{K'}(C_1, C_2), \text{ where } C_1, C_2 \in K,$$
$$C_1, C_2 \notin K'.$$

Interestingly, two chords from the same key were perceived as more closely related when the first chord was not in the harmonic core, and the second chord was in the harmonic core, than when they were heard in the reverse temporal order.[18] This asymmetry-effect was termed by Bharucha & Krumhansl (1983) as the principle of *Intrakey Asymmetry*,[19] formally stated as:

$$d(C_1, C_2) < d(C_2, C_1), \text{ where } C_1 \notin S, C_2 \in S, \text{ and } C_1, C_2 \in K.$$

When both test chords were from different keys, the highest ratings were given when there was no context. Importantly, higher ratings were given to pairs ending on a chord that belonged to the preceding tonal context compared to pairs ending on a chord that did not belong to the preceding context. That is, a pair of chords was judged as more closely related when the first chord was out of the context key (and the second chord was in the context key) than when they were heard in the reverse temporal order. This asymmetry-effect turned out to be context-dependent, since the tonal context tended to increase the perceived distance between chords belonging to different keys (the ratings virtually did not differ when there was no context). The principle describing this effect was termed as the principle of *Contextual Asymmetry* (see also Krumhansl *et al.*, 1982a), formally written as:

$$d_K(C_1, C_2) < d_K(C_2, C_1), \text{ where } C_1 \notin K, \text{ and } C_2 \in K.$$

In another experiment (Bharucha & Krumhansl, 1983), participants recognized a particular chord more easily when it was out of the context key than when there was no tonal context or when it was in the context key.[20] This effect was described

---

[18]   As mentioned above, however, it is important to remember that chords of the core were always major chords, and chords not in the core always either minor chords, because the tendency for preference of a minor–major progression over a major–minor progression, and thus for a progression from less consonant to more consonant, presumably contributed to that effect.

[19]   This asymmetry was reduced in magnitude when the chords did not belong to the context key.

[20]   Each trial consisted of two chord sequences. These two sequences were either identical, or differed in one single chord (that chord was either diatonic or nondiatonic). Moreover, sequences were either composed in a way that all chords were in-key (except possibly a target chord), and a musical context was built up towards the end of the sequences (tonal condition), or they consisted of randomly ordered chords from different keys. Participants (musically trained) had to make same/different judgments about the chord sequences presented in each trial. Recognition errors were taken to reflect the perceived relatedness between the target chords.

by the sixth principle: *Contextual Identity* (Bharucha & Krumhansl, 1983; see also Krumhansl *et al.*, 1982a), formally written as:

$$d_K(C_1, C_1) < d(C_1, C_1), \text{ and } d_K(C_1, C_1) < d_{K'}(C_1, C_1),$$
$$\text{where } C_1 \in K, \text{ and } C_1 \notin K'.$$

The six principles proposed by Bharucha & Krumhansl (1983) as governing the perceived distances between chords were taken to indicate that the internal representation of harmonic relationships was highly regular and structured. Some of the principles hold in the absence of an established tonal context, whereas other principles describe relationships that are altered by a tonal context. Similarly to the studies by Krumhansl (1979) and Krumhansl & Kessler (1982b), the study by Bharucha & Krumhansl (1983) demonstrated that the perception of chords, like the perception of single tones, is 'subject to influences of the tonal context in which they are embedded' (ibid.).

It is important to note that a major finding of the study by Bharucha and Krumhansl (1983) was that when a tonal context was introduced, the representations of in-key chords became more stable, whereas those of chords containing out-of-key notes became less stable (a feature that is relevant for chord sequence paradigms used to investigate the processing of musical syntax; see below and Chapter 9). Generally, in Western tonal music 'the more stable tones appear more frequently, in prominent positions, and with rhythmic stress' (Bharucha & Krumhansl, 1983). According to Bharucha and Krumhansl (1983), harmonically stable chords function as cognitive reference points for the system as a whole. Moreover, the perception of structure of music (which is a prerequisite for building a representation of a musical context and thereby a prerequisite for the understanding of music) relies strongly on the perceiver's ability to organize the individual musical events in terms of the hierarchies of harmonic stability reflected by the six principles.

3.6   Musical expectancies

The perception of musical relations within a hierarchy of tonal stability enables a listener to perceive and appreciate tension and release. For example: Moving away from a tonal centre to unstable chords (or keys) is perceived as tensioning, returning to the stable tonal centre as relaxing. Likewise, dissonance as well as a tone (or chord) that is harmonically unrelated to a musical context can produce or enhance tension. The interplay between expectancies, as they unfold over time, and the varying degrees to which they are fulfilled or violated are broadly considered as one fundamental aspect of the appreciation of tonal music (e.g., Meyer, 1956; Schönberg, 1969; Bharucha, 1984; Jones, 1981, 1982;

Bharucha & Stoeckig, 1986, 1987; Huron, 2006).[21] In his *Tonal Pitch Space* theory, Fred Lerdahl (2001b) suggests algorithms to quantify sequential and harmonic tension (in particular rise and fall in tension in the course of a musical sequence). This model has also been used in studies comparing the modelled tension–resolution patterns with the tension–resolution patterns perceived by listeners (Bigand *et al.*, 1996; Lerdahl & Krumhansl, 2007).

The generation of musical expectancies while listening to major–minor tonal music considerably relies on the representation of a hierarchy of harmonic stability in the brain of a listener. In a study from Bharucha & Stoeckig (1986) it was shown that a harmonic context primes the processing of chords that are related to this context (relative to chords that are unrelated to this context).[22] Subjects (whether or not musically trained) responded faster, and more accurately, to harmonically related chords compared to unrelated chords. This result was taken to reflect that a chord generates expectancies for related chords to follow. Bharucha & Stoeckig (1987) also found evidence for the hypothesis that harmonic expectancies for chords are generated, at least in part, on a cognitive level (rather than only on a sensory level), by activation spreading through a network that represents harmonic relationships (for similar results obtained with the presentation of chord sequences see Bigand & Pineau, 1997; Bigand *et al.*, 1999). The brain's ability to expect musical events to a higher or lower degree is one reflection of a psychological reality of musical syntax.

3.7   Chord sequence paradigms

Numerous behavioural studies have investigated the processing of chord functions using chord sequence paradigms (reviewed in Tillmann, 2009; Bigand & Poulin-Charronnat, 2006). In these experiments, a prime context (e.g., a chord sequence) is typically followed by a target chord. The target is either a tonic chord, or a subdominant, with the idea that the tonic is perceived as harmonically more stable than the subdominant.

For example, Bigand *et al.* (2003) used eight-chord piano sequences, with the last chord (target) either being the tonic chord, or a (less 'stable') subdominant chord (Figure 3.8). Note that, when looking at the last two chords in isolation, the succession of these two chords forms a local authentic cadence (V–I). Therefore, it is the preceding harmonic context that modifies the harmonic functions of the final two chords, making them a V–I, or a I–IV ending (see 'final tonic chord' and

---

[21]   In tonal music, the expectancy of an acoustic event is not just determined by harmonic features, but by several syntactic features (melody, metre, rhythm, harmony, intensity, instrumentation, and texture). This will be elaborated in Chapter 9, as will be the importance of differentiating between acoustical and musical expectancies.

[22]   In a priming paradigm, subjects had to make a speeded true/false decision about a chord following a prime chord to which it was harmonically either closely or distantly related.

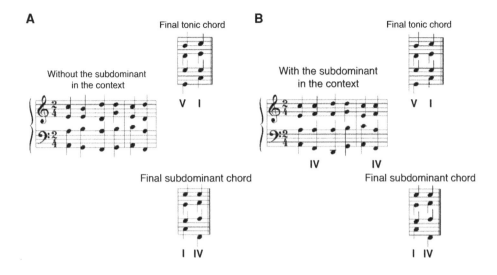

Figure 3.8 Illustration of a chord sequence paradigm used, e.g., in the study from Bigand *et al.* (2003). This paradigm employed two types of priming contexts, either without the subdominant (left of A) or with the subdominant (left of B) occurring in the context. The priming context is immediately followed by two chords, that represent either a dominant – tonic ending ('V–I,' upper panel), or a tonic – subdominant ending ('I–IV,' lower panel). Reprinted with permission from Poulin-Charronnat *et al.* (2006).

'final subdominant chord' in Figure 3.8). For example, when the prime context is in *C* major, then the endings shown in the upper panel of Figure 3.8 represent V–I endings, and the endings shown in the lower panel represent I–IV endings. When the prime context is in *F* major, then the endings shown in the *lower* panel of Figure 3.8 represent V–I endings, and the endings shown in the *upper* panel represent I–IV endings. This design rules out the possibility that any behavioural effects between conditions are simply due to an acoustical difference between the endings shown in the upper panel compared to those shown in the lower panel of Figure 3.8.[23]

In addition, Figure 3.8 also illustrates one possibility to disentangle sensory and cognitive processes evoked by harmonic priming. In musical priming, knowledge-driven processes (referred to as *cognitive priming*) are thought to prevail over sensory-driven processes (i.e., *sensory priming*). The former is a result of the activation of listeners' knowledge of the regularities of major–minor tonal syntax, the latter of the acoustical overlap, or similarity, between prime and target

[23] However, also note that the V–I ending of such sequences is not the most regular ending (because the most regular ending would require that the tonic chord occurs on a heavy beat). On the other hand, the final subdominant is a *regular* chord function at that cadential position, though not a regular chord function to terminate the sequence. This matters in terms of the ERP effects evoked by such chords, which will be discussed in Chapter 9.

(see also Chapter 9). In the sequences shown in Figure 3.8, the occurrence of the subdominant chord in the previous context was manipulated. In one type of priming context, neither the tonic nor the subdominant chord occurred in the previous context, whereas in another type of priming context, a subdominant occurred one or two times (whereas the stable tonic never occurred in the context). Such a manipulation allowed the investigators to test whether behavioural differences between the responses to the I–IV and the V–I endings are simply due to the fact that the target chord already occurred in the previous context (i.e., simply due to sensory priming), or due to a syntactic irregularity (i.e., due to cognitive priming).

To investigate behavioural effects due to the processing of the target chords, paradigms such as the one shown in Figure 3.8 typically employ an additional condition in which the last chord is acoustically quite different than all other chords (e.g., being more dissonant, or louder, or having a different timbre; for reviews see Tillmann, 2009; Bigand & Poulin-Charronnat, 2006). Participants are then asked to respond as fast as possible to the standard (e.g., consonant) chords by pressing one button, and to the deviant (e.g., dissonant) chords by pressing another button. One reason for doing so (instead of simply asking participants

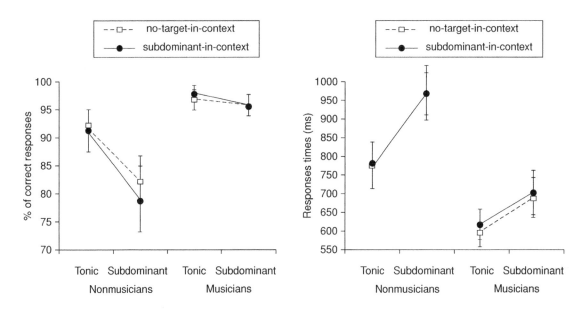

Figure 3.9 Percentage of correct responses (left) and response times of correct responses (right) for tonic and subdominant targets observed in the experiment from Bigand *et al.* (2003), using the paradigm shown in Figure 3.8 (presentation time of each context chord was 660 ms, and each final chord was presented for 2000 ms). Participants responded faster (and made fewer errors) to the more stable tonic chords than to the less stable subdominants. Note that effects were observed independently of whether or not the subdominant was presented in the priming context. Error bars represent standard errors. Reprinted with permission from Bigand *et al.* (2003).

to press one button for the tonic chords and another one for the subdominants) is that data can be obtained even from non-musicians (who have no explicit, conceptual knowledge about 'tonic' and 'subdominant').

Results usually show that participants respond faster (and make fewer errors) to the more stable tonic chords compared to the less stable subdominants (see Figure 3.9). Behavioural studies investigating differences between musicians and nonmusicians with harmonic priming paradigms have found only few (Bigand *et al.*, 2003) or no differences (Poulin-Charronnat *et al.*, 2005; Bigand *et al.*, 2001) between both groups of participants. These studies corroborate the hypothesis that knowledge about structural regularities of the Western tonal system can be acquired implicitly through passive learning processes, such as exposure to tonal music in everyday life (for a review; see Tillmann *et al.*, 2000).

4

From Electric Brain Activity to ERPs and ERFs

The human cerebral cortex is involved in perception, cognition, and emotion, including attention, action, volition, awareness, reasoning, decision making, learning and memory. Due to its modulatory influence on subcortical structures, the cerebral cortex is potentially involved in all psychological and physiological phenomena originating in the brain. The cortex contains different types of nerve cells that can be divided into two major classes: Pyramidal and nonpyramidal cells (based, e.g., on morphology and neurotransmitter content). The pyramidal cells represent about 80 to 85% of cortical neurons. They are thought to use mainly glutamate as excitatory, and γ-aminobutyric acid (GABA) as inhibitory neurotransmitter. Cortical pyramidal cells are generally oriented in parallel to each other (perpendicular to the surface of the cortex).

When a neuron is activated by an excitatory neuron, the neuron produces an excitatory postsynaptic potential (EPSP). The presynaptic activation evokes an inflow of cations and a release of anions at (and in the region around) the postsynaptic membrane; that is, the presynaptic activity leads to a local depolarization of the membrane potential. Therefore, this region is now extracellularly more negative[1] than other extracellular regions of this neuron, and intracellularly less negative[2] than other intracellular regions of the neuron. Because of these potential gradients, cations move extracellularly in the direction of this region, and intracellularly away

---

[1] Due to inflow of cations into, and release of anions by, the neuron.
[2] Due to fewer anions, and more cations.

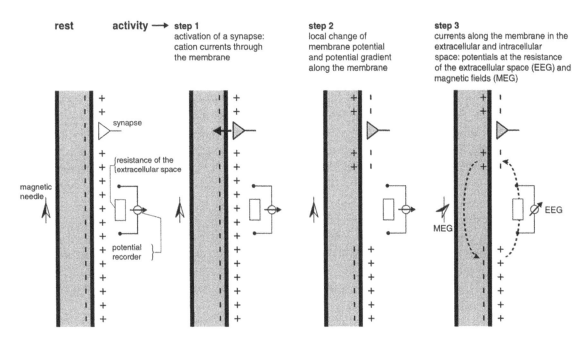

Figure 4.1 Mechanisms underlying the generation of EEG potentials (and of the corresponding magnetic fields) in the extracellular space of the central nervous system. See main text for details. Reprinted with permission from Niedermeyer & Da Silva (2005).

from this region (towards more negative intracellular regions), producing electric currents along the membrane in the extra- and intracellular space (Figure 4.1; for details see Niedermeyer & Da Silva, 2005). In the case of inhibitory postsynaptic potentials (IPSPs), these processes take place with the opposite polarity of current flows.[3]

Importantly, both EPSPs and IPSPs develop over a time period of some milliseconds (EPSPs have a decay of about 10 to 30 ms, IPSPs of about 70 to 150 ms). Therefore, when neurons of a population of pyramidal neurons[4] produce EPSPs (or IPSPs), these potentials can effectively summate (given similar geometric orientation of the neurons). The resulting electric potentials can be measured extracellularly, even from some centimetres distance.[5] The recording of electric potential changes originating from brain activity using electrodes situated

---

[3] The summation of EPSPs (i.e., decrease, or depolarization, of the membrane potential) can lead to the production of one or more action potentials, whereas summation of IPSPs (i.e., increase of the membrane potential) inhibits the production of action potentials.

[4] Such a population of neurons is sometimes also referred to as a *cell assembly*.

[5] Given that the population of neurons producing a potential is large enough, that is, presumably at least several tens of thousands of neurons.

on the scalp is called the *electro-encephalogram* (EEG).[6] The measurement of such potentials with electrodes placed on the pia mater of the brain is referred to as electrocorticography (ECoG). Another method of intracranial recording to measure such potential differences is the use of small electrodes inserted directly into cortical tissue, this method is referred to as *depth EEG*.[7]

If EPSPs are generated near the apical end of the neuron (i.e., more at the surface of the cortex), the extracellular current flows away from the soma towards the apical end of the neuron, thus a negative potential is generated at the surface electrode. IPSPs, on the other hand, produce a positive potential at the scalp surface when generated near the apical end of the neuron. If EPSPs are generated near the basal end of the neuron (i.e., deeper in the cortex), the extracellular current flows away from the soma towards the basal end of the neuron, thus a positive potential is generated at the surface electrode. IPSPs, on the other hand, produce a negative potential at the scalp surface when generated near the apical end of the neuron. Thus, a current flow (and the corresponding surface potential) generated by superficial EPSPs is similar to the current flow generated by deep IPSPs (and vice versa; for details see Niedermeyer & Da Silva, 2005). However, whether EPSPs or IPSPs contribute equally to the ongoing EEG is not known. Note that the action potentials (the potentials with the largest amplitudes generated by neurons) are actually thought to contribute only little to surface EEG potentials, because the action potentials can (due to their short latency) hardly summate over time as effectively as the (slower) EPSPs and IPSPs.[8]

The transmission of sensory information from the peripheral sensory system through the sensory pathways is also capable of producing measurable electric potentials, though considerably smaller than those originating from (neo)cortical activity. For auditory stimuli, for example, *auditory brainstem responses* (ABRs) that originate from the activation of various nuclei in the brainstem (and are thus associated with the transmission of sensory auditory information) can also contribute to the EEG, and thus to event-related brain potentials (see also Section 5.2; for detailed descriptions of brainstem responses in the auditory modality see Rowe, 1981; Picton *et al.*, 1994; Näätänen, 1992).

However, a considerable amount of neural activity is not measurable with the EEG. Neural activity might for example be insufficiently synchronous. Moreover, in several brain structures (e.g., the thalamus or the amygdala) neurons, even when associated in an assembly, do not have similar geometric orientations. Hence, their

---

[6] The term EEG was first introduced by Hans Berger (during his time at the Friedrich Schiller University Jena) in 1929 for the electric measurement of human brain activity (for a historical account see La Vaque, 1999).

[7] The extracellular recording of action potentials in animals from a single cell is referred to as *single-unit recording*.

[8] Glia cells probably also contribute to local field potentials, for details see Niedermeyer & Da Silva (2005).

activity can hardly be recorded with distant electrodes (for details on open and closed fields see Niedermeyer & Da Silva, 2005; Nunez, 1981; Rugg & Coles, 1995).

4.1 Electro-encephalography (EEG)

To measure the EEG, at least two electrodes have to be used: An active electrode which is placed over a site of neuronal activity, and an indifferent electrode which is placed at some distance from this site and serves as reference electrode. Though reduced by the electrical resistance of brain tissue, meninges, skull, cerebrospinal fluid, and skin, many potentials originating from the brain are measurable even outside the head. In both clinical applications and scientific research, numerous active electrodes are usually situated over different parts of the scalp (and connected to a single reference electrode). The EEG is mainly applied as a non-invasive method (though intracranial recordings may be used in clinical applications). The active electrodes are often placed according to conventional schemes, e.g., the 10–20 system (see below). The frequencies of EEG that are the subject of most investigations vary between about 1 and 45 Hz, although EEG frequencies of up to several hundreds of Hz can be measured as well (such high frequencies, however, usually have relatively small amplitudes). The amplitudes of a scalp-recorded EEG normally vary between about 1–100 μV.[9]

A major advantage of EEG as compared with functional neuroimaging techniques such as functional magnetic resonance imaging (fMRI) or positron emission tomography (PET) is that EEG reflects brain activity with a high temporal resolution. With a sampling rate of 1000 Hz, for example, changes of brain potentials can be investigated in 1 ms time-intervals. In addition, EEG is by far not as expensive as MEG, fMRI, or PET, and EEG data are relatively easy to obtain.

4.1.1 The 10–20 system

The locations of electrodes for scalp-recorded EEGs are usually described with reference to the extended 10–20 system (Figure 4.2, Sharbrough, 1991).[10] In this system, electrode positions are specified with regard to their proximity to particular brain regions (F: Frontal, C: Central, P: Parietal, O: Occipital, T: Temporal), and to their locations in the lateral plane (odd numbers: Left, even numbers: Right, subscript z for midline). The principal electrode locations are defined with regard

---

[9] μV stands for microvolt, i.e., one millionth of one Volt.

[10] The extended 10–20 system is an extended (i.e., specifying more electrode locations), and slightly modified version of the 10–20 system introduced by Jasper (1958). The modifications include a more consistent labelling of the electrode positions. The positions T7/T8 of the extended 10–20 system were labelled as T3/T4 in the original 10–20 system, and the positions P7/P8 were labelled T5/T6 (see also black circles in Figure 4.2).

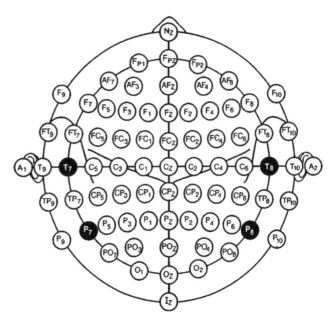

Figure 4.2 Electrode positions according to the extended version of the 10–20 system. A1 and A2 are electrodes placed on the earlobes. M1 and M2 (not depicted here) are situated behind the ears on the mastoidal bones, close to the T9 and T10 locations. Reprinted with permission from Sharbrough (1991).

to the relative distances along the anterior-posterior axis (from nasion over the vertex to inion), and the coronal axis (from the left post-auricular point over the vertex to the right post-auricular point). On each of these axes, the distance between two adjacent electrodes is 10% of the length of the axis. The other locations are defined in relation to these principal locations.

Figure 4.2 shows the electrode positions according to the extended 10–20 system. The outer circle is drawn at the level of the nasion and the inion. The inner circle represents the temporal line of electrodes. Tracing along the anterior-posterior line from nasion to inion, after 10% of the length of this line the electrode Fpz is located, after another 10% the AFz electrode is located, etc. It is also common to apply electrodes behind the ears over both left and right mastoidal bones.[11] These electrodes are referred to as *M1* (left) and *M2* (right), and they are often used as reference electrodes (A1 and A2 refers to electrodes placed on the left or right earlobe). Note that the locations specified by the extended 10–20 system do not imply that in every EEG study all electrode locations have to be used; the number of electrodes for each study may vary (depending, for example, on which brain-electric effect is investigated, whether source localization is intended, etc.). Moreover, electrodes can be placed wherever they are useful: Placement of nasopharyngal or sphenoidal electrodes, for example, enhances detection of activity in medial temporal lobe structures such as the hippocampus.

---

[11]   That is, on the mastoid part of the temporal bone, behind the left and right ear.

4.1.2   Referencing

For the analysis of neural generators located in the superior auditory cortex it is often useful to use an electrode placed on the nose tip as reference, or to use a *common average reference*.[12] One reason to use such references is that numerous cell assemblies of the auditory cortex are located in the supratemporal cortex, i.e., on the upper surface of the superior temporal gyrus. The apical dendrites of the pyramidal neurons of these assemblies are oriented perpendicular to the surface of the superior temporal plane. That is, they lie roughly on a horizontal plane passing through the centre of the head, and when active, they generate potentials that spread out with one polarity in superior direction (towards the top of the head), and with the opposite polarity in inferior direction. To observe maximal amplitudes of both negative and positive potentials over both the left and the right hemisphere, a reference electrode is ideally placed onto the anterior-posterior line within the plane that draws through the generator (i.e., just below the Sylvian fissure). By doing so, a clear polarity inversion can be observed, e.g., with negative potentials at Fz and positive potentials at the mastoid electrodes. Such a polarity inversion indicates that the neural generator of the potential is located on, or near the anterior-posterior line within the plane that roughly draws through the Sylvian fissure. Therefore, neural generators located in the auditory cortex perpendicular to this plane will produce such a polarity inversion. Note, importantly, that it is a logical fallacy to assume that a polarity inversion at electrodes located above and below the Sylvian fissure indicates that the underlying neural generator is located in the auditory cortex, or within the temporal lobe: Neural generators located in the upper bank of the Sylvian fissure (such as parietal or frontal opercular areas), or neural generators located anywhere on the anterior-posterior line within the plane that roughly draws through the Sylvian fissure, will produce such polarity inversions as well.

4.2   Obtaining event-related brain potentials (ERPs)

A *spontaneous EEG* (recorded without experimental stimulation) over three seconds, obtained from the vertex electrode, is shown in Figure 4.3. Though no

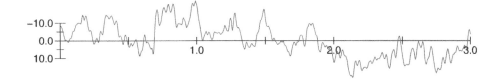

Figure 4.3   Spontaneous EEG with 3 seconds duration, recorded from Cz.

---

12   That is, to reference each EEG channel to the mean of all (scalp-recorded) EEG channels.

stimulation was present, distinct electric potential changes with amplitudes from around −20 to 20 μV are observable. This activity is called 'noise', and generated in part (a) by a vast number of operations permanently, or occasionally, performed by a living brain, (b) by potentials produced by the muscles of the face, jaws, neck etc., and (c) technical noise (such as line noise). Thus, the brain potentials recorded during the presentation of an experimental stimulus are composed of brain activity related to the processing of the stimulus, brain activity which is not related to the presentation of the stimulus, muscular activity, and technical noise. The brain potentials which correlate in time with the presentation of a stimulus are referred to as *signal*.

The signal of one single trial (i.e., from a single stimulus presentation) is usually not larger than the spontaneous EEG itself, and can thus not be distinguished from the noise (see top left of Figure 4.4). However, when the same stimulus (or a similar stimulus) is presented repeatedly, the signal in the brain potentials following each trial systematically correlates with the presentation of the stimuli,

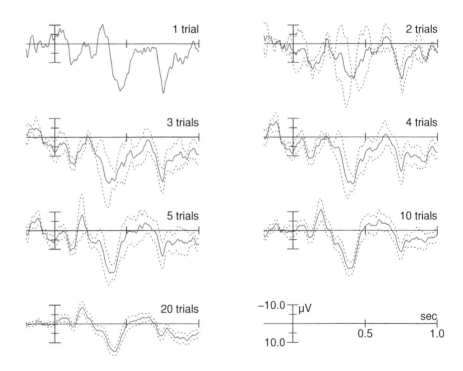

Figure 4.4 ERP of a single trial (top left), and ERPs of 2–20 averaged similar trials (solid lines). A 250 ms pre-stimulus baseline was used for averaging. The standard error of mean (SEM) is indicated by the dotted lines. The position of the y-axis indicates the onset of the stimulus. Note the SEM decrease with increasing amount of trials. Also note that potentials approximate the x-axis during the pre-stimulus ('baseline') interval with increasing amount of trials.

whereas the noise does not. Therefore, it is possible to extract the signal from the noise by averaging the potentials of all collected trials (thereby increasing the *signal-to-noise ratio*, SNR). That is, for each corresponding sampling point of all trials, arithmetic mean (and standard deviation etc.) of the electric potentials can be calculated. By doing so, the stimulus-correlated signal becomes apparent, whereas the uncorrelated noise is averaged out. A brain response that becomes visible by the averaging of trials (time-locked to the presentation of a stimulus) is called an *event-related potential* (ERP).

Figure 4.4 illustrates how the brain activity before the onset of the stimulus (indicated by the vertical line) approximates the zero-line with increasing number of averaged trials, reflecting that brain activity that did not correlate with the presentation of the stimuli was averaged out of the ERP-data. Moreover, the figure illustrates how the standard error of mean of the averaged event-related potential decreases with increasing number of averaged trials.

ERPs are usually characterized by their polarity (positive or negative), their latency (that is, by their amplitude maximum), by their scalp distribution, and by their functional significance.[13] In this regard, three points are important to consider. *Firstly*, the latency of an ERP component does not indicate that the cognitive process related to the generation of this potential has the same latency – it is well possible that the cognitive process happened earlier, and that the ERP reflects (a) that information is propagated to other regions, and/or (b) that inhibitory loops terminate previous or ongoing processes. *Secondly*, the polarity of a potential is dependent on the referencing of the electrodes. *Thirdly*, even though the electric potential of a component is often largest at a particular electrode site, an amplitude maximum at a certain electrode does not indicate that the neural generator of this component is located below this electrode; if a generator is not oriented radially to the head surface, then the maximum amplitude of the generated potential will emerge at a region over the scalp different from the one located directly above the generator. Even if one generator is oriented radially, it is possible that the ERP potential was not generated by a single neural source, but by a set of generators located in different brain areas. This is presumably the case for most (if not all) ERP-components with latencies longer than about 50 ms, and the spatial distance of sources generally increases with increasing latency (see also Näätänen & Picton, 1987; Scherg & Picton, 1991). The waveforms of all generators superimpose linearly at the scalp, that is, each electrode receives activity from each source (to a greater or lesser extent, depending on the distance to the source).[14] Moreover, the spatial distribution of the electric potentials over the scalp is determined by the orientations of the neural sources (e.g., whether they are oriented more tangentially or more radially with respect to the surface of the head).

---

[13] For the description of some ERP components see Chapter 5.
[14] For a mathematical description see, e.g., Scherg & Picton (1991).

4.3   Magnetoencephalography (MEG)

A major task of brain research is the investigation of the functional significance of circumscribed brain regions. This can be achieved by localizing the neural generators of processes underlying perception, cognition, emotion, etc. EEG measurements are strongly influenced by the volume conducting properties of the whole head, because the electric currents produced by the brain potentials have to pass the brain tissues and the skull to reach the skin surface. The low conductivity of the skull causes severe attenuation of the potential values, as well as blurring of their spatial distribution (e.g., Elbert, 1998; Knösche, 1997). Electric activity of the brain, however, also produces magnetic fields (similarly to a magnetic field produced by an electric current passing through a wire), and therefore the electric ERPs are thought to have magnetic equivalents in event-related magnetic fields (ERFs; for discussion see Näätänen, 1992; Hämäläinen *et al.*, 1993).[15,16] The magnetic fields are measured using superconducting quantum interference devices (SQUIDs), which contain coils of superconducting wire (the SQUID is cooled by liquid helium to sustain the superconducting state). In principle, magnetic fields generated in the head (given a particular orientation and a certain strength) produce via the SQUID a voltage proportional to the magnetic flux through the area of the detection coil (for a detailed description see, e.g., Vrba & Robinson, 2001).

A neural source is well represented by a short segment of current which is, as a model, usually referred to as an equivalent current dipole (ECD).[17] To determine the focus of electrical activity in the brain from MEG data (e.g., by calculating ECDs), the magnetic field has to be measured from a number of locations (using up to hundreds of SQUIDs). The data from all sensors can be interpolated, resulting in a topographical map representing the spatial distribution of the amplitudes and polarities of the magnetic field at a certain latency for a particular experimental condition. By doing so, it is possible to determine the loci of the extrema of the magnetic field strength. Given, for example, a dipolar magnetic field pattern, it is then possible to derive the localization of the electric

---

[15]   Whether EEG and MEG actually measure the same neuronal activity is still a matter of debate (for discussions see Näätänen, 1992, p. 89–90).

[16]   These neuromagnetic fields are extremely weak. For a rough illustration: The order of magnitude of an event-related magnetic field is around 100 femtotesla (fT), that is, 100×10^{-15} tesla. This is about one billionth of the earth's magnetic field (which has a strength of $70\,\mu$ tesla, i.e. 70×10^{-6} tesla.

[17]   The localization of a source of a magnetic field relies on the law of Biot and Savart (e.g., Bleaney & Bleaney, 1976), which specifies the contribution made by the current density at each point in space to the field at a given point of observation. According to the Biot-Savart law, the magnetic field generated from a current dipole is tangential to a circle centred on a straight-line extension of the current's direction. The field is thus parallel to a plane that is perpendicular to the dipole. The orientation of the field can be predicted by the right-hand rule.

activity.[18] One important advantage of MEG is that the magnetic field related to an electric potential is influenced mostly by the conductivity profile of the tissue surrounding the source (whereas EEG source localization is dependent on a correct volume conductor model between a source and all electrodes). Thus, an appropriate volume conductor model is more complicated to construct for the source localization with EEG compared to MEG data.

4.3.1   Forward solution and inverse problem

Given a volume conductor model and a sensor configuration, the magnetic field or electric potential that would arise from a certain source can be predicted with the Maxwell equations. This prediction is usually referred to as the forward solution. By contrast, when electric potentials or magnetic fields are measured outside the head, the measured information can be used to reconstruct the cerebral current sources underlying the (electric) brain activity. This reconstruction is usually referred to as the bioelectromagnetic inverse problem. Importantly, the solution to this problem is generally not unique, because different source configurations (e.g., each consisting of a different number of dipoles) can give rise to the same distribution of magnetic fields (or electric potential) measured over the head surface. A reasonable solution of the inverse problem thus depends on a reliable forward solution which, in turn, depends on both an appropriate volume conductor model and an assumption of a plausible source configuration (especially with regard to the number of sources; for detailed descriptions see e.g., Elbert, 1998; Vrba & Robinson, 2001; Mattout *et al.*, 2006). However, 'reasonable source configuration' is a matter of subjectivity, and a potential source of bias introduced by the individual performing the source analysis (but see also Mattout *et al.*, 2006).

4.3.2   Comparison between MEG and EEG

As mentioned above, MEG data can lead to more accurate localizations of the sources of brain activity compared to EEG data. In addition, depending on the EEG system used, MEG measurements can be performed more quickly, because usually only few electrodes (e.g., for measuring the electro-oculogram, EOG) have to be applied. However, there are also some advantages of EEG compared to MEG:

1. Depending on the MEG device, the orientation of the SQUIDs placed over the head may only register magnetic fields that pass a detection coil of the SQUID (three-dimensional vector devices better support measuring radial magnetic field components). Therefore, unless a three-dimensional vector

---

[18]   The generator is located between the two maxima, the distance between the two maxima determines the distance of the generator.

device is used, the topographical map representing the spatial distribution of the amplitudes and polarities of a magnetic field may not show focal maxima of the magnetic field if a dipole is oriented radially to the skull surface (even when a non-spherical, realistically shaped volume conductor model is used, such as a Boundary Element Model, or a Finite Element Model). This effect also contributes to the phenomenon that deeper sources are hardly measurable with MEG, because with increasing depth (i.e., when moving towards the centre of the volume conductor) neural sources become radially oriented with regard to the head surface (Hämäläinen *et al.*, 1993; Näätänen, 1992; Elbert, 1998).

2. In EEG data, the presence of multiple generators being active at the same time usually leads to a clear difference in the ERPs of different experimental conditions, whereas magnetic fields easily cancel out each other (so that no magnetic field is measurable outside the head). Therefore, using MEG, early cognitive processes can best be investigated (within approximately 250 ms after the onset of a stimulus). Physiologically, the number of neural activation foci increases with time after stimulus onset, which is why potentials with longer latencies (such as latencies >300 ms) are increasingly difficult to capture and localize with MEG.[19] They may, however, still be localized with an extremely high signal-to-noise ratio.

3. MEG is more expensive than EEG.

4. EEG data can easily be grand-averaged across subjects (due to the application of electrodes on the head surface, the electrode configuration is individually scaled; in addition, the electric potentials blur to a greater degree over the head compared to the magnetic fields). A grand-average often leads to significant qualitative differences between the ERPs elicited in different experimental conditions. By contrast, it is not appropriate to simply average ERFs across subjects: Subjects differ with regard to their head size, and their positioning within the sensor. The magnetic fields (which are more focal compared to the widely distributed electric potentials) thus hardly overlap across subjects.

5. Therefore, considerably more trials have to be collected per subject using MEG to obtain reliable results per subject. This may lead to repetition effects, or experimental sessions with inconvenient duration. To obtain a signal-to-noise ratio reasonable for dipole-fitting, the signal-to-noise ratio for a two-dipole solution should ideally exceed a value of at least 10 (and considerably more for solutions employing more than two dipoles). As a rule-of-thumb, for fitting of dipoles, the amount of trials obtained with a single subject with MEG should therefore equal the amount of trials obtained across several subjects in an EEG-experiment. For current source density analysis, the SNR does not have to be that high, because source resolutions can be averaged across subjects (and, thus, fewer trials per subject may be sufficient).

---

[19]   For example, whereas we could localize the ERAN (which usually has peak latencies <200 ms) with MEG using dipole models (Maess *et al.*, 2001), we were not able to localize the N5 (which usually has peak latencies >500 ms), presumably due to a number of generators contributing to the generation of the N5.

5

ERP Components

This chapter provides a brief overview of a few event-related brain potentials (ERPs) that can emerge during the processing of music or language. Technically, ERPs are the same as *evoked potentials* (EPs). However, for historic reasons, the term 'event-related potential' is rather used for later potentials (such potentials are related to more cognitive processes), whereas the term 'evoked potential' is more used with regard to early potentials elicited by perceptual processes.

5.1 Auditory P1, N1, P2

In general, a predominant classification divides ERP components into exogenous and endogenous components (Donchin *et al.*, 1978). With regard to exogenous components, the presence of an external stimulus is regarded as necessary, as well as sufficient, condition to elicit exogenous components. By contrast, the presence of an external stimulus is not a necessary, though sometimes a sufficient, condition for the elicitation of an endogenous component. Associated with the concept of exogenous and endogenous components is the notion that latencies and amplitudes of exogenous components are determined mainly by external stimulus characteristics, and that latencies and amplitudes of endogenous components are determined mainly by endogenous factors such as attention, intention, and volition. However, because endogenous factors considerably influence the processing

Brain and Music, First Edition. Stefan Koelsch.
© 2013 John Wiley & Sons, Ltd. Published 2013 by John Wiley & Sons, Ltd.

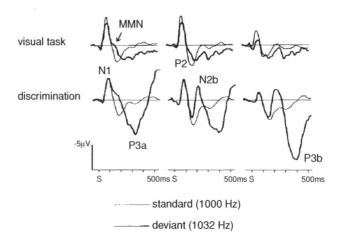

Figure 5.1 ERPs of auditory standard and deviant stimuli while performing a demanding visual task (top row), or while trying to discriminate deviant stimuli among standard stimuli (bottom row). ERPs were recorded from Fz (left column), Cz (middle), and Pz (right). Adapted with permission from Näätänen (1990).

of external stimulus characteristics (and thus latencies and amplitudes of the so-called exogenous components), and vice versa, the distinction between exogenous and endogenous ERP components is not a fortunate one.

The earliest ERP responses elicited by an auditory stimulus are the auditory brainstem responses, which occur within the first few milliseconds (ms) after the onset of a stimulus (for review see Näätänen, 1990, 1992). These responses appear to be automatic, that is, unaffected by attentional factors (Woldorff *et al.*, 1987). The brainstem responses are followed by the so-called middle-latency responses which are presumably generated in the primary auditory cortex.[1] Their latency is from about 9 to 50 ms after stimulus onset (Picton, 1980; Celesia & Puletti, 1971). The largest of these potentials is the P1, which is usually maximal at around 50 ms. Effects of attention on the ERP waveforms already occur in the midlatency range (Woldorff *et al.*, 1987), but the P1 is usually nevertheless considered as an exogenous component. The auditory P1 is followed by the N1 (a negative potential elicited around 100 ms after stimulus onset), and usually by a P2 (a positivity occurring around 200 ms; see also Figure 5.1).[2,3] These later components have considerably larger amplitudes compared to the early and middle-latency responses, and they are clearly modulated by attentional factors.

---

[1] These responses are also categorized as exogenous components.

[2] For reviews see Näätänen & Picton (1987), Näätänen (1990).

[3] The N1, and sometimes even the P2, are also classified as exogenous components. Note that N1-like responses can also be evoked in the absence of an auditory stimulus, e.g., (Janata, 2001).

The N1 can be elicited by both the onset and the ending of a stimulus, as well as by a change in tonal frequency or intensity of a continuous auditory stimulation (for review see Näätänen & Picton 1987; Näätänen, 1992). That is, the N1 is evoked by relatively abrupt changes in the level of energy impinging on the sensory receptors (Clynes, 1969), and therefore related in part to the detection of transients. A study by Pantev *et al.* (2001) showed that the N1 is modulated by experience; in that study, trumpet players showed a larger N1 amplitude in response to trumpet tones than to violin tones, whereas violinists showed a larger N1 amplitude in response to violin (compared to trumpet) tones.

The functional significance of the P2 has remained elusive (several studies reported P2 effects in association with emotion words, with the P2 being larger for emotion words compared to neutral words; for an overview see Kanske & Kotz, 2007). N1 and P2 are presumably generated by different sets of neural generators, although both N1 and P2 appear to be generated in the auditory cortex (Näätänen, 1992; Näätänen & Picton, 1987; Scherg, 1990).

5.2   Frequency-following response (FFR)

The frequency-following response (FFR) is an oscillatory potential (often with gradients of the oscillatory frequency) that resembles the periodicity of an acoustic signal. FFRs are assumed to reflect the synchronized activities of neuronal potentials generated by populations of neurons in the auditory brainstem (probably located in the inferior colliculus, although contributions from the lateral lemniscus are also discussed; Smith *et al.*, 1975; Hoormann *et al.*, 1992). It is assumed that the temporal encoding of frequencies at the level of the auditory nerve and the brainstem is represented by discharge patterns of single neurons, as well as by synchronous neuronal activity of populations of neurons. The pattern of neural discharge (which is related to the phase-locking of auditory nerve neurons) is modulated by the temporal structure of the eliciting sound. Therefore, if the periodicity of a sound changes, the discharge pattern of auditory brainstem neurons changes accordingly, which can be measured as a periodicity-following, or 'frequency-following' response, even with relatively remote scalp electrodes (such as Cz, referenced to an earlobe or mastoid electrode; for a detailed review and tutorial see Skoe & Kraus, 2010).

For example, in the temporal structure of oscillations of melodic speech sounds, the periodicity corresponds to the fundamental (F0) frequency. This periodicity is reflected neurally in interspike intervals that occur in synchrony with the period of the F0 (as well as some of its harmonics). The temporal structure of oscillations of the evoking stimulus is represented with remarkably high fidelity in the FFR (sometimes even so high that it is possible to play the FFR back through a loudspeaker, and recognize the original signal), and the fidelity of an FFR can be compared statistically, for example, between two groups (such as musicians and non-musicians; Hall, 1979; Galbraith *et al.*, 2000; Kraus & Nicol, 2005; Johnson *et al.*, 2005).

Wong *et al.* (2007) measured FFRs to three Mandarin tones in Mandarin speech that differed only in their F0 contours. Participants were amateur musicians and non-musicians, and results indicated that musicians had more accurate encoding of the pitch contour of the phonemes (as reflected in the FFRs) than nonmusicians. These findings corroborate the notion that the auditory brainstem is involved in the processing of pitch contours of speech information (vowels), and they show that the correlation between the FFRs and the properties of the acoustic information is modulated by musical training. Similar training effects on FFRs elicited by syllables with a dipping pitch contour have also been observed in adult native English speakers (non-musicians) after a training period of 14 days (with eight 30 minute sessions; Song *et al.*, 2008). The latter results show that effects of language learning are reflected in FFR responses (in adults).[4]

Note that in a typical experimental session in which an FFR is measured, a single stimulus (such as a phoneme) is repeated several thousand times. Therefore, it is highly likely that the auditory cortex is involved in shaping the FFRs (by virtue of top-down projections to the inferior colliculus). That is, the effects of musical (or language) training on the fidelity of FFR responses is most likely due to plastic changes in both the auditory cortex and the brainstem. How much the auditory cortex, and how much the brainstem contributes to the fidelity of the FFRs remains to be specified, but given the importance of the auditory cortex for the high resolution of spectral information (see Chapter 1), the contribution of the auditory cortex to the fidelity of the FFRs is likely to be considerable.

5.3   Mismatch negativity

The mismatch negativity (MMN) is elicited by deviant auditory stimuli in a repetitive auditory environment of discrete *standard stimuli* (for reviews see Näätänen *et al.*, 2005, 2007, 2010). Typically, experimental paradigms used to investigate the MMN consist of frequent so-called standard stimuli, and rarely occurring *deviant stimuli* (for illustration see Figure 5.2). Such experimental paradigms are therefore also referred to as *auditory oddball paradigms* (with the deviant stimuli being the 'oddballs'). The generation of the MMN is intimately linked to auditory sensory (or 'echoic') memory operations; the human brain permanently encodes physical features of the auditory environment into neural memory representations that are stored for a few seconds in the auditory sensory memory (the memory traces storing such information are thought to decay gradually; see also p. 97). The auditory cortex automatically detects changes in regularities inherent

---

[4]   For a study using FFRs to investigate effects of musical training on audiovisual integration of music as well as speech see Musacchia *et al.* (2007); for a study on the development of auditory brainstem responses to speech in 3- to 12-year-old children see Johnson *et al.* (2008).

Figure 5.2 Examples of stimuli evoking a physical (frequency) MMN (top), and an abstract feature MMN (bottom). The frequent stimuli are usually referred to as 'standard stimuli', and the infrequently occurring stimuli as 'deviants' (indicated by arrows). Note that the whole notes (occurring in every third bar) do not elicit MMNs (although their overall probability is lower than that of the standard half notes), because the three-bars-pattern (with four half notes followed by a whole note) is perceived as a pattern (as long as the tempo is not too slow, for details see also Sussman, 2007). Also note that the frequency deviants (i.e., the deviants shown in the upper panel) will not only elicit an MMN, but also a larger N1 due to refractoriness effects (see text for details). Adapted from Koelsch *et al.* (2001).

to the acoustic input on the basis of information buffered in the auditory sensory memory, and such detection is thought to be reflected electrically in the MMN. Therefore, the MMN 'appears to provide a physiological measure, although an indirect one, of the actual sensory information processed in the brain' (Näätänen, 1992; see also Figure 5.1).

The latency of the MMN is usually around 90 to 150 ms, thus overlapping in time with the N1 and the P2 waves (which are elicited by both standard and deviant stimuli). The MMN has a fronto-central, often right-hemispheric preponderant scalp distribution, and inverts polarity at mastoidal sites when a nose-reference or common average reference is used. Note that the frequency deviants shown in the upper panel of Figure 5.2 do not only elicit an MMN response, but also a (slightly) larger N1 response than the standard stimuli, due to refractoriness effects of neurons stimulated by the standard tones: When a standard tone is frequently presented, the neurons tuned to the frequency of the standard stimulus become refractory, whereas when a deviant is presented, 'fresh' neurons (with a lower state of refractoriness) tuned to the frequency of the deviant are elicited, resulting in a larger N1 elicited by deviants than by standards. Thomas Jacobsen and Erich Schröger (Jacobsen *et al.*, 2003) have developed experimental paradigms suited to investigate 'true' MMN responses, i.e., MMN responses without N1 contributions due to refractoriness effects (for a review see Schröger, 2007).

Initially, the MMN was investigated with stimuli containing physical changes, such as changes in frequency (e.g., Sams *et al.*, 1985), intensity (e.g., Näätänen *et al.*, 1987), spatial location (Paavilainen *et al.*, 1989), stimulus duration (Näätänen *et al.*, 1989), phonetic features (Näätänen *et al.*, 1997), and timbre (Tervaniemi *et al.*, 1997b). In the early 1990s, a study from Saarinen *et al.*

(1992) showed that a brain response reminiscent of the MMN can also be elicited by changes of abstract auditory features. In that study, standard stimuli were tone pairs with frequency levels that varied across a wide range, but were always rising in pitch, whereas deviants were tone pairs falling in pitch (see bottom of Figure 5.2). By introducing the concept of an 'abstract feature MMN' (afMMN), Saarinen and colleagues implicitly changed the previous concept of the MMN as a response to a physical deviance within a repetitive auditory environment (henceforth referred to as phMMN) to a concept of the MMN as a negative event-related potential (ERP) response to mismatches in general, i.e., to mismatches that do not necessarily have to be physical in nature (for other studies reporting abstract feature MMNs see, e.g., Paavilainen et al., 1998, 2001, 2003, 2007; Korzyukov et al., 2003; Schröger et al., 2007).

It is important to understand that the generation of both phMMN and afMMN is based on an on-line establishment of regularities – that is, based on representations of regularities that are extracted on-line from the acoustic environment. The on-line establishment of regularities must not be confused with long-term experience or long-term representations that might influence the generation of the MMN. For example, pitch information can be decoded with higher resolution by some musical experts (leading to a phMMN to frequency deviants that are not discriminable for most non-experts; Koelsch et al., 1999), or the detection of a phoneme is facilitated when that phoneme is a prototype of one's language (leading to a phMMN that has a larger amplitude in individuals with a long-term representation of a certain phoneme compared to individuals who do not have such a representation; Näätänen et al., 1997; Winkler et al., 1999; Ylinen et al., 2006). In this regard, long-term experience has clear effects on the processing of physical oddballs (as reflected in the MMN). However, in all of these studies (Koelsch et al., 1999; Näätänen et al., 1997; Winkler et al., 1999; Ylinen et al., 2006), the generation of the MMN was dependent on representations of regularities that were extracted on-line from the acoustic environment; for example, in the classical study by (Näätänen et al., 1997), the standard stimulus was the phoneme /e/, and one of the deviant stimuli was the phoneme /õ/, which is a prototype in Estonian (but not in Finnish). This deviant elicited a larger phMMN in Estonians than in Finnish subjects, reflecting that Estonians have a long-term representation of the phoneme /õ/ (and that Estonians were, thus, more sensitive to detect this phoneme). However, the regularity for this experimental condition ('/e/ is the standard and /õ/ is a deviant') was independent of the long-term representation of the phonemes, and this regularity was established on-line by the Estonian subjects during the experiment (and could have been changed easily into '/õ/ is the standard and /e/ is the deviant'). That is, the statistical probabilities that make up the regularities in such an experimental condition are learned within a few moments, and the representations of such regularities are not stored in a long-term memory format. This is an important difference compared to music- or language-syntactic regularities, which are stored in a long-term memory format (and often refer to long-distance dependencies involving hierarchical syntactic organization). This issue will be dealt with in detail in Chapter 9.

The phMMN most presumably receives its main contributions from neural sources located within and in the vicinity of the primary auditory cortex, with additional (but smaller) contributions from frontal cortical areas (Alho *et al.*, 1996; Alain *et al.*, 1998; Giard *et al.*, 1990; Opitz *et al.*, 2002; Liebenthal *et al.*, 2003; Molholm *et al.*, 2005; Rinne *et al.*, 2005; Schonwiesner *et al.*, 2007); for a review see Deouell (2007). Likewise, it appears that the main generators of the afMMN are located in the temporal lobe (Korzyukov *et al.*, 2003).

The processes underlying the generation of the MMN may be evoked even by unattended deviant stimuli (for detailed discussion see Sussman, 2007). Therefore, the MMN seems to reflect auditory feature encoding and mismatch-detection processes which operate largely automatically (or 'pre-attentively', i.e., independently of attention). However, note that several studies have shown that the MMN amplitude can be reduced in some cases by attentional factors (reviewed in Sussman, 2007), and it has been suggested that such modulations can possibly be attributed to effects of attention on the formation of representations for standard stimuli, rather than to the deviant detection process (Sussman, 2007). That is, the deviant detection mechanism itself is possibly largely unaffected by attentional modulations (Grimm & Schroger, 2005; Sussman *et al.*, 2004; Gomes *et al.*, 2000).

5.3.1   MMN in neonates

MMN-like responses can be recorded even in the foetus (Draganova *et al.*, 2005; Huotilainen *et al.*, 2005), and several studies showed MMN-like discriminative responses in newborns (although sometimes with positive polarity; Ruusuvirta *et al.*, 2003, 2004; Winkler *et al.*, 2003; Maurer *et al.*, 2003; Stefanics *et al.*, 2007) to both physical deviants (e.g., Alho *et al.*, 1990; Cheour *et al.*, 2002a,b; Kushnerenko *et al.*, 2007; Winkler *et al.*, 2003; Stefanics *et al.*, 2007; Háden *et al.*, 2009; Winkler *et al.*, 2009b) and abstract feature deviants (Ruusuvirta *et al.*, 2003, 2004; Carral *et al.*, 2005). Cheour *et al.* (2000) reported that, in some experiments, the amplitudes of such ERP responses are only slightly smaller in infants than the MMN usually reported in school-age children (but for differences see also, e.g., Maurer *et al.*, 2003; Kushnerenko *et al.*, 2007; Friederici, 2005). With regard to pitch perception, Háden *et al.* (2009) reported that already neonates can generalize pitch across different timbres. Moreover, reliable pitch perception has been shown in neonates for complex harmonic tones (Ceponiene *et al.*, 2002), speech sounds (Kujala *et al.*, 2004), environmental sounds (Sambeth *et al.*, 2006), and even noise (Kushnerenko *et al.*, 2007). Neonates are less sensitive than adults, however, for changes in the pitch height of sinusoidal tones (Novitski *et al.*, 2007).

5.3.2   MMN and music

As mentioned above, the generation of the MMN is based on auditory sensory memory operations – these operations are indispensable for music perception.

Therefore, practically all MMN studies are inherently related to music processing, and relevant for the understanding of music processing. Numerous MMN studies have contributed in particular to this issue, for example (a) by investigating different response properties of the auditory sensory memory to musical and speech stimuli (e.g., Tervaniemi *et al.*, 1999, 2000), (b) by using melodic and rhythmic patterns to investigate auditory Gestalt formation (e.g., Sussman, 2007; see p. 12 for explanation of the term 'Gestalt'), and/or (c) by studying effects of long- and short-term musical training on processes underlying auditory sensory memory operations (reviewed in Tervaniemi & Huotilainen, 2003; Tervaniemi, 2009). Especially the latter studies have contributed substantially to our understanding of neuroplasticity,[5] and thus to our understanding of the neural basis of learning. A detailed review of these studies goes beyond the scope of this chapter (for reviews see Tervaniemi & Huotilainen, 2003; Tervaniemi, 2009). Here, suffice it to say that MMN studies showed effects of long-term musical training on the pitch discrimination of chords (Koelsch *et al.*, 1999),[6] on temporal acuity (Rammsayer & Altenmüller, 2006), on the temporal window of integration (Rüsseler *et al.*, 2001), on sound localization changes (Tervaniemi *et al.*, 2006a), and on the detection of spatially peripheral sounds (Rüsseler *et al.*, 2001). Moreover, using MEG, an MMN study from Menning *et al.* (2000) showed effects of three-weeks auditory musical training on the pitch discrimination of tones.

Auditory oddball paradigms were also used to investigate processes of melodic and rhythmic grouping of tones occurring in tone patterns (such grouping is essential for auditory Gestalt formation; see also Sussman, 2007), as well as effects of musical long-term training on these processes. These studies showed effects of musical training on the processing of melodic patterns (Tervaniemi *et al.*, 1997a, 2001; Fujioka *et al.*, 2004; von Zuijen *et al.*, 2004; in these studies, patterns consisted of four or five tones), on the encoding of the number of elements in a tone pattern (von Zuijen *et al.*, 2005), and effects of musical training on the processing of patterns consisting of two voices (Fujioka *et al.*, 2005).

Finally, several MMN studies investigated differences between the processing of musical and speech information. These studies report larger right-hemispheric responses to chord deviants than to phoneme deviants (Tervaniemi *et al.*, 2000, 1999), and different neural generators of the MMN elicited by chords compared to the MMN generators of phonemes (Tervaniemi *et al.*, 1999). Similar results were also shown for the processing of complex tones compared to phonemes (Tervaniemi *et al.*, 2006, 2009; the latter study also reported effects of musical expertise on chord and phoneme processing).

---

[5]   That is, to changes in neuronal structure and function due to experience.
[6]   But note that superior attentive pitch discrimination accuracy is not always reflected in the MMN (Tervaniemi *et al.*, 2005; that study used complex tones as stimuli).

5.4   N2b and P300

During active oddball paradigms, that is (only) when participants detect occasionally presented target stimuli among a series of standard stimuli, the MMN is followed by an N2b (for an exception see Näätänen *et al.*, 1982).[7] The N2b is typically maximal over central scalp electrodes and does not invert polarity at mastoid electrodes when nose- or common average reference is used (Näätänen & Gaillard, 1983; Näätänen, 1990; see Figure 5.1). The N2b is usually followed by a positive-going ERP-component which is maximal around 300 ms and has a frontally predominant scalp distribution (for an exception see Knight, 1990). This component is called the P3a (Squires *et al.*, 1975; Ritter & Ruchkin, 1992; Näätänen, 1992).

A P3a might occur without a preceding N2b, in the case where a deviant stimulus is not task-relevant, but nevertheless attracts the attention of a subject. The amplitude of the P3a is related to physical stimulus deviation (rather than to the dimension of task-relevancy, i.e., whether or not a stimulus has to be detected; Näätänen, 1992). When the deviant sound is a complex environmental sound (usually referred to as 'novel' sound), additional cognitive processes might be involved in the generation of a P3a. Therefore, the frontally predominant ERP-deflection elicited by novel sounds in the time-window of the P3a is often referred to as 'Novelty P3' (e.g., Courchesne *et al.*, 1975; Cycowicz & Friedman, 1998; Spencer *et al.*, 1999; Opitz *et al.*, 1999a,b). The generation of both P3a and Novelty P3 is observable under both attend and ignore conditions.

The N2b-P3a-complex is usually followed by another positive ERP-component with a latency slightly longer than the latency of the P3a and with a parietal amplitude maximum, the P3b (often just 'P300', or 'P3'; see also Figure 5.1). The P3b reflects the decisional processes during the conscious recognition and detection of a target-stimulus (Donchin *et al.*, 1978).[8]

5.5   ERP-correlates of language processing

In the following, a few language-related ERPs will be described, especially with regard to the processing of semantic (content) and syntactic (structural) information. Note that in many of the studies mentioned in the following, language perception was investigated with visual word stimuli (that is, only a few of these experiments used auditory stimuli).

---

[7]   N2b may also be elicited without a preceding MMN.
[8]   Though the P3 does not necessarily reflect conscious processing (Donchin & Coles, 1998). For an interpretation of the P3 as an index of processes underlying context updating see Donchin & Coles (1988).

5.5.1   Semantic processes: N400

The N400 ERP is an electrophysiological index of the processing of meaning information, particularly conceptual/semantic processing or lexical access, and/or post-lexical semantic integration (Friederici & Wartenburger, 2010; Lau *et al.*, 2008). The N400 (or just 'N4') has negative polarity, slight right-hemispheric preponderance, and is generally maximal over centro-parietal electrode sites when elicited by visual, or over central to centro-frontal sites when elicited by auditory stimuli. It usually emerges at about 250 ms after the onset of word stimulation and reaches its maximal amplitude at around 400 ms (note that the term N400 is used to refer to the functional significance of this ERP component, rather than to its peak latency, which might differ depending on the experimental paradigm). The N400 elicited by words is highly sensitive to manipulations of semantic relations, being attenuated for words that are preceded by a semantically congruous context, compared to when preceded by a semantically incongruous context (Kellenbach *et al.*, 2000). That is, when a word is preceded by a semantic context, the amplitude of the N400 is inversely related to the degree of semantic fit between the word and its preceding semantic context (for an example see Figure 5.3, and upper panel of Figure 10.1).

In a classic study by Kutas & Hillyard (1980), subjects were required to read sentences consisting of about seven words, each word being presented individually at a rate of one word per second. Infrequently, the final word was either semantically incongruous (but syntactically correct), or semantically correct but larger in letter size. Semantically incongruous final words elicited a negative component that was maximal around 400 ms after stimulus onset and broadly distributed over the posterior part of both hemispheres (the N400; see also Figure 5.3). By

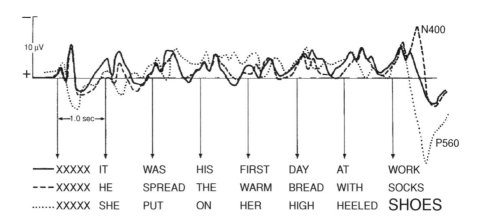

Figure 5.3   ERPs to sentences ending with a non-anomalous, semantically anomalous, or physically anomalous word. Semantically anomalous words elicit an N400. Reprinted with permission from Kutas & Hillyard (1980).

contrast, words presented in larger font ('physical deviants') elicited a P3b (maximal at around 560 ms). Neither of these components was evident when a sentence ended on a word that was both semantically and physically congruous with the preceding words.

The amplitude of the N400 is sensitive to the semantic expectancy built up by the preceding context for a given word. Kutas *et al.* (1984) showed that semantically anomalous words had a smaller N400 when they were related to the expected ending than when they were not. While the semantically congruous final word '*eat*' presented after the words *The pizza was too hot to . . .* showed no N400, the word '*drink*' elicited a smaller N400 than the semantically unrelated word '*cry*' (see also Kellenbach *et al.*, 2000).

Fischler *et al.* (1983) suggested that the N400 is sensitive to the associative strength between entries in the mental lexicon, rather than to the propositional content of a statement. Subjects were asked to verify a set of simple semantic propositions (e.g., *A robin is a not a car* or *A robin is not a bird*). The truth of a statement did not affect the N400, but the association between the two content words on the other hand did.[9] However, the N400 can be elicited by false statements that are easily interpretable: In another experiment by Fischler *et al.* (1985), subjects had to learn a set of statements such as *Matthew is a lawyer*. False statements such as *Matthew is a dentist* presented a day after the practice session elicited an N400. These findings were corroborated by the observation that the N400 can in fact be elicited by a false negation under circumstances in which the negation is normally used (and thus easily understandable; Nieuwland & Kuperberg, 2008).

The N400 seems to be a controlled, rather than an automatic, process. In a semantic priming paradigm, Chwilla *et al.* (1995) observed that different levels of processing (assessed by changing the task demands) affected the amplitude of the N400 priming effect (which was only present in a lexical decision task compared to a physical task). These results were taken to indicate that an N400 priming effect is only evoked when the task performance induces the semantic aspects of words to become part of an episodic trace of the stimulus event.[10]

The N400 can not only be elicited by semantically incongruous information (such as the final word of the sentence 'He spread the warm bread with socks'), but can also be elicited by semantically correct words that require more semantic processing resources than other (semantically correct) words: Van Petten & Kutas

---

[9]   Presumably because the word 'robin' automatically primed the word 'bird' more strongly than the word 'car'.

[10] The hypothesis of the N400 being a controlled process was supported by a study from Gunter & Friederici (1999). Similar to the study from Chwilla *et al.* (1995), an N400 was present under a grammatical judgment task, but distinctly attenuated under a physical task. An interesting finding of the study from Gunter & Friederici (1999) was that the N400 was (somewhat unexpectedly) elicited by *syntactic* violations (verb inflection and word category violation). Nevertheless, according to the authors, the N400 was still triggered by semantic expectancies (Gunter & Friederici, 1999).

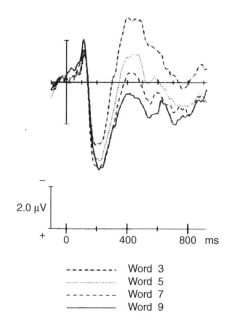

2.0 μV

Figure 5.4 Amplitude-decline of the N400 elicited by semantically correct open class words during sentence comprehension (from Van Petten & Kutas, 1990).

- - - - - - - -   Word 3
....   Word 5
- - - - - - -   Word 7
─────────   Word 9

(1990) showed that the amplitude of the N400 elicited by open class words (i.e., nouns, verbs, etc.) is inversely correlated with the word's ordinal position in relatively simple English sentences (Figure 5.4). That is, correct words in sentences also elicit N400 potentials, the amplitude of the N400 being larger in response to words presented at the beginning of the sentence compared to the N400 elicited by words presented in the middle of the sentence, and being smallest for words presented at the end of the sentence. Commonly, this reduction of the N400 is interpreted as the reflection of semantic context build-up during sentence comprehension, and as a result of the fact that words occurring at the beginning of a sentence require more semantic processing resources than words occurring at the end of a sentence (where the semantic context is already established, and the range of possible words is narrower than at the beginning or middle of the sentence).

During the last years, the scope of investigations using the N400 has been expanded to, for example, development of semantic processing in children (Friederici, 2005; Friedrich & Friederici, 2006),[11] interactions between discourse knowledge and discourse context (Hald *et al.*, 2007), interactions between words and gestures (Holle & Gunter, 2007; Holle *et al.*, 2008; Obermeier *et al.*, in press), interactions between social context and language comprehension (Van Berkum *et al.*, 2008), as well as interactions between semantics and affective prosody (Wambacq & Jerger, 2004; Hayashi *et al.*, 2001; Schirmer & Kotz, 2006).

---

[11]   Showing, e.g., that the N400 is observable in 14-, but not 12-month-old children (Friederici, 2005).

Notably, the N400 can not only be elicited by words (contrary to what was initially believed Holcomb & Neville, 1990), but N400-like effects were found in priming studies to pairs of related and unrelated pictures (Barrett & Rugg, 1990; Holcomb & McPherson, 1994; McPherson & Holcomb, 1999; Hamm *et al.*, 2002; Willems *et al.*, 2008), environmental sounds (Van Petten & Rheinfelder, 1995, Cummings *et al.*, 2006; Orgs *et al.*, 2006, 2007), faces (e.g., Debruille *et al.*, 1996; Jemel *et al.*, 1999; Bentin & Deouell, 2000; Boehm & Paller, 2006), gestures (Gunter & Bach, 2004), actions (Bach *et al.*, 2009), odours (Grigor *et al.*, 1999), and during the retrieval of object forms (taken to reflect involvement of conceptual semantic integration processes; see Mecklinger, 1998). With regard to the present book, it is particularly interesting that the N400 can also be elicited in response to musical information such as short musical excerpts (Koelsch *et al.*, 2004a; Daltrozzo & Schön, 2009b; Frey *et al.*, 2009; Goerlich *et al.*, 2011), chords (Steinbeis & Koelsch, 2008a, 2011), and the timbre of sounds (Grieser-Painter & Koelsch, 2011; Schön *et al.*, 2010). Moreover, studies with chord sequence paradigms identified an ERP component somewhat reminiscent of the N400, the N500 (or N5), which appears to be related, at least in part, to the processing of musical meaning. Chapter 10 will provide a detailed review of studies investigating the processing of musical meaning with the N400 and the N5.

5.5.2 Syntactic processes: (E)LAN and P600

It is generally agreed that sentence comprehension also requires an analysis of constituent structure, that is, an analysis of the relative ordering of words in the sentence and of the grammatical roles played by these words. Syntactic processes are reflected electrically either in early (left anterior) negativities, and/or in a later positivity with posterior scalp distribution (the P600).

Early Negativities With regard to the left anterior negativities described in the literature for the processing of sentences, it has been suggested to distinguish the *left anterior negativity* (LAN) with a latency of about 300 to 500 ms, and the *early left anterior negativity* (ELAN) with a latency of about 100 to 300 ms (for a review see Friederici, 2004).

The LAN is typically elicited by morpho-syntactic violations of tense, sub-categorization information, number agreement,[12] by agreement errors in pseudoword combinations, and in association with syntactical-relational processes such as verb-argument structure information (Coulson *et al.*, 1998; Friederici *et al.*, 1993; Gunter *et al.*, 1997; Osterhout & Mobley, 1995; Osterhout & Holcomb, 1993; Rösler *et al.*, 1993; Münte *et al.*, 1997; Kutas & Hillyard, 1983; Gunter

---

[12] E.g., 'He kisses a girls' (number disagreement), or 'Er trinkt den kühlen Bier' ('He drinks the$_{masc}$ cool$_{masc}$ beer$_{neuter}$'; gender disagreement).

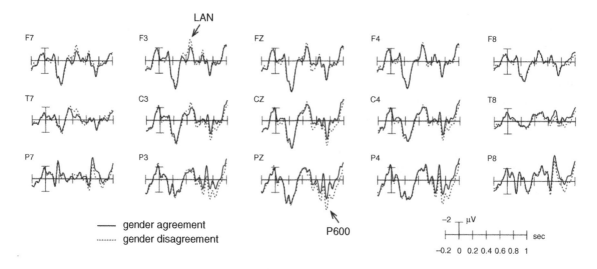

Figure 5.5   ERPs elicited by a morpho-syntactic violation (gender disagreement, dashed line) compared to ERPs elicited by syntactically correct words (solid line). An example for a correct sentence is *Sie bereist das Land...* (She travels the_neuter land_neuter...), an example for an incorrect sentence is *Sie bereist den Land...* (She travels the_masculine land_neuter...). Morpho-syntactic violations elicit an LAN followed by a P600. Reprinted with permission from Gunter *et al.* (2000).

et al., 2000). The LAN usually has a centro-frontal or frontal maximum with a left-hemispheric preponderance, and is normally followed by a late positivity, the so-called P600 (see Figure 5.5).[13]

During the processing of language, early syntactic processes reflected in the LAN and semantic processes reflected in the N400 are presumably carried out in parallel. In a 2-by-2 design employed in a study by Gunter *et al.* (1997), both LAN and N400 were found to become significant around the same time window (260 ms) but did not show an interaction (see also Friederici & Weissenborn, 2007).[14]

The ELAN elicited during the presentation of connected speech was first described by Friederici *et al.* (1993). In that experiment, word category violations (e.g., *Der Freund wurde im besucht / The friend was in the visited*) evoked an ELAN that was maximal at around 180 ms, followed by a second negativity between 300 and 500 ms (Figure 5.6). The ELAN has so far been observed for the processing of phrase structure violations and closed class elements only.[15]

---

[13]   For the observation of the LAN in sentences whose processing required working memory see Kluender & Kutas (1993).

[14]   An interaction between syntax and semantics was found, however, in the time window of the P600.

[15]   The first experiment reporting such an effect using visual word presentation was published by Neville *et al.* (1991). In that experiment, the violation of phrase structure elicited an early left anterior

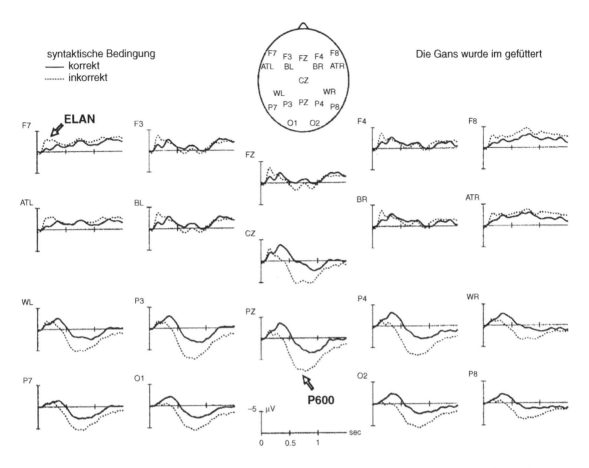

Figure 5.6 ERPs elicited by syntactically incongruous words (phrase structure violations, dotted line) compared to ERPs elicited by syntactically correct words (solid line). The ELAN is best to be seen at F7 around 180 ms. See text for example sentences. Reprinted with permission from Hahne (1999).

The ELAN appears to be quite robust against attentional influences. In a recent study (Maidhof & Koelsch, 2011), this issue was investigated using a selective attention paradigm, employing one condition in which participants were required to attend to musical stimuli and ignore simultaneously presented sentences.[16] In this condition, phrase structure violations elicited an ELAN (even though participants shifted their attention away from the speech stimuli, onto the music stimuli), and the amplitude of the ELAN was only slightly (and statistically not significantly) smaller than in a condition in which the speech stimuli were to

---

negativity (around 125 ms) which was followed by a left temporo-parietal negativity between 350 and 500 ms. The left anterior negativity was evoked by a word category error, e.g., *Max's of proof the theorem.*

[16]   Both musical and speech stimuli were presented auditorily.

be attended (and the music was to be ignored). These results indicate that the syntactic processes reflected by the ELAN operate (at least) partially automatically. This notion is consistent with studies using auditory oddball paradigms to elicit the ELAN (in those studies, the early negativity elicited by syntactic irregularities is referred to as *syntactic MMN*, e.g., Pulvermüller & Shtyrov, 2006; Pulvermüller *et al.*, 2008; Hasting *et al.*, 2007; Hasting & Kotz, 2008).

Notably, the latter two studies (Hasting *et al.*, 2007; Hasting & Kotz, 2008) showed that, in auditory oddball paradigms using word pairs, morpho-syntactic violations can also elicit negative potentials in the ELAN time-range. This suggests that the timing of early ERP effects reflecting specific syntactic processes is influenced to a large degree by methodological (and, thus, not only by linguistic) factors.

Both ELAN and LAN bear a remarkable resemblance with ERPs related to the processing of musical syntax, such as the early right anterior negativity (ERAN), and the right anterior-temporal negativity (RATN). This is one piece of evidence suggesting that the neural resources underlying the processing of linguistic and musical syntax overlap. This issue will be dealt with in detail in Chapter 9.

P600   The P600 is a relatively late positivity that can be elicited by words that are difficult to integrate structurally into meaningful sentences. Its amplitude is maximal over parietal leads, and typically maximal between around 500 and 1000 ms after the onset of the critical word (see Figures 5.5 & 5.6), although the latency of the P600 can vary considerably. The P600 has been found to be elicited by a variety of syntactic anomalies such as garden-path sentences and other syntactically non-preferred structures (Friederici *et al.*, 1996; Hagoort *et al.*, 1993; Mecklinger *et al.*, 1995; Osterhout & Holcomb, 1992, 1993; Osterhout *et al.*, 1994), agreement violations (Coulson *et al.*, 1998; Friederici *et al.*, 1993; Gunter *et al.*, 1997; Hagoort *et al.*, 1993; Osterhout & Mobley, 1995), outright phrase structure violations (Friederici *et al.*, 1996; Neville *et al.*, 1991; Osterhout & Holcomb, 1992, 1993), and subjacency violations (Neville *et al.*, 1991; McKinnon & Osterhout, 1996). During the last years, several studies showed that the P600 can also be elicited in sentences whose interpretation is difficult due to a conflict between semantic (potential agency of a noun) and syntactic information (word position of the noun in the sentence; for reviews see Bornkessel-Schlesewsky & Schlesewsky, 2008; Friederici & Wartenburger, 2010). Moreover, it was recently shown that a P600 can also be elicited by wrong stress patterns within metrically structured sentences (Schmidt-Kassow & Kotz, 2009).[17]

[17]   In that study, sentences with a trochaic pattern, such as *'Vera 'hätte 'Christoph 'gestern 'morgen 'duzen 'können*, were contrasted with sentences containing a word that represented an auditory 'stress-oddball', and was – in addition – wrongly stressed: *'Vera 'hätte 'Christoph 'gestern 'morgen 'duZEN 'können* (for further conditions employed in that experiment see Schmidt-Kassow & Kotz,

The P600 component is assumed to reflect processes of reanalysis and repair that require attention, and occur later than the (fairly automatic) parsing processes reflected in the early negativities described in the previous section (for reviews see Kaan *et al.*, 2000; Friederici & Wartenburger, 2010). It seems likely that the P600 is a type of P3b (Osterhout & Holcomb, 1995), calling somewhat into question the hypothesis of the P600 reflecting a pure *syntactic positive shift* (SPS, Hagoort *et al.*, 1993). Evidence supporting the hypothesis that the P600 rather belongs to the P3-family was provided by a study from Gunter *et al.* (1997). In that study, the P600 was affected by the probability of violation (25% vs. 75%). This finding was taken as evidence that the P600 resembles the P3b (but see also Osterhout, 1999).[18] It has also been reported that the P600 is related to understanding of the meaning of ironic sentences, suggesting that the P600 can also reflect pragmatic interpretation processes (Regel *et al.*, 2011). In addition, Van Herten *et al.* (2005) reported that the P600 can be elicited by semantic anomalies (such as 'The cat that fled from the mice ...'). Van Herten *et al.* (2005), therefore, suggested that the P600 is not a purely syntactic component, but that the P600 might reflect a monitoring component that is associated with checking whether one's perception of sequential information was veridical.

It thus seems plausible that a P600 is not specific for the processing of language, as demonstrated by an experiment from Patel *et al.* (1998): As will be described in more detail in Chapter 6, that experiment reported that both linguistic and musical structural incongruities elicited positivities that were statistically indistinguishable, though the positive component peaked earlier during the processing of music (600 ms) compared to language processing (900 ms). This finding was suggested by the authors to indicate that the neural processes reflected in the P600 are not uniquely linguistic, but index more general cognitive operations involved in the processing of structural relations in rule-governed sequences.[19] Mireille Besson also described P600-like positivities in experiments using musical stimuli, and introduced the term *late positive component* (LPC) for these effects (e.g., Besson & Faita, 1995; Regnault *et al.*, 2001; Magne *et al.*, 2006; see also Chapter 6).

5.5.3   Prosodic processes: Closure Positive Shift

During comprehension of a spoken sentence, listeners not only rely on semantic and syntactic information, but also on suprasegmental phonologic information.

2009); such double violations elicited early negative effects (possibly MMN- and LAN-effects), and a late P600 effect.

[18]   However, studying the P600 is still capable of providing valuable information about language processing (cf. Osterhout & Holcomb, 1995).

[19]   Note that the P3b is also larger in amplitude for unexpected compared to expected targets (cf. Pritchard, 1981; Ritter & Ruchkin, 1992), and that P600 potentials might sometimes be late P3b potentials.

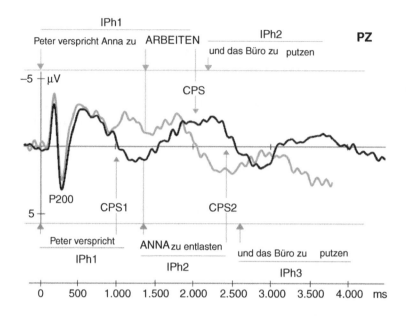

Figure 5.7 Closure Positive Shift elicited at intonational phrase boundaries (IPh) during sentences such as *[Peter verspricht Anna zu arbeiten] IPh1 [und das Büro zu putzen] IPh2* (Peter promises Anna to work, and to clean the office), and *[Peter verspricht] IPh1 [Anna zu entlasten] IPh2 [und das Büro zu putzen] IPh3* (Peter promises, to support Anna, and to clean the office). Reprinted with permission from Steinhauer *et al.* (1999).

This suprasegmental information is part of the *prosody*, and contains important information such as the F0 contour (i.e., the speech melody), as well as rhythmic information in terms of metric patterns and stress patterns (other prosodic features include emotional information, information about the speaker identity, the gender of the speaker, etc.; for reviews see Heim & Alter, 2006; Friederici & Wartenburger, 2010). Prosodic information is important for marking so-called intonational phrase boundaries (IPh). Functionally, an IPh is a syntactic boundary (e.g., the boundary between a main clause and a subordinate clause) and, therefore, particularly relevant for the parsing of spoken sentences. Acoustically, an IPh is realized by (a) a rise in the pitch contour (F0 contour) along with (b) a pause and (c) the pre-final lengthening of the syllable/s preceding the pause.

The processing of IPhs has been found to correlate with a positive shift following the end of an IPh, referred to as the *Closure Positive Shift* (CPS, Steinhauer *et al.*, 1999; see Figure 5.7). The CPS is also present when the pause at the IPh is deleted and the boundary is only indicated by the F0 variation and the pre-final lengthening. The CPS can be elicited even by hummed sentences (that is, in the absence of segmental information; Pannekamp *et al.*, 2005), indicating that purely prosodic processes can elicit the CPS (see also Li & Yang, 2009). Notably, the CPS can also be elicited by the processing of musical phrase boundaries (Knösche *et al*., 2005; Neuhaus *et al*., 2006; Nan *et al*., 2006). This issure is dealt with in more detail in Chapter 6.[20]

---

[20]   Using fMRI, results of a study by Meyer *et al*. (2004) suggest that the processing of the prosodic aspects of the sentences involves premotor cortex of the (right) Rolandic operculum, (right) auditory cortex locted in the planum temporale, as well as the anterior insula (or perhaps deep frontal operculum) and the striatum bilaterally. Sentences used in that study were those used in the study by Steinhauer *et al*. (1999), but stripped of phonological information by filtering, sounding like hummed sentences. A very similar activation pattern was also observed for the processing of the melodic contour of spoken main clauses (Meyer *et al*., 2002). That is, similar to the right-hemispheric weighting of activations observed in functional neuroimaging studies on music perception (reported in detail in Chapter 9), a right-hemispheric weighting of (neo-cortical) activations is also observed for the processing of speech melody.

6

A Brief Historical Account of ERP Studies of Music Processing

6.1 The beginnings: Studies with melodic stimuli

The first ERP studies on music processing investigated ERP effects of melodic deviance using familiar melodies (such as *Happy birthday to you*) that consisted either of the original tones, or in which the pitch of one tone was modified (e.g., *d* or *d sharp* instead of *c*), rendering that tone unexpected and music-syntactically less regular. In 1986; Mireille Besson and Françoise Macar published a study attempting to determine whether the N400 could be elicited by other than only linguistic incongruences (Besson & Macar, 1986). Sentences, sequences of geometric patterns, scales and short melody-excerpts (containing only the first few bars of familiar melodies) were presented to the participants. Melodies ended in 25% of all trials on a 'wrong note' (p. 111). An N400 was elicited only by language stimuli (i.e., by semantically incongruous words), taken to support the hypothesis that the N400 indexes the processing of lexico-semantic incongruity. Note that the incongruous endings of scales and melodies were structural violations, rather than violations of semantic expectancy, and would therefore not necessarily be expected to elicit an N400 potential. Auditory (and geometric) stimuli also elicited a P3b, due to the unexpectedness of the less probable incongruous stimuli.

Notably, already the data by Besson & Macar (1986) showed a clear 'longer and larger N1' (p. 111) that was elicited by incongruous endings of melodies, and in another article presenting the same data (Besson & Macar, 1987) it was reported that a 'negative component peaking between 150 and 200 ms ... could be seen in some of the individual averages' (p. 23). In light of present knowledge,

Brain and Music, First Edition. Stefan Koelsch.
© 2013 John Wiley & Sons, Ltd. Published 2013 by John Wiley & Sons, Ltd.

this early negative potential can be interpreted as (early component of) the early right anterior negativity (ERAN) reflecting the processing of a musical expectancy violation (the ERAN is described in detail in Chapter 9). In the original article, this negativity was only briefly discussed in terms of 'mismatch detection based upon complex information' (Besson & Macar, 1986, p. 114).

A study by Verleger (1990) used melodies that ended not only in the midst of the phrases (on a regular or irregular tone), but also at the end of a melodic phrase (with the final tone either having its proper or a deviant pitch). As in the study by Besson & Macar (1987), no N400 was found in any condition. Instead, P3-effects were found and suggested to correlate independently (a) with deviance and (b) with ending, the former reflecting arousal, the latter reflecting subjects' expectancy of the closing stimulus. As in the study by Besson & Macar (1987), deviant tones elicited early negative effects with frontal maximum (denoted as N150 by the authors), which were not further discussed.

Paller *et al.* (1992) conducted a similar experiment, but allowing additional time for expectations to develop for the terminal note. As in the studies by Besson & Macar (1987) and Verleger (1990), deviant (final) notes did not elicit N400s, but P3-effects. Even in a condition in which the P3-amplitude was minimized (used to investigate whether the N400 was possibly overlapped by a P3), no N400 was evident. Similar to the studies by Besson & Macar (1987) and Verleger (1990), deviant notes also elicited early negative effects with frontal maximum (denoted as N120 by the authors), briefly discussed by the authors as a possible attention effect.

In a prominent study by Besson & Faita (1995), familiar and unfamiliar melodies with either congruous or incongruous endings (diatonic, non-diatonic, and rhythmic) were presented to both musicians and non-musicians. In one experiment of that study, participants had to detect the incongruous endings, and in another experiment, participants 'were told to listen to the phrases carefully to be able to answer questions at the end of each block' (Besson & Faita, 1995, p. 1288). Because incongruous endings were task-relevant in both experiments, a P3b (reflecting the decisional processes connected to the detection of the incongruity) is to be expected to be present in the ERPs. Incongruous endings elicited positivities from around 300 ms post-stimulus onset and with a parietal maximum. These positivities were referred to by Besson & Faita as *late positive components* (LPCs). Besson & Faita (1995) also reported negative components in the 200–600 ms range for incongruous endings (diatonic and non-diatonic), which differed neither between hemispheres, nor along the anterior-posterior dimension. Their functional significance could, however, not be specified, because they were largest for incongruous endings of unfamiliar musical phrases.

In the experiment in which incongruous endings had to be detected, the LPCs showed a greater amplitude and a shorter latency for musicians than for non-musicians, presumably because the melodies were more familiar for musicians than for non-musicians (and, thus, incongruities were easier to detect for musicians). In both experiments, the LPCs had a larger amplitude for familiar melodies than for novel melodies (presumably because incongruous endings of familiar phrases were easier to detect), and for non-diatonic than for diatonic endings (Figure 6.1). Diatonic incongruities terminating unfamiliar melodies did not elicit an LPC (presumably because they are hard to detect), whereas non-diatonic incongruities did (they were detectable for participants by the application of tonal rules). Results from that study indicated that the LPC is at least partly connected to the detectional processes of a music-structural violation (detectable mainly through specific, memory-driven expectations), rather than to a genuine processing of music. This hypothesis is supported by the finding that the LPC was significantly larger in the first experiment (where the incongruous endings were to be detected).

Note that in all of the mentioned studies (Besson & Macar, 1986; Verleger, 1990; Paller *et al.*, 1992; Besson & Faita, 1995), as well as in three later ERP-studies on melody processing (Besson *et al.*, 1998; Miranda & Ullman, 2007; Peretz *et al.*, 2009), the melodic dimension of the stimuli (although not necessarily the unexpected tones) was somehow task-relevant (except Experiment 2 of Paller *et al.*, 1992). Consequently, in all of these studies (except Experiment 2 of Paller *et al.*, 1992), unexpected tones elicited P300/LPC components. This suggests that, as soon as participants have any kind of melody-related task in experimental paradigms using familiar melodies, unexpected notes elicit P300/LPC potentials. Accordingly, a study by Brattico *et al.* (2006) reported that a direct comparison between a passive listening condition (ignoring the melodies and watching a silent movie), and an active condition (judging the correctness of tones) showed a P300/LPC (peaking between 500–700 ms after tone onset) in the active, but not in the passive condition. Likewise, when participants have a timbre-detection task (and thus do not attend to the melodic dimension of the stimulus), irregular tones in melodies do not evoke such positive potentials (Koelsch & Jentschke, 2010).

It is also remarkable that in *all* of the mentioned studies on melodies (Besson & Macar, 1986; Verleger, 1990; Paller *et al.*, 1992; Besson & Faita, 1995; Besson *et al.*, 1998; Miranda & Ullman, 2007; Brattico *et al.*, 2006; Peretz *et al.*, 2009), as well as in another study by Koelsch & Jentschke (2010), the unexpected notes also elicited a frontal negative ERP emerging around the N1 peak (i.e., around 100 ms) and peaking between 120 to 180 ms after stimulus onset (see also arrows in Figure 6.1). This early component had a peak latency of around 150 ms in the study by Verleger (1990), around 120 ms in the study by Paller *et al.* (1992), around 125 ms in the study by Koelsch & Jentschke (2010), and around 180 ms in the passive condition (elicited by out-of-key tones) in the study by Brattico *et al.* (2006) (the other studies did not report peak latencies). This ERP effect resembles the ERAN (or, more specifically, an early component of the ERAN; for details see Section 9.3.5), but it is important to note that this effect was presumably overlapped by a subsequent N2b in those studies in which the melodic dimension was task-relevant (Besson & Macar, 1986; Verleger, 1990; Paller *et al.*, 1992; Besson & Faita, 1995; Besson *et al.*, 1998; Miranda & Ullman, 2007; Peretz *et al.*, 2009).[1] An ERAN elicited by melodies without overlapping N2b has so

---

[1]   The N2b is centrally maximal and often precedes the P300; see p. 49.

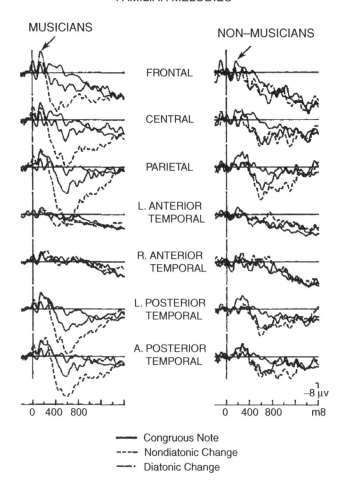

FAMILIAR MELODIES

MUSICIANS   NON—MUSICIANS

FRONTAL

CENTRAL

PARIETAL

L. ANTERIOR
TEMPORAL

R. ANTERIOR
TEMPORAL

L. POSTERIOR
TEMPORAL

A. POSTERIOR
TEMPORAL

0 400 800   0 400 800   m8

—— Congruous Note
---- Nondiatonic Change
—— Diatonic Change

Figure 6.1 ERPs elicited by congruous and incongruous melody-endings, separately for musicians and non-musicians (from Besson & Faita, 1995). Participants were instructed to listen to the sequences and to answer infrequently asked questions about the melodies. Arrows (added to the original figure) show early negative potentials elicited by diatonic changes. Reprinted with permission from Besson & Faita (1995).

far been reported by three studies (Miranda & Ullman, 2007; Brattico *et al.*, 2006; Koelsch & Jentschke, 2010) for out-of-key notes occurring in unfamiliar melodies, and it seems that a similar potential was elicited by out-of-key notes occurring in unfamiliar melodies in musicians in the study by Besson & Faita (1995, p. 1284).

Presenting the scores of unfamiliar melodies (without acoustic presentation of the tones), Schon & Besson (2005) also showed that unexpected notes elicit

ERAN potentials. Finally, a study by Herholz *et al.* (2008) reported a similar response to unexpected tones of familiar melodies (presented auditorily) after participants imagined the six tones preceding these tones (in the absence of auditory information).

6.2   Studies with chords

In 1995; Petr Janata published the first ERP-study with multi-part stimuli (i.e., chords). In that study (Janata, 1995), chord sequences consisting of three chords were terminated equiprobably either by the tonic (i.e., the most expected chord function), or a less regular submediant (e.g., an *a* minor chord in *C* major), or a major chord based on the tritone (IV#) of the original key (this chord is harmonically only distantly related, and thus perceived as unexpected by listeners familiar with major–minor tonal music). Musicians had to judge whether or not a chord sequence ended on 'the best possible resolution' (Janata, 1995).

In general, the degree of expectancy violation of a chord terminating a chord-sequence was reflected in the amplitude of positive ERP-peaks in two temporal regions: A P3a with a latency of 310 ms reflecting attentional processes, and a P3b peaking at 450 ms reflecting decisional processes. A larger degree of violation was reflected in larger P3-peaks.

To test the language-specificity of the P600, Patel *et al.* (1998) compared ERPs elicited by 'syntactic incongruities' in language and music, whereby harmonic incongruities were taken as grammatical incongruity in music. Target chords within homophonic musical phrases were manipulated, so that the targets were either within the key of a phrase, or out-of-key (from a key that was either closely, or only distantly related in terms of the circle of fifths).

Both musical and linguistic structural incongruities elicited positivities with a latency of about 600 ms, which were maximal at posterior sites and did not significantly differ from each other. Moderate and high degrees of structural anomaly differed in amplitude of the elicited positivities. Therefore, the results indicated that the P600 reflects more general knowledge-based structural integration during the perception of rule-governed sequences.

Additionally, a negative music-specific ERP with a latency of around 350 ms and an anterior, right-hemispheric lateralization was observed. This *right anterior-temporal negativity* (RATN) was elicited by out-of-key target chords (Fig. 6.2). The RATN was taken to reflect the application of music-syntactic rules and working memory processes.

In 2000, the first article describing the ERAN and the N5 using a chord sequence paradigm was published (Koelsch *et al.*, 2000; for a detailed review see Chapter 9). Using these ERP components, numerous subsequent studies investigated, for example, the development of music processing in children (Koelsch *et al.*, 2003e; Jentschke *et al.*, 2008a; Jentschke & Koelsch, 2009), effects of attention (Koelsch *et al.*, 2002; Loui *et al.*, 2005, Maidhof & Koelsch, 2011),

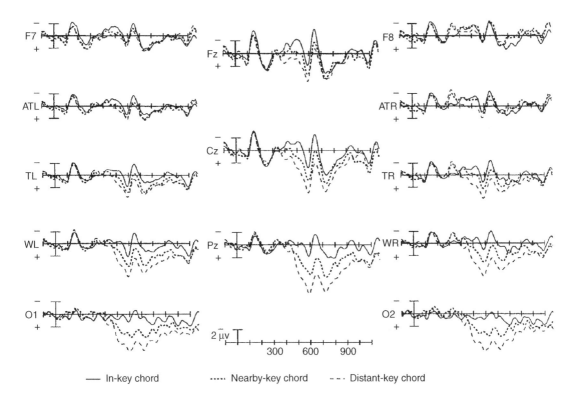

Figure 6.2   ERPs elicited by the three target chord types used in the study by Patel *et al.* (1998). Onset of the following chord is 500 ms after target onset. Reprinted with permission from Patel *et al.* (1998).

effects of long- and short-term experience (Koelsch *et al.*, 2002c; Koelsch & Jentschke, 2008), differences between music-syntactic processing and the processing of acoustic oddballs (Koelsch *et al.*, 2001, 2005b; Leino *et al.*, 2007; Koelsch *et al.*, 2007a), learning of artificial grammar (Carrión & Bly, 2008), emotional responses of music-syntactic processing (Steinbeis *et al.*, 2006; Koelsch *et al.*, 2008b), aesthetic aspects of music perception (Müller *et al.*, 2010), music-syntactic processing of melodies (Miranda & Ullman, 2007; Koelsch & Jentschke, 2010), music-syntactic processing during the reading of scores (Schon & Besson, 2005), gender differences (Koelsch *et al.*, 2003c), music processing under anaesthesia (Heinke *et al.*, 2004; Koelsch *et al.*, 2006b), music perception in cochlea implant users (Koelsch *et al.*, 2004b), processing of tonal modulations (Koelsch *et al.*, 2003b), and oscillatory correlates of music-syntactic processing (Herrojo-Ruiz *et al.*, 2009b).

6.3   MMN studies

Also around the year 2000, the first MMN studies were published investigating effects of musical training on pre-attentive pitch discrimination of chords

(Koelsch *et al.*, 1999) or melodies (Menning *et al.*, 2000), on the formation
of cortical memory traces for melodic patterns (Tervaniemi *et al.*, 2001), and on
the processing of sound omissions (Rüsseler *et al.*, 2001). Moreover, the hemispheric weighting of the MMN response to chords and phonemes was investigated
using MEG (Tervaniemi *et al.*, 1999) as well as PET (Tervaniemi *et al.*, 2000). [2]

6.4   Processing of musical meaning

In 2004, a study using ERPs to investigate the processing of musical meaning was
published (Koelsch *et al.*, 2004). Using a semantic priming paradigm, musical
excerpts were used to prime the meaning of a target word. Target words that were
semantically unrelated to a preceding short musical excerpt (about ten seconds
long) elicited an N400 compared to target words that were semantically related to
a preceding musical stimulus. Subsequent studies showed that an N400 can even
be elicited by musical target stimuli following word prime stimuli (i.e., not only
by word targets following a music prime; Daltrozzo & Schön, 2009b; Goerlich
et al., 2011). Other studies showed that affective qualities of chords (Steinbeis
& Koelsch, 2008a, 2011), and timbres of tones with iconic or indexical sign
quality (Grieser-Painter & Koelsch, 2011; Schön *et al.*, 2010) can modulate the
processing of word meaning (in these studies, the N400 elicited by target words
was modulated by the semantic relatedness to a preceding chord or a tone).
The latter studies (Grieser-Painter & Koelsch, 2011; Schön *et al.*, 2010) also
showed that N400 potentials can even be elicited by chords (Steinbeis & Koelsch,
2008a, 2011), or tones (Grieser-Painter & Koelsch, 2011) that are semantically
unrelated to a preceding target word (a detailed account on this topic is provided in
Chapter 10).

Moreover, a study by Miranda & Ullman (2007) showed that in-key deviant
notes in familiar well-known melodies elicit an N400, and a recent study by
Gordon *et al.* (2010) showed that melodic variations of words affect the processing
of word meaning. The latter study (Gordon *et al.*, 2010) used sung tri-syllabic
nouns as primes (each syllable was sung on a different tone). Each prime stimulus
was followed by a target that consisted either of (a) the same word sung on the
same melody, (b) a different word sung on the same melody, (c) the same word
sung on a different melody, and (d) a different word sung on a different melody.
When a different word was sung on the same melody, the authors observed
an N400 response (as expected). Importantly, an N400 was also elicited when
the target was the same word sung with a different melody (with the N400
being followed by a late positive component), and the N400 elicited by the word
meaning appeared to interact with the N400 elicited by the different melodies.[3]

---

[2]   For further information see also Section 5.3.2 on MMN and music.

[3]   This was indexed by a non-additive N400 in the condition in which both the word and the
melody were different.

These results are remarkable, because they show that the processing of word meaning interacts with the processing of the melody with which a word is spoken, and that the melody is important for the message conveyed by speech (see also Fernald, 1989; Papoušek, 1996).

6.5   Processing of musical phrase boundaries

In 2005, a study by Knösche *et al.* (2005) showed that the processing of musical phrase boundaries elicits a *Closure Positive Shift* (CPS). This (music-evoked) CPS was strongly reminiscent of the CPS evoked by the processing of intonational phrase boundaries during language perception (see also Chapter 5.5.3), except perhaps a slightly different latency. That study (Knösche *et al.*, 2005) investigated musicians only, but a subsequent study also reported a CPS in non-musicians (although with considerably smaller amplitude than in musicians; Neuhaus *et al.*, 2006). The latter study also reported larger CPS amplitudes for longer pauses as well as longer boundary tones (which make the phrase boundary more salient). Both studies (Knösche *et al.*, 2005; Neuhaus *et al.*, 2006) used EEG as well as MEG, showing that the CPS can be observed with both methods. In a third study, Chinese and German musicians performed a cultural categorization task with Chinese and Western music (with both groups being familiar with Western music, but only the Chinese group being familiar with Chinese music; Nan *et al.*, 2006). Both groups showed CPS responses to both types of music (with no significant difference between groups).

6.6   Music and action

During the last few years, five ERP studies have begun to investigate action processes with music by obtaining ERPs from pianists playing the piano (Katahira *et al.*, 2008; Maidhof *et al.*, 2009; Herrojo-Ruiz *et al.*, 2009a; Maidhof *et al.*, 2010; Herrojo-Ruiz *et al.*, 2010). These studies investigated error-related potentials (e.g., the *error-related negativity*) that were evoked either when the pianists played a wrong note, or when they were provided with false feedback of a wrong note (these studies are described in detail in Chapter 11). Interestingly, the data of self-produced errors show error-related potentials even *before* the erroneous movement was fully executed, that is, before the musicians heard the (wrong) tone. Such studies have the potential to investigate action processes that are considerably more complex, and ecologically more valid, than those investigating action in experiments using 2 or 3 buttons on a response box (a typical setting in this field). Moreover, such studies open the precious opportunity to investigate interactions between action and emotion.

Currently, there is a strong interest in the neural mechanisms underlying a possible neural entrainment to music, as well as in neural correlates of synchronizing to an external beat, and of inter-individual synchronization during music making and music listening. Studies investigating electrophysiological correlates of inter-individual synchronization and music making in groups can also provide important contributions to the investigation of the influence of social factors on emotional processes, as well as on cognitive processes such as action monitoring and error processing. Moreover, such studies can contribute to the investigation of the neural correlates of joint action, shared intentions, and the nonverbal communication of music performers.

7

Functional Neuroimaging Methods: fMRI and PET

In functional magnetic resonance imaging (fMRI), time series of magnetic resonance (MR) images are recorded. The intensity in an image is a representation of the chemical environment of hydrogen isotopes in the brain (particularly in water and fat). These isotopes are subjected to a strong magnetic field, and excited by suitable radiowave pulses. MR images mainly reflect brain structure, but also contain small contributions from blood flow (on the order of 2% of maximum intensity), which is a delayed function of brain cell activity: If neurons are active, their metabolic rate, including the consumption of oxygen, increases. This causes changes in the local metabolite composition, which in turn modulates blood vessel diameters, so that an over-compensation of oxygenated blood concentration (and a higher local cerebral blood volume) in the milieu of the preceding nerve cell activity is established. The peak of this haemodynamic response usually has a latency of several seconds. Oxygenated blood has a higher ratio of oxyhaemoglobin (i.e., haemoglobin-carrying oxygen) to deoxyhaemoglobin (i.e., haemoglobin that has dispensed its oxygen), than deoxygenated blood. Importantly, oxyhaemoglobin and deoxyhaemoglobin have different magnetic properties – oxyhaemoglobin is diamagnetic and exerts little influence on the local magnetic field, whereas deoxy-haemoglobin is paramagnetic and causes signal loss in the corresponding volume elements of the image. This blood-oxygen-level dependent (BOLD) image contrast is most commonly measured with fMRI, and provides an indirect measure of preceding neural activity in the brain (for a more detailed account see, e.g., Faro & Mohamed, 2006).

Brain and Music, First Edition. Stefan Koelsch.
© 2013 John Wiley & Sons, Ltd. Published 2013 by John Wiley & Sons, Ltd.

Functional MRI does not employ radioactive substances (unlike PET; see below), or ionizing radiation (unlike Computed Tomography, CT), and is therefore regarded as a harmless and safe procedure, if safety precautions are strictly followed. These include that individuals with ferromagnetic metal implants and cardiac pacemakers are prevented from having MRI scans (due to the high magnetic field), and that no ferromagnetic metals are brought into the scanner room, especially not while an individual is scanned (the magnetic forces of the scanner can easily turn ferromagnetic objects into projectiles).

Image acquisition relies on supplementary magnetic fields to encode spatial information, which is achieved by means of dedicated magnetic gradient coils. Their rapid switching requires strong electric currents, which interact with the main magnetic field. The resulting minute vibrations of the coil materials propagate into the surrounding air, in which they are audible as fairly loud noise. The sound intensity of such noise can reach up to 130 dB, which can damage the hair cells in the inner ear. Hence, the appropriate use of ear protection is mandatory during fMRI. Both the loud scanner and the requirement for ear protection make it challenging to perform experiments with auditory stimulation in the fMRI scanner. Using ISSS or 'sparse temporal scanning' designs (see below) can help to overcome such problems.

Another functional imaging technique is positron emission tomography (PET).[1] As mentioned above, activity of nerve cells correlates with the regional flow of blood, and this regional cerebral blood flow (rCBF) can also be measured using a radioactive tracer, which decays by emitting a positron (e.g., oxygen-15, O-15). Thus, after injection of such a tracer, areas with increased brain activity are associated with areas of high radioactivity. When the radioisotope emits a positron, the positron encounters, and annihilates with, an electron (the antimatter counterpart of a positron), producing a pair of annihilation (gamma) photons moving in opposite directions. These two photons are detected by the scanning device, and registered when they reach the detector at the same time (i.e., within a few nanoseconds). Because most gamma photons are emitted at almost 180 degrees to each other, it is possible to localize the source of a pair of gamma photons that reach the detector at the same time. O–15 has a short half-life of (around 2 minutes), hence this tracer must be injected directly before the beginning of an experimental block (for a more detailed account see, e.g., Phelps, 2006).

Besides the obvious disadvantage of using radioactive substances, PET has also a lower spatial resolution than fMRI. However, PET has also important advantages compared with fMRI. Firstly, PET is practically silent. Secondly, tracers can be used that distribute in tissues by partially following the metabolic pathways of their natural analogues, or that bind with specificity in tissues containing a particular receptor protein. Hence, PET can be used to trace the biological

---

[1]   Although fMRI is used more widely now as a functional neuroimaging technique, PET was actually used before fMRI in neuroscience.

pathways of specific compounds, or the distribution of compounds (e.g., dopamine) in response to an experimental stimulus.

7.1   Analysis of fMRI data

Functional MRI data can be analyzed in numerous ways. In commonly applied statistical analyses, the time course of blood oxygen level dependent (BOLD) response is correlated with an independent variable, which may for example relate to different tasks (such as listening to music-syntactically correct chord functions, or to music-syntactically incorrect chord functions), or a difference in perceived pleasantness of the stimulus categories, or different subject groups (such as non-musicians and musicians). Figure 7.1 shows the time-line of intensity in a voxel located in the auditory cortex (indicated by the cross-hair in Figure 7.2) during three music listening trials; each music listening trial is followed by a pause (in which the scanner noise is still present due to continuous image acquisition). Note that the BOLD response is relatively slow (taking about 6 to 8 seconds to reach the maximum response after stimulus onset), and that the signal variation is rather small in relation to the average voxel intensity. That is, the BOLD response shown here is less than the tip of an iceberg, and relatively small in relation to the noise generated by, e.g., cardiovascular and respiratory activity.

Figure 7.2b shows the image of a scan obtained while a subject is listening to music (taken from one of the scans indicated by the black horizontal bars in Figure 7.1), and Figure 7.2c shows the image of a scan obtained during a period in which no music was presented (the cross-hair indicates the voxel of which the time-course of BOLD activity is displayed in Figure 7.1). Note that, whereas the difference in voxel intensity can be nicely seen in the time-course of Figure 7.1, this difference is hardly visible when looking at the functional images themselves. However, one way to detect correlations between brain responses and a variation of an experimental parameter (in our example presentation of music and intermediate periods without music) is to statistically test the difference between the BOLD signals measured during the two experimental conditions across numerous trials: Figure 7.2d shows such a contrast (that is, a statistical test for significance) between the music condition and the condition without music. The resulting statistical parametric map (SPM) of the contrast indicates whether differences in BOLD response between the contrasted conditions are significant. A reliably detected difference is then assumed to be due to different (cognitive or affective) processes evoked during the two conditions (in our example auditory cortex activity in response to music compared to a period in which no music is presented). Thus, such SPMs do not directly show neuronal activity (as often suggested by the hot red colours prominent in publications of fMRI data), although the BOLD signal strongly correlates with neuronal activity. That is, the SPMs show nothing but a statistical comparison, here between the two

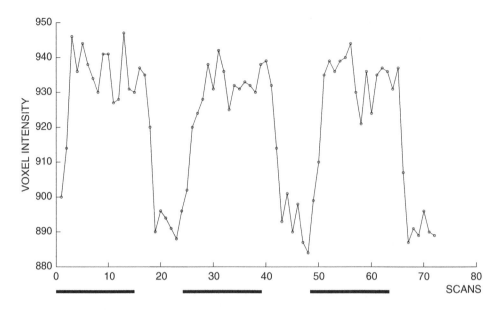

Figure 7.1 Time-line of intensity in a voxel located in the auditory cortex, during three music listening trials (the three black horizontal bars at the bottom of the figure indicate the time intervals in which music was presented). The x-axis represents the scan-number (time of repetition was 2 sec, that is, every two seconds a functional image of the whole brain was acquired), each music listening trial had a duration of 30 sec (corresponding to 15 scans). Note that the BOLD response is relatively slow: It takes about 6 to 8 seconds to reach the maximum response after stimulus onset (and considerably longer to return to baseline). Also note that the signal variation is rather small in relation to the average voxel intensity (less than about 0.5 percent; see scale of y-axis).

experimental conditions for each measured voxel with regard to the changes in blood oxygenation.

To investigate correlations between spatially remote neurophysiological events, connectivity analyses can be applied. In this correlational approach (referred to as *functional connectivity analysis*), the time-course of BOLD signal in one region (often referred to as the 'seed-region') is correlated with the time-course of BOLD signals in all other voxels of the brain. This provides a measure for BOLD response synchrony, and it is assumed that brain structures showing such synchrony are functionally connected with each other. Functional connectivity analyses have been developed to investigate functional networks in the brain. Note, however, that functional connectivity does not provide information about anatomical connections, only about BOLD time-course synchrony. Another method for investigating functional connectivity with fMRI is the *psycho-physiological interaction analysis* (PPI). In PPI, functional connectivity is statistically compared between experimental conditions (Friston *et al.*, 1997). This is relevant because the BOLD synchrony between brain regions may differ between different experimental conditions.

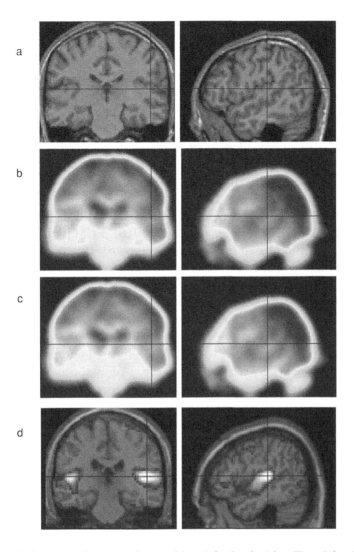

Figure 7.2   (a) Anatomical image of one subject (obtained with a T_1-weighted scan, the left image shows a coronal view, the right image a sagittal view). (b) and (c) show T_2-weighted scans, (b) is one scan obtained while the subject was listening to an auditory stimulus, (c) was obtained while the subject was not presented with an auditory stimulus. Note that such single scans do not inform about functional differences between two experimental conditions, and contain 'noise' in addition to the signal due to the experimental stimulation, just as the spontaneous EEG shown in Figure 4.3. To obtain a reliable signal from such images, several images are acquired for each condition in one experimental session (similar to obtaining several epochs in ERP experiments; see also Chapter 4). A statistical comparison of several images results in statistical parametric maps (SPMs) such as the one shown in (d), indicating a statistically significant difference of BOLD signal between the two conditions in the auditory cortex bilaterally.

Unlike functional connectivity analyses, *effective connectivity analyses* aim to investigate directional influences of one neuronal system on another. Such methods comprise *dynamic causal modelling* (DCM), *Granger causality mapping* (GCM), and *Structural Equation Modelling* (SEM), methods which will not be described further in this chapter (for detailed explanations; see Friston *et al.*, 2003; Goebel *et al.*, 2003; McIntosh & Gonzalez-Lima, 1994; Lohmann *et al.*, 2011).

A further method for the analysis of BOLD signals is the pattern classification analysis, which can be applied in an attempt to read out psychological states from fMRI data. For this kind of analysis, pattern recognition algorithms are trained through the pairing of a psychological variable and patterns of BOLD response (for details; see Haynes & Rees, 2005).

Finally, a recently developed method for analyzing fMRI data is the *Eigenvector Centrality Mapping* (ECM, Lohmann *et al.*, 2010). ECM is an assumption- and parameter-free method based on *eigenvector centrality* (a particular form of node centrality). Eigenvector centrality attributes a value to each voxel in the brain such that a voxel receives a large value if it is strongly correlated with many other nodes that are themselves central within the network (Google's page-rank algorithm is a variant of eigenvector centrality, this algorithm assigns high ranks to web-pages to which many other web-pages point). ECM yields information about an intrinsic neural architecture on a voxel-wise level, based on linear correlations or on spectral coherences between fMRI time series (thus also allowing one to draw conclusions of connectivity patterns in different spectral bands). ECM can be applied to the entire cerebrum, or to a set of regions of interest. After the voxels of the computational 'hubs' have been identified, the coordinates of these voxels can be used for functional connectivity or PPI analysis. ECM is included in the LIPSIA software package.[2]

Due to the correlational analysis, experimental trials are recommended to be quite long for the application of ECM (about 2 to 3 minutes, although also trials with a length of 30 seconds can lead to good results, perhaps with neocortical structures being more easily detected than, e.g., subcortical structures). However, if one is interested in looking at longer trial periods (which is often the case in experiments on mood and emotion), or if it is possible to present stimuli in blocks of about 2 to 3 minutes length, then ECM has the advantage that just one trial per condition is sufficient per subject. ECM can also nicely be used for the comparison of different states measured in different experimental sessions (such as being hungry or satiated, or pre-/post-training, or pre-/post-treatment), or for between-subject comparisons.

7.2   Sparse temporal sampling in fMRI

Using fMRI, the loud scanner noise makes it difficult to perform experiments with auditory stimulation. One possibility to make the musical stimulus better

---

[2]   LIPSIA can be downloaded from http://neuro.debian.net/pkgs/lipsia.html

audible in an fMRI experiment is to apply a sparse temporal scanning design (Gaab *et al.*, 2003, 2007; Hall *et al.*, 1999), in which functional MR images are not acquired continuously, but with a pause of a few seconds between two image acquisitions, so that an auditory stimulus can be presented in the pause between two acquisitions. The pause between two acquisitions can vary between experiments, from about 4 to up to 14 seconds (depending, e.g., on how long the stimuli are). When activations within the auditory cortex are the focus of the investigation, then a sparse temporal sampling design is indispensable (e.g., Gaab *et al.*, 2007). In fMRI experiments on emotion, participants might find the loudness related to continuous scanning distressing, particularly in experiments using devices operating at or above 3 T (tesla). Therefore, 1.5 T devices (which are not as loud as 3 T devices) are, for example, better suited when a high signal-to-noise ratio and minimized distress for participants are more important than high spatial resolution.

7.3   Interleaved silent steady state fMRI

Another option is interleaved silent steady state (ISSS) imaging (Schwarzbauer *et al.*, 2006; ISSS is only an option if the required sequences can be made available for the MR scanner used for the experiment). The ISSS imaging design is also a sparse temporal sampling design, but here several images (not just one) are acquired after a 'silence' period of no scanning. Importantly, the images are acquired more rapidly than usual, and the T1-related signal decay during the silence periods is avoided by maintaining a steady state (longitudinal) magnetization with a train of relatively silent slice-selective excitation pulses during the silence period, resulting in a better signal-to-noise ratio of the fMRI images acquired after the presentation of the stimulus. However, maintaining the steady state magnetization also produces noise (although it is considerably less loud than the echoplanar imaging noise), and the ISSS design is intended for stimuli that do not exceed a duration of a few seconds. For such stimuli, this scanning protocol is very useful. Whether this scanning protocol is suited for block designs with block durations of several tens of seconds remains to be investigated.

7.4   'Activation' vs. 'activity change'

In research using fMRI or PET, a signal increase is often referred to as 'activation' (and a signal decrease as 'deactivation'). The word 'activation' is semantically associated with excitatory, rather than inhibitory, processes, and in unimodal cortical areas, it seems quite plausible that the BOLD signal correlates mainly with excitatory postsynaptic activity. However, for heteromodal, and particularly for limbic/paralimbic structures with complicated receptorarchitectonics (such as the amygdala and the hippocampus), an increase in BOLD signal does not necessarily originate from excitatory (post)synaptic activity. It is also possible that such signal changes are the result of inhibitory synaptic processes (Buxton, 2002; Lauritzen, 2008; Shibasaki, 2008). Therefore, one has to be careful about simply speaking of 'activation' of a brain structure. The more accurate terminology is often 'activity change.'

With regard to the interpretation of functional neuroimaging data, this is an important distinction. For example, an activity change in a particular nucleus of the amygdala (or other regions such as hippocampus, temporal poles, or parahippocampal gyrus) in response to a fear-inducing stimulus might reflect three different things. Firstly, it might simply reflect inhibition (or down-regulation) of processes related to positive emotions; the emergence of a negative emotion usually also means decrease of a positive emotion, and it is possible that only the decrease of positive, but not the emergence of negative, emotion is actually reflected in the BOLD or rCBF signals. Secondly, the signal change might actually reflect activation of a fear network. Thirdly, due to restricted spatial resolution, the signals might reflect both excitatory and inhibitory activity. However, with the measurement techniques usually employed with PET and fMRI, one can often not tell which of the three possibilities the data reflect.

Part II

Towards a New Theory of Music Psychology

8

Music Perception: A Generative Framework

This chapter introduces a neuro-cognitive model of music perception (Figure 8.1). The model was developed by Koelsch & Siebel (2005), and serves as a basis to develop several aspects of music perception and production in the following chapters. Therefore, several sections of this chapter are rather brief introductions into topics that will be dealt with in more detail in the following chapters. The model synthesizes different areas of neuroscientific research in the field of music perception into a common theoretical framework. It describes different stages, or dimensions, of music perception, introducing research findings that provide information about the time course of the processes of these different stages (as reflected in electrophysiological indices), and about where in the brain these processes are located. These stages are not thought of as exclusively serving music processing, but also, at least in part, the processing of (spoken) language.[1,2] In fact, as will be shown later in this book, the model can largely be adopted for spoken language processing.

Auditory feature extraction   Music perception begins with the decoding of acoustic information. As described in Chapter 1, decoding of acoustic information involves the auditory brainstem and thalamus, as well as the auditory cortex

---

[1]   For a model assuming music- and language-specific modules see Peretz & Coltheart (2003).

[2]   In previous articles, I have used the term *module*, which was sometimes misunderstood by readers who thought that I wanted to express that the described processes serve music processing only. This was, however, never meant. For a discussion on the term *modularity* see, e.g., Fodor *et al.* (1991).

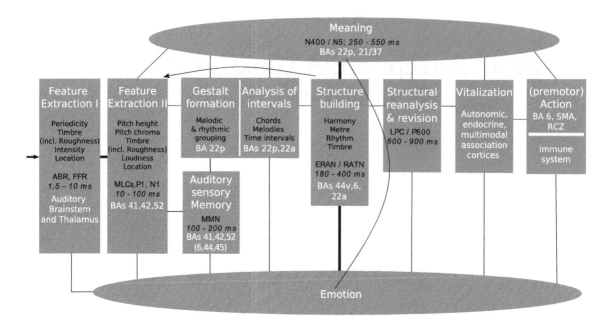

Figure 8.1 Neurocognitive model of music perception. ABR: Auditory Brainstem response; BA: Brodmann area; ERAN: Early right anterior negativity; FFR: Frequency Following Response; LPC: Late positive component; MLC: Mid-latency component; MMN: Mismatch Negativity; RATN: Right anterior-temporal negativity; RCZ: Rostral cingulate zone; SMA: Supplementary motor area. Italic font indicates peak latencies of scalp-recorded evoked potentials.

(mainly auditory core and belt regions, corresponding to BA 41, 42, and 52). It was emphasized that one role of the auditory cortex is the transformation of acoustic features (such as periodicity and intensity) into percepts (such as pitch height, pitch chroma, and loudness; see *Feature Extraction I* and *II* in Figure 8.1). Chapter 5 described how ABRs and FFRs can be used to measure neural activity reflecting processes of auditory feature extraction in the brainstem, and that on the level of the cortex, the mid-latency components (including the P1) as well as the N1 (possibly also the P2) reflect in part neural activity related to feature extraction.

Echoic memory and Gestalt formation While auditory features are extracted, the acoustic information enters the auditory sensory memory (or 'echoic memory'), and auditory Gestalten are formed (e.g., representations comprising several perceptual elements are integrated over time). Echoic memory and processes of Gestalt formation appear to be mediated mainly by the auditory cortex, in the middle and posterior region of the superior temporal gyrus (STG) including Heschl's gyrus and the planum temporale. As described in Section 5.3, operations of the auditory sensory memory appear to be reflected, at least partly, in the mismatch negativity (MMN). The MMN receives main contributions from the auditory cortex, with additional contributions from frontal areas (possibly BA 6, 44, and 45).

The formation of auditory Gestalten entails processes of perceptual separation, as well as processes of melodic, harmonic, metric, rhythmic, timbral and spatial grouping. Such processes have been summarized under the concepts of auditory scene analysis and auditory stream segregation (Bregman, 1994). Grouping of acoustic events follows Gestalt principles such as similarity, proximity and continuity (for acoustic cues used for perceptual separation and auditory grouping see Darwin, 1997, Darwin, 2008). In everyday life, such operations are not only important for music processing, but also, for instance, for separating a speaker's voice during a conversation from other sound sources in the environment. That is, these operations are important because their function is to recognize acoustic objects, and thus to establish a cognitive representation of the acoustic environment. Knowledge about neural mechanisms of auditory stream segregation, auditory scene analysis, and auditory grouping is still relatively sparse (for reviews see Griffiths & Warren, 2002, 2004; Carlyon, 2004; Nelken, 2004; Scott, 2005; Shinn-Cunningham, 2008; Winkler *et al.*, 2009a). However, it appears that the planum temporale (BA 22p, i.e., part of the auditory association cortex) is a crucial structure for auditory scene analysis and stream segregation (Griffiths & Warren, 2002), particularly due to its role for the processing of pitch intervals and sound sequences (Patterson *et al.*, 2002; Zatorre *et al.*, 1994).

Minute interval analysis Presumably closely linked to the stage of auditory Gestalt formation is a stage of a more minute analysis of intervals, which might include: (1) A more detailed processing of the pitch relations between the tones of a melody, or between the tones of a chord (e.g., to determine whether a chord is a major or minor chord, or whether a chord is played in root position, in inversion, etc.),[3] and possibly (2) a more detailed processing of time intervals. Melodic and temporal intervals appear to be processed independently, as suggested by the observation that brain damage can interfere with the discrimination of pitch relations, but spare the accurate interpretation of time relations, and vice versa (Peretz & Zatorre, 2005; Di Pietro *et al.*, 2004).

Correspondingly, identification of minute pitch interval relations appears to involve auditory cortical areas, whereas identification of minute time intervals (within an isochronous tactus) appears to also involve the basal ganglia and the cerebellum. As mentioned in Chapter 1, it has been suggested that the analysis of the contour of a melody (which is part of the auditory Gestalt formation) relies particularly on the posterior part of the right STG (BA 22p), whereas the processing of more detailed interval information appears to involve both posterior

---

[3] To illustrate this: Try to determine whether a chord is played in major or minor, in root position, or in inversion, while listening to a piece of tonal music, such as a chorale by J.S. Bach. Such a task is surprisingly difficult, and even experienced listeners of classical music often have huge problems simply telling on-line whether successive chords are major or minor chords.

and anterior areas (BA 22p & BA 22a) of the supratemporal cortex bilaterally (Peretz & Zatorre, 2005; Liegeois-Chauvel *et al.*, 1998; Patterson *et al.*, 2002). Processing of time relations, on the other hand, appears to involve the planum polare, supplementary and premotor areas, as well as the basal ganglia and the cerebellum (e.g., Grahn & Brett, 2007).

When intelligence comes into play: Syntactic structure building   All kinds of music involve forms of local dependencies between sounds. However, in tonal music, as well as in many other kinds of music, single musical elements (such as tones, or chords) can also be arranged into hierarchical structures involving long-distance dependencies on a phrase structure level (context-free grammar). Such hierarchically organized syntactic structures are particularly powerful in communicating intra-musical meaning, and processing of such structures can evoke emotions. Chapter 9 will outline a theoretical basis of musical syntax, emphasizing that tonal music has several syntactic features (melody, metre, rhythm, harmony, intensity, instrumentation, and texture). However, so far neuroscientific research has mainly focussed on the processing of melodies and chord functions (i.e., on the processing of harmony).

As will be described in detail in Chapter 9, the disruption of syntactic structure building by a music-syntactically irregular event is reflected electrically in relatively early anterior, or anterior-temporal, negativities such as early right anterior negativity (ERAN), or the right anterior-temporal negativity (RATN). The ERAN receives main contributions from the inferior pars opercularis (BA 44v),[4] presumably with additional contributions from the ventrolateral premotor cortex (BA 6), and the anterior STG (planum polare, BA 22a).

The strongest evidence for the assumption that the ERAN reflects music-syntactic processing stems from data showing that the generation of the ERAN is influenced by the simultaneous processing of language-syntactic irregularities (but not by language-semantic irregularities). This indicates an intersection of cognitive operations of music-syntactic and language-syntactic processing (and of neural populations mediating music-syntactic and language-syntactic processing). That is, these findings indicate that music-syntactic and language-syntactic processing relies on at least partly shared cognitive and neural resources. Beyond music- and language-syntactic processing, this intersection probably also includes the processing of hierarchically organized actions and mathematical formulae (or, on a more general level, processing of hierarchically organized structures involving long-distance dependencies). Notably, the ability to process phrase-structure grammar is available to all humans (as evidenced, e.g., by the use of music and language), whereas non-human primates are apparently not able to master such grammars (Fitch & Hauser, 2004). Thus, it is highly likely that only humans can adequately process music-syntactic information on the phrase-structure level.

---

[4]   According to the anatomical labelling of Broca's area proposed by Amunts *et al.* (2010).

Structural reanalysis and revision   Later stages of syntactic processing may occur when structural reanalysis and revision is required. During the syntactic processing of a sequence of elements, perceivers usually tend to structure the elements in the most likely way (for example, based on thematic role assignment, minimal attachment, late closure, etc.).[5] However, it may also happen that a listener recognizes that an established hierarchical model has to be revised. That is, the headedness of branches, the assignment of elements to branches, etc. may have to be modified. An example from language is the 'garden-path' sentence effect,[6] an example for music is that a tonal key which was initially presumed to be the home key turns out to be not the home key.[7]

It appears that these processes are reflected electrically in P600/LPC potentials (Chapters 5 & 6). As described in Chapter 6, a study by Patel *et al*. (1998) compared ERPs elicited by structural events that required syntactic reanalysis and/or revision, in both music and language. The linguistic incongruities were 'garden-path' sentences[8] or ungrammatical sentences. Musical incongruities were out-of-key chords (either from a nearby, or from a distant key) presented in sequences of successive chords. Both musical and linguistic structural incongruities elicited P600 components, taken to reflect processes of knowledge-based structural integration during the perception of rule-governed sequences.

The neural generators of the music-generated LPC/P600 are not known. During language perception, processes of re-analysis and revision (as reflected in the P600) appear to involve posterior superior and middle temporal cortex (Grodzinsky & Friederici, 2006; Bahlmann *et al*., 2007), as well as basal ganglia (Kotz *et al*., 2009). Moreover, perhaps due to increased working memory demands, the generation of the P600 also appears to involve (dorsolateral) frontal and prefrontal regions (Bahlmann *et al*., 2007; Opitz & Kotz, 2011; Wassenaar & Hagoort, 2007) as well as the inferior parietal lobule (Bahlmann *et al*., 2007).

Vitalization: Effects of music on the autonomic and the endocrine system
Music listening, and perhaps even more so music making, can have vitalizing effects on an individual. Vitalization entails activity of the autonomic nervous system (i.e., regulation of sympathetic and parasympathetic activity) along with the conscious cognitive integration of 'musical' and 'non-musical' information

---

[5]   For details see, e.g., Sturt *et al*. (1999).

[6]   Think of 'garden-path' sentences such as 'When Fred eats food gets thrown', 'Mary gave the child the dog bit a bandaid', or 'They painted the wall with cracks'. These sentences are syntactically correct, but readers tend to build an initial hierarchical structure that turns out to be wrong towards the end of the sentence. In order to establish the correct representation of the sentence, or the correct hierarchical structure, readers have to re-analyze the sentence and to revise their structural model.

[7]   Think, for example, of the beginning of Beethoven's first symphony in *C* major, which starts in *F* major, then moves to *G* major, before reaching the actual home key of *C* major.

[8]   Such as 'Some of the senators endorsed promoted an old idea of justice' – here, the word 'endorsed' is locally ambiguous between a main-verb interpretation and a reduced-relative clause interpretation.

(an unconscious individual is hardly vitalized). For example, Chapter 10 will describe how music can evoke (non-musical) physiological effects that have meaning quality for the perceiver. The interpretation of such meaning information requires conscious awareness, and therefore presumably involves multimodal association cortices such as parietal cortex in the region of BA 7. Due to its role for conscious awareness (Block, 2005), BA 7 is probably also involved in the conscious awareness of a musical percept.

Effects of music perception on activity of the autonomic nervous system have been investigated mainly by measuring electrodermal activity, heart rate, and heart-rate variability, as well as the number and intensity of musical frissons (i.e., of intensely pleasurable experiences often involving goosebumps or shivers down neck, arms, or spine; Khalfa et al., 2002; Blood & Zatorre, 2001; Panksepp & Bernatzky, 2002; Sloboda, 1991; Grewe et al., 2007a,b; Lundqvist et al., 2009; Orini et al., 2010).

Because autonomic activity always acts in concert with endocrine activity, music perception (as well as music making) also has effects on hormonal activity. So far, most of the studies investigating effects of music on endocrine activity have measured levels of cortisol. These studies show increased cortisol levels (presumably related to the vitalization of individuals) after listening to music (VanderArk & Ely, 1992; Gerra et al., 1998; Evers & Suhr, 2000), singing (Beck et al., 2000; Kreutz et al., 2004), dancing (West et al., 2004; Quiroga Murcia et al., 2009), and music therapy (Burns, 2001; McKinney et al., 1997). On the other hand, music listening can lead to a reduction of cortisol levels in individuals with high levels of worry (e.g., due to impending clinical procedures such as colonoscopy, gastroscopy, angiography, or surgery; Escher et al., 1993; Miluk-Kolasa et al., 1994; Schneider et al., 2001; Uedo et al., 2004; Nilsson et al., 2005; Leardi et al., 2007; Nilsson, 2009; Koelsch et al., 2011). However, as a note of caution it should be mentioned that only few of the studies investigating effects of music on cortisol (in healthy or clinical populations) were randomized controlled trials, or used a control-group design (for a critical review see Koelsch & Stegemann, in press).

Effects of music on the immune system   Activity of the (autonomic) nervous system as well as of the endocrine system has, in turn, effects on the immune system. Effects of music processing on the immune system have been assessed mainly by measuring variations of (salivary) immunoglobulin A (IgA) concentrations (e.g., Hucklebridge et al., 2000; Beck et al., 2000; McCraty et al., 1996; Kreutz et al., 2004; Quiroga Murcia et al., 2009).[9] In all of these studies, IgA concentrations increased, and in all of these studies the effects of music on the immune system

---

[9]   Immunoglobulins are antibodies. Antibodies can identify and neutralize alien objects such as bacteria and viruses. Immunoglobulins are produced by the humoral immune system, specifically by so-called B cells (a type of lymphocyte), that is, by a class of white blood cells (or 'leukocytes').

were related to an increase in positive mood (music-evoked emotions will be dealt with in detail in Chapter 12).

Premotor processes evoked by music perception   Neural activities of the late stages of perception overlap with those related to the early stages of action (such as premotor functions related to action planning; Rizzolatti & Craighero, 2004; Janata *et al.*, 2002b). For example, music perception can interfere with action planning in musicians (Drost *et al.*, 2005a,b), and premotor activity can be observed during the perception of music (a) in pianists listening to piano pieces (Haueisen & Knösche, 2001), (b) in non-musicians listening to song (Callan *et al.*, 2006), and (c) in non-musicians who received one week of piano training, and subsequently listened to the trained piano melody (Lahav *et al.*, 2007). Details on neural correlates of perception-action mediation during music listening will be described in detail in Chapter 11.

Movement induction by music perception in the way of tapping, dancing or singing along to music is a very common experience (Panksepp & Bernatzky, 2002). Such movement induction has social relevance, because synchronized movements of different individuals represent coordinated social activity (the social functions engaged by music are described in Chapter 12). Action induction by music perception is accompanied by neural impulses in the reticular formation (in the brainstem; for example, for the release of energy to move during joyful excitement). It is highly likely that the reticular formation, in turn, projects to the auditory brainstem (as well as to the auditory cortex; Levitt & Moore, 1979), and that the neural activity of the reticular formation, therefore, also influences the processing of (new) incoming acoustic information.

Processing meaning in music   Music can transfer meaningful information, and is an important means of communication. Chapter 10 will describe dimensions of meaning information, arguing that meaning information conveyed by music can be divided into three fundamentally different classes: Extra-musical, intra-musical, and musicogenic meaning. A number of studies indicate that processing of extra-musical meaning is reflected in N400 ERP responses (Steinbeis & Koelsch, 2008a; Daltrozzo & Schön, 2009b; Steinbeis & Koelsch, 2011; Grieser-Painter & Koelsch, 2011; Goerlich *et al.*, 2011). In these studies, N400 responses were elicited by musical information that was semantically unrelated to the meaning of a preceding (prime-) word. Processing of intra-musical information appears to be reflected in part in the N5. For example, a study by Steinbeis & Koelsch (2008b) showed that the generation of the N5 (elicited by irregular chord functions) can be modulated by the semantic properties of a simultaneously processed word. These findings suggest that processing of musical meaning information is reflected in (at least) two different brain-electric responses: The N400 reflecting processing of extra-musical meaning, and the N5 reflecting processing of intra-musical meaning.

With regard to the N400, two studies (Koelsch *et al.*, 2004a; Steinbeis & Koelsch, 2008a) suggest that the processing of extra-musical meaning, as

Table 8.1   Systematic overview of basic processes involved in music perception.

music perception	feature extraction	Gestalt formation	interval analysis	structure building	structural re-analysis	vitalization	premotor, immune system

reflected in N400 potentials, involves neural resources located in posterior superior temporal (BA 22p), and posterior middle temporal regions (BA 21/37). Neural generators of the N5 have remained elusive. Because the N5 interacts with the N400 (for details see Chapter 9), it seems likely that the neural generators of the N5 overlap at least in part with those of the N400. However, the scalp distribution of the N5 is also considerably more frontal than that of the N400. Therefore, it seems highly likely that the N5 receives additional contributions from frontal areas. Such additional contributions possibly originate from Broca's area, the inferior frontal sulcus, dorsolateral prefrontal cortex, and premotor cortex (possibly due to operations such as working memory, semantic selection and semantic integration, although this remains to be specified).

The theoretical basis   The different stages of music perception described in this chapter are summarized in Table 8.1. In the next chapters, which deal with neurophysiological correlates of the processing of musical syntax, musical semantics, music and action, as well as music-evoked emotions, further theoretical considerations will be added step by step to this table. Thus, Table 8.1 will serve as a basis for the construction of a new systematic theory of music psychology. With regard to Table 8.1, and the expansion of this table in the next chapters, this theory is organized in a way that *if* a phenomenon A is a necessary condition for a phenomenon B in a row (in the sense of $\neg A \rightarrow \neg B$, or $B \rightarrow A$), *then* it is listed *left* of this phenomenon (that is, A is listed then left of B). Note, however, that simply the appearance of a phenomenon left of another phenomenon does not require that the former is a necessary condition for the latter (nor does it require that the former is a sufficient condition for the latter).

For example, *extraction of acoustic features* is a necessary condition for the other processes of *music perception* and, therefore, listed left of all other phenomena in that row. *Gestalt formation* is a prerequisite for syntactic *structure building* and, therefore, listed left of structure building. Syntactic *structure building* requires (in tonal music) the distinction between major and minor, or between different inversions of chords (therefore minute *interval analysis* is listed left of structure building). Processes of syntactic *reanalysis* and revision follow processes of structure building. Note that the fact that *vitalization* (including autonomic and endocrine hormonal activity) is listed left of *premotor* processes does not mean that it is a necessary condition for premotor processes (which it is not: It is possible that, for example, a single tone activates premotor representations). It is also important to note that processes can feed back into those located left of them. For example, all processes of music perception located right of the *extraction of acoustic features* can feed back into, and thus modulate, processes of acoustic feature extraction.

   In the following chapters, further rows will be added to this table. Henceforth, I will refer to the outermost left cell of a row as *domain*, and to the cells right of each domain as *dimensions*. For example, Table 8.1 summarizes the domain of *music perception*, and the dimensions of music perception are auditory *feature extraction*, *Gestalt formation*, etc.

9

Musical Syntax

9.1 What is musical syntax?

This chapter deals with neurophysiological correlates of the syntactic processing of major-minor tonal music, particularly with regard to the syntactic processing of chord functions. Although the syntax of tonal music comprises several structural aspects (melody, metre, rhythm, harmony, intensity, instrumentation, and texture), investigations on the neural correlates of music-syntactic processing have so far mainly focussed on the processing of harmony. One possible theoretical description of the syntax of harmony is the classical theory of harmony as formulated, e.g., by Jean-Philippe Rameau (1722), as well as their continuation and expansion by, e.g., Gottfried Weber (1817), Moritz Hauptmann (1873), Hugo Riemann (1877/1971) and Siegfried Karg-Elert (1931). These descriptions primarily deal with the derivation of chord functions, and the construction of cadences, rather than with how cadences are generated, and how they are chained into pieces.[1]

Heinrich Schenker was the first to see an entire movement as an enlarged cadence and to provide a systematic analysis of principles underlying the structure of entire pieces (e.g., Schenker, 1956).[2] Schenker's analysis included his thoughts on the *Ursatz*, which implicitly assume a hierarchical structure such as the large-scale tonic-dominant-tonic structure of the sonata form. Moreover, his analysis implies

---

[1] For an account on the term *cadence* see Caplin (2004).
[2] Similarly, Arnold Schoenberg (1978) noted that 'we can consider the chorale, as well as every larger composition, a more or less big and elaborate cadence' (p. 290).

Brain and Music, First Edition. Stefan Koelsch.
© 2013 John Wiley & Sons, Ltd. Published 2013 by John Wiley & Sons, Ltd.

recursion (although Schenker himself never used this term),[3] and that certain chord functions can be omitted without destroying the Ursatz, whereas others cannot.[4,5] The *Generative Theory of Tonal Music* (GTTM) by Lerdahl & Jackendoff (1999) began to formalize Schenker's thoughts (although the GTTM differs in many respects from Schenker's theory). Note that GTTM is an analytical (not a generative) model.[6] Lerdahl's *Tonal Pitch Space* theory (TPS; Lerdahl, 2001b) attempts to provide algorithms for this analytical approach which are, however, often not sharp and precise enough to operate without subjective corrections. Nevertheless, the primary advance of the TPS theory (and GTTM), lies in the description of Schenker's reductions as tree-structure (in this regard, GTTM uses terms such as *time-span reduction* and *prolongation*).[7] These tree-structures parallel the description of linguistic syntax using tree structures (note, however, that Lerdahl & Jackendoff's GTTM at least implicitly assumes that musical structure and linguistic syntax have only little to do with each other, except perhaps with regard to certain prosodic features). TPS also provides the very interesting approach to model tension-resolution patterns from tree-structures (e.g., Lerdahl & Krumhansl, 2007), and aims to provide a theoretical framework that applies not only to tonal music, but (with suitable modifications) also to other kinds of hierarchically organized music.

A more recent approach to model tonal harmony with explicit generative rules according to tree-structures is the *Generative Syntax Model* (GSM) by Martin Rohrmeier (2007) (see also Rohrmeier, 2011). In contrast to GTTM (and TPS), GSM does not (yet) address rhythm or melody, and GSM is up to now specified for tonal music only (whereas GTTM can relatively easily be generalized to other kinds of music with hierarchical organization). Rohrmeier's GSM distinguishes between four structural levels: Phrase-structure level, functional-structural level, scale-degree structure level, and a level with surface structure (see also Figure 9.1). Each of these levels is described with generative rules of a phrase-structure grammar, the generation beginning on the highest (phrase-structure) level, and propagating information (in the process of derivation) through the sub-ordinate levels. This opens the possibility for the recursive derivation of complex sequences on both a functional and a scale-degree structural level (e.g., 'dominant of the dominant of the dominant'). The actual tonality of a tree is propagated from the head into the

---

[3]   The elaborative and transformational principles of his theory apply recursively at different levels (from his deep structure to the musical surface). This is similar, for example, to a sequence in one key which is embedded in another sequence in a different key, which is embedded in another sequence with another different key, etc. Note, however, that Schenker's theory does not primarily deal with key relations.

[4]   Note that his theory dealt primarily with the principle that tonal music can be reduced to a fundamental Ursatz. In this regard, harmony was not as central for his theory as it was, e.g., for Riemann.

[5]   For a summary of Schenker's theory see, e.g., Lerdahl (2001b).

[6]   Despite its name, GTTM does not provide any generative rules.

[7]   For an introduction and explanation of terms see Lerdahl (2009).

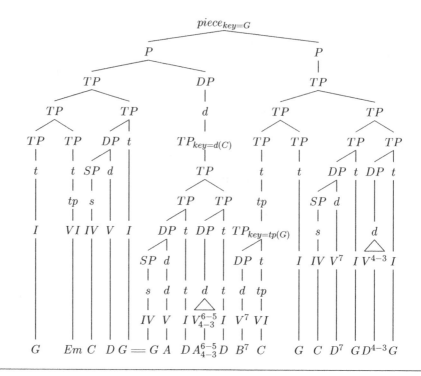

Figure 9.1 Analysis of Bach's chorale 'Ermuntre Dich, mein schwacher Geist', mm. 1–4. The = sign indicates that both instances of the *G* chord refer to the identical surface pivot chord; the triangle is an abbreviation symbol for omitted derivations. The phrase-structure level (top) is indicated by the upper-case symbols (*P, TP, DP, SP*), the functional-structural level is indicated by the lower-case letters (*t, s, d, tp, sp, dp, tcp*), the scale-degree structure level by Roman numeral notation, and the surface structure by the chord symbols. *TP*: Tonic phrase; *DP*: Dominant phrase; *SP*: Subdominant phrase. Figure reprinted with permission from Rohrmeier (2011).

sub-ordinate branch(es). The tree-structure reflects (a) the formal arrangement of the piece, (b) the phrase structure, (c) the functional aspects of partial phrases and chords, (d) the relation of key-regions, and (e) the degree of relative stability (which is determined by the sub-ordination rules between the branches).

The GSM combines three principles, as follows. (1) Given (a) that musical elements (e.g., a tone or a chord) always occur in relation to other musical elements (and not alone), and (b) that each element (or each group of elements) has a functional relation either to a preceding or to a subsequent element, the simplest way of representing these relations graphically is a tree-structure. This implies that relations of elements that are functionally related, but not adjacent, can be represented by a tree-structure. For example, in a sequence which consists of several chords and which begins and ends with a tonic chord, the first and the last tonic are not adjacent (because there are other chord functions between them), but the last tonic picks up (and 'prolongates') the first tonic. This can be represented by a tree structure in a way that the first and the last tonic chords build beginning- and end-points of the tree, and that the relations of the other chord functions can be represented by ramification of branches ('prolongating' chords can extend the tonal region, and 'progressing' chords determine the progression of tonal functions). Of course this also implies that adjacent elements do not necessarily have a direct structural relation (for example, an initial tonic chord can be followed by a secondary dominant to the next chord: The secondary dominant then has a direct relation to the next chord, but not to the preceding tonic chord). It is not possible that a chord does not have a functional relation to a preceding nor subsequent chord; if in our example the secondary dominant would not be followed by an (implicit) tonic, the sequence would be ungrammatical (with regard to major–minor tonal regularities). Compared to the classical theory of harmony, this GSM approach has the advantage that chord functions are not determined by their tonality (or the derivation from other chords), but by their functional position within the sequence.

(2) The tree-structure indicates which elements can successively be omitted in a way that the sequence still sounds correct. For example, secondary dominants can be omitted, or all chords between first and last tonic chord can be omitted, and a sequence still sounds correct (though less interesting). Assume a less trivial case, in which a dominant occurs in a local phrase between two tonic chords (and several chords between each tonic and dominant chord); then, the tree structure indicates that on the next higher level all chords except the two tonic chords and the dominant chord can be omitted. This principle is a consequence of the 'headedness,' in which each node in the syntax tree dominates sub-ordinate branches. That is, the tree-structure provides a weighting of the elements, leading to an objectively derivable deep structure (another advantage compared to the classical theory of harmony).[8]

---

[8]   With regard to the terms *cadence* and *phrase*, Caplin (2004) argued that 'many of the problems associated with cadence (and indeed with phrase) can be dispelled when the two concepts are entirely disengaged. Cadence can then be viewed as a manifestation of formal functionality, whereas phrase can be used as a functionally neutral term for grouping structure (embracing approximately four measures of music)' (p. 59). However, Caplin also notes that by 'separating cadence from phrase it is possible to describe more clearly which phrases have cadential closure and which do not.' For example, both half cadence and authentic cadence usually represent a phrase, but whereas the half cadence requires another phrase to follow, the authentic cadence brings a phrase, or a group of

(3) Chord functions are a result of their position within the branches of the tree (notably, the rules of the GSM specify the positions of the functions in the tree). For example, a pre-dominant is the sub-ordinate branch preceding a dominant, or there are stable and instable dominants which can be differentiated based on how deep they are located in the syntax-tree. Similarly, a half-cadence is a case in which a dominant is reached as stable local endpoint of a phrase, which must be followed by a phrase that ends on a tonic, thus resolving the open dominant (with regard to the deep structure).

Up to now, no neurophysiological investigation has tested whether individuals perceive music cognitively according to tree-structures. Similarly, behavioural studies on this topic are extremely scarce (but see Cook, 1987; Bigand *et al.*,1996; Lerdahl & Krumhansl, 2007). That is, it is still an open question whether tree-structures have a psychological reality in the cognition of listeners of (major–minor tonal) music. As will be described later in this chapter, studies investigating neurophysiological correlates of music-syntactic processing have so far mainly utilized the classical theory of harmony (according to which chord functions are arranged within harmonic sequences according to certain regularities).[9] Before such neurophysiological correlates are described in more detail (in Section 9.3), cognitive processes involved in the syntactic processing of (tonal) music will be enumerated.

9.2 Cognitive processes

Processing of musical syntax comprises several (sub-)processes. Based on the processes underlying music perception described in Chapter 8, these processes are enumerated in the following (for a summary see Table 9.1). Although formulated here with regard to the processing of tonal music, the logic of this enumeration is supposed to be applicable to other kinds of music as well. Note that not all of these processes are necessarily involved in different domains, or different kinds of music: For example, the music of the Mafa people (see p. 130 and Fritz *et al.*, 2009) does not seem to have a hierarchically organized syntactic structure, therefore cognitive operations related to the processing of hierarchical structures do not appear to be applicable to their music. Also note that the ordering of the enumerated processes is not intended to reflect a temporal order of music-syntactic processing (that is, the processes may partly happen in parallel).

The syntax of tonal music comprises several structural aspects: Melody, metre, rhythm, harmony, and timbral structure (including intensity, instrumentation, and texture). So far, processing of such aspects has been investigated mainly

phrases, to an end that does not necessarily require the continuation of another phrase. Moreover, whereas a cadence is usually also a phrase, a phrase does not need to be a cadence (for example, a phrase may consist of tones that belong to only one chord function).

[9] To my knowledge, Hugo Riemann (1877/1971) was the first to refer to such regularity-based arrangement as *musical syntax* (Riemann used the term *musikalische Syntaxis*).

Table 9.1   Systematic overview of syntactic processes (bottom row), in relation to processes of music perception. Note that syntactic *element extraction* is related to auditory *feature extraction*. *Knowledge-free structuring* is based on operations on the auditory sensory memory and grouping, and therefore related to *Gestalt formation*. The establishment of (long-term) knowledge about the intricate statistical probabilities in order to form *musical expectancies* involves minute *interval analysis* to determine probabilities for chord-transitions between major and minor chords, and to recognize how the position of a chord (e.g., root position, sixth, or six-four chord) influences the probabilities of chord-transitions. The pleasure and relaxation arising from the *integrated* representation of the (simultaneous) operation of syntactic features is related to processes of music-evoked *vitalization*. Finally, the systemic nature of *large-scale structuring* parallels the systemic nature of *immune* activity as an effect of autonomic and endocrine activity during music perception and production (note that both span relatively long time intervals, in contrast to the other processes).

music perception	feature extraction	Gestalt formation	interval analysis	structure building	structural re-analysis	vitalization	premotor, immune system
syntactic processing	element extraction	knowledge-free structuring	musical expectancy formation	structure building	structural re-analysis	syntactic integration	large-scale structuring

with regard to harmony, other aspects have been investigated only sparsely.[10] Therefore, a particular emphasis is put here on the syntactic processing of harmony (future conceptualizations, however, might elaborate on other aspects as well).

(1) Element extraction.   Elements such as tones and chords (or words in language) are extracted from the continuous stream of auditory information. In homophonic and polyphonic music, a representation of a current melodic and harmonic event is established (with the harmonic event colouring the melodic event). With regard to the temporal structure, a *tactus* is extracted.[11] Because a tactus is not an auditory feature, I use the term *syntactic element extraction* here (instead of *auditory feature extraction*; see also Table 9.1).

(2) Knowledge-free structuring.   A representation of structural organization is established on-line (on a moment-to-moment basis) without obligatory application of long-term knowledge. Consider, for example, a sequence like:

.....ˉ...ˉ......ˉ....ˉ.. etc.

Without any prior knowledge, it is possible to establish a representation of '.' being a high-probability standard, and 'ˉ' being a low-probability deviant. Listening

---

[10]   For investigations of the syntactic processing of metre and rhythm see Tomic & Janata (2008) and Winkler *et al.* (2009b).
[11]   The tactus is represented by the most salient time periodicity with which musical elements occur, corresponding to the rate at which one might clap, or tap to the music. The colloquial expression is *beat*.

to a musical passage in one single key, an individual can establish a representation of the tones of a key, and to detect out-of-key tones (thus also enabling them to determine key membership) based on information stored in the *auditory sensory memory*: The in-key tones become standard stimuli, and any out-of-key tone (such as any black piano key producing a tone within a sequence of *C* major) represents a deviant stimulus ('auditory oddball'). Such auditory (acoustic) oddballs elicit an *MMN* (see Chapter 5). Models that can represent such regularities, only considering information available in the auditory sensory memory (without application of long-term memory information) may be referred to as *on-line minimal fragment models*.[12]

The processes underlying the establishment of such models include *grouping*. For example, in a sequence like:

-...-...-...-...-... etc.

four elements are grouped together (most likely to '-...' or '...-') based on *Gestalt principles*.[13] With regard to the melodic structure of a piece, grouping is required to assemble single tones to a melodic contour. With regard to the temporal structure, grouping serves the extraction of the *metre* of a piece, as well as of rhythmic patterns.[14]

Moreover, based on the (Gestalt) principles of contextual identity, contextual distance, and contextual asymmetry (see Chapter 3), relations between tones and chords can be established that give rise to a tonal order, or a *hierarchy of stability* (Bharucha & Krumhansl, 1983; Krumhansl, 1990). According to the hierarchy of stability, the configuration of chords forms a tonal structure. For example, within the tonal hierarchy of stability the tonic chord is the most 'stable' chord, followed by the dominant and the subdominant, whereas chords such as the supertonic represent less stable chords.[15]

The Gestalt principles that give rise to a hierarchy of stability are grounded (a) in acoustic similarity and representations of tones in auditory sensory memory,[16]

---

[12]   The term *on-line minimal fragment models* was suggested to me by Martin Rohrmeier (personal communication).

[13]   Within such a sequence, a group like '-....' will also elicit an MMN if the subject realizes the grouping of the tones (reviewed in Sussman 2007).

[14]   Metre refers to the grouping of the basic isochronous temporal units into regular measures, e.g., the 4/4 ('March') metre, the 3/4 ('Waltz') metre, or the (more interesting) 5/4, 7/8, etc. metres.

[15]   I refer to chord functions ('tonic', 'dominant' etc.) to illustrate the hierarchy of stability. This does not mean that at this stage a representation of a chord function is already fully established. Here, psychoacoustic principles determine the functional roles of chords (and tendencies to move from less stable to more stable tones and chords, or to move from more dissonant to more consonant tones and chords). In the next stage (musical expectancy formation), the functional role of chords is also determined based on convention-based regularities, such as the dominant-character of a six-four chord.

[16]   For detailed accounts see Huron & Parncutt (1993) and Leman (2000).

as well as (b) in representations of major–minor tonal key-, chord-, and tone-relationships. As mentioned above, auditory sensory memory operations establish representations of standard pitches, and detect deviant pitches (the detection of deviants being modulated by the degree of acoustic dis/similarity; see Chapter 1). Leman (2000) showed that key profiles obtained by Krumhansl & Kessler (1982a) show a remarkable resemblance to those provided by an auditory sensory memory model (the model was based on echo effects of periodicity pitch patterns; for a similar account see also Huron & Parncutt, 1993). Leman (2000) even went so far as to claim that 'the probe-tone experiments provide no evidence for the claim that listeners familiar with Western music have abstracted tonal hierarchies in a long-term memory' (p. 508). This is relevant because, although such key profiles describe representations of the hierarchy of tonal functions of single tones (not chords), such a tonal hierarchy is taken as basis for a hierarchy of stability of chord functions, and thus the mentioned contextual principles.[17] Corroboratingly, Bharucha & Krumhansl (1983) reported in their initial study that 'no regular differences appeared between the responses of listeners with and without backgrounds in music theory' (p. 84). On the other hand, acculturation may shape such establishment (as suggested by a developmental trajectory of such establishment during childhood; for a review see Krumhansl & Cuddy, 2010). Therefore, long-term knowledge about the relations of tones and chords is not a *necessary condition* for the establishment of a tonal hierarchy, but a *sufficient condition* for the modulation of such establishment (that is, the establishment of such a hierarchy is shaped by cultural experience).

The tonal centre of the harmonic hierarchy corresponds to the root of a tonal key (and is perceived as the best representative of a tonal key; see Krumhansl & Kessler, 1982a). In terms of harmonic function, the tonal centre is also the root of the tonic chord, and thus the reference point for the tonal hierarchy or chord functions. The process of establishing a representation of a tonal centre is normally an iterative process (Krumhansl & Kessler, 1982a), and this process has to be engaged each time the tonal key changes.[18] Listeners (at least those familiar with tonal music) tend to interpret the first chord (or tone) of a sequence as the

---

[17]   Krumhansl & Kessler (1982a) stated that the 'alterations in judgments, which are summarized in three context-dependent principles, are a function of the distance between the instantiated context key and the key in which the chords play harmonic roles. As with single tones, then, the perception of chords depends significantly on their functions within the established key. This suggests that if invariant relationships are found in music, they are neither at the level of single tones nor at the level of chords but instead may be found at the more abstract level of musical keys or tonal centres.' (p. 337). On p. 363, it is stated that the observation of a hierarchy of tones within an instantiated key 'raises the possibility that the internalization of interkey structure might likewise be derivative of the perceptual assignment of tones to positions within the hierarchy of tonal functions.' On p. 366, the authors state that 'at the most basic level … relations between keys and between individual chords and keys are all mediated through the internalized hierarchy of tonal functions at the level of individual tones.'
[18]   For an overview of approaches to modelling perception of tonal space see Janata (2007).

tonic (that is, as the tonal centre; Krumhansl & Kessler, 1982a). In case the first chord is not the tonic, listeners have to modify their initial interpretation of the tonal centre during the perception of subsequent tones or chords (Krumhansl & Kessler, 1982a).[19]

(3) Musical expectancy formation.   The 'on-line minimal fragment models' described in the last section are established on a moment-to-moment basis without long-term knowledge of rules or regularities. By contrast, music-syntactic processing also involves representations of regularities that are stored in a long-term memory format (for example, probabilities for the transition of musical elements such as chord functions). Such representations can be modelled as *fragment models*, e.g., n-gram, Markov, chunking, or PARSER models (see, e.g., Servan-Schreiber & Anderson, 1990; Perruchet & Vinter, 1998).[20]

For example, with regard to tonal music, Rohrmeier (2005) found in a statistical analysis of the frequencies of diatonic chord progressions occurring in Bach chorales, that the supertonic was five times more likely to follow the subdominant than to precede it (a Markov-table of the transition probabilities of chord functions in this corpus is reported in Rohrmeier & Cross, 2008). Such statistical properties of the probabilities for the transitions of chord functions are implicitly learned during listening experience (see also Tillmann, 2005; Jonaitis & Saffran, 2009) and stored in a long-term memory format. The mathematical principles from which the probabilities for chord transitions within a tonal key might have emerged are under investigation (see, e.g., Woolhouse & Cross, 2006, for the interval cycle-based model of pitch attraction), and it appears that many of these principles represent abstract, rather than physical (or acoustical) features (Woolhouse & Cross, 2006).

With regard to the model of music perception (Chapter 8 and Table 9.1), the establishment of knowledge about the intricate statistical probabilities of tonal music involves minute *interval analysis* to determine probabilities for chord-transitions between major and minor chords, and to recognize how the position of a chord (e.g., root position, sixth, or six-four chord) influences the probabilities of chord-transitions. For example, six-four chords often have dominant character,[21] a dominant that does not occur in root position is unlikely to indicate cadential arrival, a tonic presented as sixth chord is unlikely to be the final chord of a chord sequence

---

[19]   For a conception of key identification within the tonal idiom see also the intervallic rivalry model from Brown *et al.* (1994).

[20]   Computationally, 'on-line minimal fragment models' can also be represented as n-gram models, but the important difference between *knowledge-free structuring* and *musical expectancy formation* is that the former is based on psychoacoustic principles and information stored the auditory sensory memory, whereas the latter is based on long-term memory. Importantly, this does not exclude that during listening to a piece, experience based on *knowledge-free structuring* is immediately memorized and used throughout the musical piece.

[21]   In C major, the chord g – c – e is not heard as a tonic, but as a dominant. This chord is usually referred to as cadential 6/4 chord, i.e., dominant with 6/4 suspension.

(the same holds for a tonic in root position with the third in the top voice), etc. (for further examples see also Caplin, 2004). In this sense, a *chord function* parallels a *lexeme*,[22] and the different versions of a chord function parallel word inflections.

On the metrical and tonal grid established due to *knowledge-free structuring*, *musical expectancies* for subsequent structural elements are formed on the basis of *implicit knowledge*. Note that such musical expectancies are different from the expectancies (or predictions) formed due to *knowledge-free structuring*, because the latter are formed on the basis of acoustic similarity, acoustic regularity, and Gestalt principles, without long-term memory representations of statistical probabilities being required. Importantly, processing of information based on fragment models can only represent *local dependencies* (not long-distance dependencies); processing of long-distance dependencies is dealt with in the next paragraph.

(4) Structure building.   Tonal music (as well as many other kinds of music) is hierarchically organized, giving rise to building and representing structures that involve long-distance dependencies on a phrase-structure level (that is, structures based on a *context-free grammar*). Such *hierarchical* structures may involve *recursion*, and they can best be represented graphically as a tree-structure (see, for example, Figure 9.1). The representation of such structures requires (auditory) *Working Memory*.[23]

Musically unexpected syntactic elements (such as an irregular chord function), that is, elements that contradict predictions based on *musical expectancy formation*, disrupt structure building.

(5) Structural reanalysis and revision.   During the syntactic processing of a sequence of elements, perceivers often tend to structure the elements in the most likely way (for example, based on thematic role assignment, minimal attachment, late closure, etc.).[24] However, it may also happen that a listener recognizes that an established hierarchical model has to be revised. That is, the headedness of branches, the assignment of elements to branches, etc. may have to be modified. To give an example with regard to language comprehension, the beginning of 'garden-path' sentences suggests a certain hierarchical structure, which turns out to be wrong (Townsend & Bever, 2001, provide examples such as 'The cotton clothing is made of grows in Mississippi'). Note that while building a hierarchical structure, there is always ambiguity as to which branch a new element might belong, whether a new element belongs to a left- or a right-branching part of a tree, whether the functional assignment of a node to which the new element

---

[22]   And a chord in root position parallels a lemma.
[23]   For fMRI studies investigating tonal and verbal working memory see Schulze *et al.* (2011b), Koelsch *et al.* (2009), Hickok *et al.* (2003), and Schulze *et al.* (2011a). For behavioural studies see Williamson *et al.* (2010a) and Williamson *et al.* (2010b).
[24]   For details see, e.g., Sturt *et al.* (1999).

Figure 9.2   Interaction between function of chords and metre: The left sequence is a regular cadence. The chord functions of the middle sequence are identical to those of the left sequence, but due to the tonic occurring on a light beat, the sequence sounds incorrect. If the penultimate dominant seventh chord is repeated (right sequence), thus not altering the harmonic structure of the chord sequence, the tonic occurs on a heavier beat and the sequence sounds correct again (though not closing as strongly as the left sequence).

belongs is correct, etc. However, once a branch has been identified with reasonable certainty, it is represented as a branch (not as single elements). If this branch (or at least one node of this branch), however, does subsequently turn out to not fit into the overall structure (e.g., because previously assumed dependencies break), the structure of the phrase, or sentence, has to be re-analyzed and revised.

(6) Syntactic integration.   As mentioned in the beginning of this section, several syntactic features constitute the structure of a sequence (e.g., in tonal music, melody, harmony, metre etc.). Therefore, *integration* of these features by a listener is required to establish a coherent representation of the structure, and thus to understand the structure. Figure 9.2 illustrates the interaction between the function of chords and metre: The four-gram $I - VI - IV - V^7$ shown in the left sequence of Figure 9.2 yields a very high probability for a tonic chord being the next chord function. Note that this four-gram is in a 4/4 metre, starting on the first beat, with each chord having the duration of a quarter note. The same sequence of chord functions sounds incorrect, for example, when starting on the second beat (middle sequence of Figure 9.2), or when played in a 5/4 metre. Thus, whether a listener perceives a chord sequence as correct or incorrect is not only determined by the succession of chord functions, but by the integrated representation of harmonic, metric, rhythmic, etc. information.

Note that in many listeners, the simultaneous operation of melody, metre, rhythm, harmony, intensity, instrumentation, and texture evokes feelings of pleasure. After the closure of a cadence, and particularly when the closure resolves a previous breach of expectancy, and/or previous dissonances (for details see Chapters 10 & 12), the integrated representation of the (simultaneous) operation of all syntactic features is perceived as particularly pleasurable and relaxing. This relates processes of syntactic integration to processes of music-evoked vitalization during music listening and music making (see also Table 9.1).

(7) Large-scale structuring.   The previous sections dealt with the processing of phrase structure (i.e., processing of phrases that close with a cadence). However, musical pieces usually consist of numerous phrases, and thus have large-scale structures. For example, verse and chorus in a song, the A–B–A(') form of a minuet, the parts of a sonata form, etc. When listening to music from a familiar

musical style with such organization, these structures can be recognized (often with the help of typical forms and transitions), and such recognition is the basis for establishing a representation of the large-scale structuring of a piece. Large-scale structures imply the spanning of relatively long time intervals (minutes to hours). With regard to the model of music perception (Table 9.1), the systemic nature of such long time-range relations parallels the systemic nature of autonomic, endocrine, and immune activity during music perception and production.[25]

9.3   The early right anterior negativity (ERAN)

This section describes the early right anterior negativity (ERAN), one electrophysiological correlate of music-syntactic processing (for descriptions of the LPC/P600, and the closure positive shift; see Chapters 5 & 6). The ERAN has been used mainly in experiments investigating the processing of chord functions with chord-sequence paradigms. The first of such experiments (Koelsch *et al.*, 2000) used chord sequences consisting of five chords each (see Figure 9.3 for examples). There were three sequence types of interest: (1) sequences consisting of music-syntactically regular chords, (2) sequences with a music-syntactically irregular chord at the third position of the sequence, and (3) sequences with a music-syntactically irregular chord at the fifth position of the sequence (see Figure 9.3a). The irregular chords shown in the examples of Figure 9.3a are referred to here as *Neapolitan sixth chords* (for explanation of the Neapolitan sixth chord see Section 2.3), although, historically, Neapolitan sixth chords only occur in minor contexts. Note that the Neapolitans are consonant chords when played in isolation, but that they are functional-harmonically only distantly related to the preceding harmonic context. Moreover, the Neapolitans contain out-of-key notes. Hence, the Neapolitan chords sound highly unexpected when presented at the end of a chord sequence (right panel of Figure 9.3a). The same chords presented in the middle of the chord sequences (middle panel of Figure 9.3a), however, sound much less unexpected because Neapolitan sixth chords are similar to subdominants (which are music-syntactically regular at that position of the sequence). However, due to their out-of-key notes, the Neapolitans at the third position still sound less expected than a regular subdominant (for acoustic irregularity and dissimilarity see next section).

Figure 9.4 shows the hierarchical organization of the sequences shown in Figure 9.3 with regard to the GSM. The Neapolitans at the third position represent an unexpected change of key (because Neapolitan sixth chords occur only in minor contexts), but due to their nature as subdominant chords these chords can still be integrated into the cadential context. In addition, the Neapolitan

---

[25]   Autonomic changes are relatively fast (in the range of milliseconds to seconds), the concerted endocrine changes are slower (in the range of seconds to minutes), and the immune system effects are in the time range of minutes to hours.

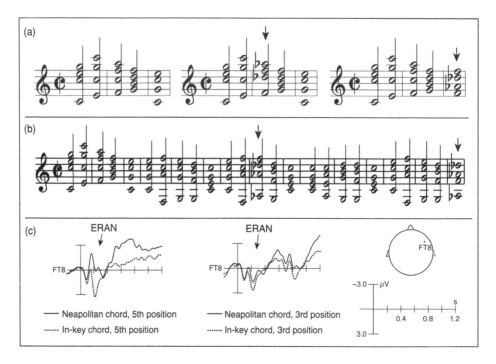

Figure 9.3 (a) Examples of chord sequences containing in-key chords only (left panel), a 'Neapolitan sixth chord' at the third position (middle panel), and at the fifth position of the sequence (right panel). All together, the stimulus pool comprised of 172 different chord sequences. Presentation time of chords 1 to 4 was 600 ms, and the fifth chord was presented for 1200 ms. (b) In the experiment, sequences were presented in direct succession; 10% of the sequences contained a chord that was played by an instrument other than piano (e.g., harpsichord, celesta, marimba). Participants were only informed about the deviant instruments, not about the Neapolitan sixth chords or their nature, and the task of participants was to ignore the harmonies and to count the deviant instruments. Sequences were presented in blocks of about 35 sequences each, in which all sequences were presented in the same tonal key. After each block, participants were asked about the number of deviant instruments (and to report their answer by pressing a response button). The next block was in a different tonal key than the preceding block. (c) Grand-average ERPs of in-key chords (solid line) and Neapolitan sixth chords (dotted line). Compared to regular in-key chords, the music-syntactically irregular Neapolitan chords elicited an ERAN. Note that when Neapolitans were presented at the fifth position of a chord sequence (where they are music-syntactically highly irregular), the ERAN has a larger amplitude compared to when Neapolitan chords are presented at the third position of the sequences (where they are music-syntactically less irregular than at the fifth position). Neapolitans at the third and at the fifth positions are comparable in terms of their acoustic irregularity (because sequences were presented in the same key within each block), therefore the difference between the ERAN amplitude elicited at the third and fifth positions provides a reasonable estimate for the 'true' ERAN amplitude. Reprinted with permission from Koelsch (2009).

chords presented at the fifth position represent a rupture of the tree-structure: The dominant-branch of the penultimate chord (necessarily) points toward the top-right. According to the GSM, the only possible way to integrate the Neapolitan sixth chord would be to assume another branch pointing to the top-right as well (Figure 9.4). Therefore, both branches could only be combined indirectly (and in a rather long, cumbersome way). In this regard, the Neapolitan sixth chord at the fifth position of the sequence shown in Figure 9.3a represents a 'breach' of the tree-structure.

In the experiments by Koelsch *et al.* (2000), chord sequences were presented in direct succession (reminiscent of a musical piece, Figure 9.3b), with 50% of the stimuli being regular sequences, 25% containing an irregular chord at the third, and 25% an irregular chord at the final position of the sequence. Importantly, sequences were presented in blocks with a duration of about 2 minutes, and in each block all sequences were presented in the same tonal key (the tonal key changed between blocks). This is important because by doing so, the irregular chords at the third and fifth position were comparable in terms of their acoustic deviance (mainly because they contained the same out-of-key tones), but they differed in terms of their functional-harmonic regularity.

The irregular chords elicited an early ERP effect that had negative polarity, maximal amplitude values over frontal leads (with right-hemispheric predominance), and a peak latency of about 150–180 ms (Figure 9.3c). Thus, the observed ERP effect had a strong resemblance to the MMN (for explanation of the MMN see Chapter 5). However, when Thomas Gunter and I discovered this brain response, we did not label it as 'music-syntactic MMN', but as early right anterior negativity (ERAN, Koelsch *et al.*, 2000). One reason for this terminology was that the ERAN was also reminiscent of an ERP effect elicited by syntactic irregularities during language perception: The early left anterior negativity (ELAN; see Chapter 5, section 5.5.2). Thus, labelling the ERP response to harmonic irregularities as ERAN emphasized the notion that this ERP was specifically related to the processing of musical syntax.[26]

The ERAN elicited in that study (Koelsch *et al.*, 2000) was probably due to (a) a disruption of musical *structure building*, (b) the violation of a local prediction based on *musical expectancy formation*, and (c) acoustic deviance (this aspect will be elaborated in the next section). That is, it seems likely that, in that study, processing of local dependencies, processing of (hierarchically organized) long-distance dependencies, and processing of acoustic deviance elicited early negative potentials, and that the observed ERAN-effect was a conglomerate of these potentials.

The ERAN was followed by a later negative ERP that was maximal at around 500–550 ms over frontal leads bilaterally. This ERP effect was termed the N5 (the N5 will be dealt with in more detail in Chapter 10).

---

[26]   The distinction between ERAN and MMN will be dealt with in detail in section 9.5; the overlap between neural resources of music- and language-syntactic processing will be dealt with in section 9.6.

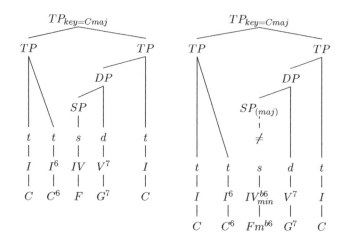

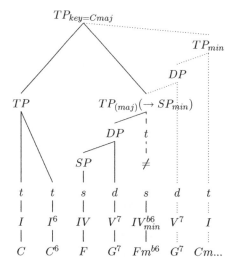

Figure 9.4 Tree structures (according to the GSM) of the sequences shown in Figure 9.3. Left: Regular sequence. Right: Sequence with Neapolitan sixth chord at the third position; the tonality of the tree is propagated from the head into the sub-ordinate branches, leading to a mismatch between expected major context, and the minor context required by a Neapolitan sixth chord (≠). *maj*: Major; *min*: Minor. Bottom: Sequence with Neapolitan sixth chord at the fifth position; dashed line: Expected structure (≠: The tonic chord is expected, but a Neapolitan sixth chord is presented); dotted lines: A possible solution for the integration of the Neapolitan. $TP(\rightarrow SP)$ indicates that the Neapolitan sixth chord can still be integrated, e.g., if the expected tonic phrase is re-structured into a subdominant phrase (note that the phrase ends then on C minor, which would additionally require a resolution to C major). As for the Neapolitans at the third position, the tonality of the tree is propagated from the head into the sub-ordinate branches, leading to an additional mismatch between the expected major context, and the minor context required by a Neapolitan.

9.3.1   The problem of confounding acoustics and possible solutions

With regard to interpreting ERPs such as the ERAN, it is important to be aware of the difficulty that music-syntactically irregular chords often also deviate in terms of acoustics from the preceding chords: They often introduce new pitches, and are acoustically less similar to preceding chords than regular chords. For example, with regard to the stimulus example shown in Figure 9.3a, the regular chords belonged to one tonal key (thus, most tones played in each experimental block belonged to this key; e.g., in *C* major all tones produced by white keys on a keyboard). By contrast, the Neapolitan sixth chords introduced tones that had not been presented in the previous harmonic context (see the flat notes of the Neapolitan chords in Figure 9.3b). These tones represent frequency-deviants (see also Section 5.3 and Figure 5.2), which potentially elicit a physical MMN (phMMN). Thus, the ERAN potentials elicited by the Neapolitan chords were presumably partly overlapped by a phMMN. In addition, the pitches of the Neapolitans were acoustically less similar to the pitches of the preceding chords (compared to regular final tonic chords). That is, the degree of sensory dissonance emerging from an overlap of the frequencies of Neapolitans with the representations of pitches of the preceding chords stored in the echoic memory was larger compared to final tonic chords. In other words, the Neapolitan chords presumably not only evoked electric brain potentials related to musical expectancy violation and disruption of syntactic structure building (as reflected in the ERAN), but also early negative potentials elicited by the processing of pitch deviants (as reflected in an MMN) and acoustic dissimilarity (sensory dissonance, presumably reflected in N1, and perhaps in P1 potentials). Particularly the overlap of ERAN and MMN potentials elicited by the Neapolitan chords is problematic, because these two potentials overlap in time.

To circumvent this problem, it is important to compare the ERAN elicited at the fifth position with the ERAN elicited at the third position of the chord sequences; as described above, the Neapolitan chords at the third position of the sequences served as acoustical control stimuli, allowing an estimate of the degree of physical mismatch responses to Neapolitan sixth chords. Figure 9.3c shows that the ERAN elicited by chords at the final position of chord sequences was considerably larger than the ERAN elicited by chords at the third position of the sequences. This showed that the effects elicited by the Neapolitan sixth chords at the final position of the chord sequences could not simply be an MMN, because an MMN would not have shown different amplitudes at different positions within the stimulus sequence.

The notion that the MMN amplitude would not differ between positions three and five was tested in a study (Koelsch *et al.*, 2001) that compared ERPs elicited by (a) regular and irregular chords (the same as shown in Figure 9.3), (b) frequency deviants (eliciting a phMMN), and (c) abstract-feature deviants suited to elicit an abstract-feature MMN (afMMN; examples of the stimuli used for the phMMN- and afMMN-blocks are shown in Figure 5.2 on p. 47). All sequences (i.e., chord sequences and 'MMN sequences' with auditory oddballs) were presented with the same timing, and with the same probabilities for irregular events at the third and at the fifth position. The results of that study (Koelsch *et al.*, 2001) indicated that neither the amplitude of the phMMN, nor the amplitude of the afMMN differed between positions three and five. By contrast, the ERAN had a larger amplitude at position five than at position three. This shows that the amount of MMN potentials contributing to the ERAN in the data shown in Figure 9.3 was not larger than part of the early effect elicited by Neapolitans at the third position. In other words, the amplitude difference of the ERAN elicited by Neapolitans at positions five and three is a reflection of music-syntactic processing (and not simply due to the acoustical deviance of the Neapolitan sixth chords). This also indicates that the ERAN elicited by Neapolitans at the fifth position reflects, to a significant amount, music-syntactic processing.

Corroborating these findings, a study by Leino *et al.* (2007) showed that the amplitude of the ERAN, but not the amplitude of an MMN elicited by mistuned chords, differed between different positions within chord sequences. A nice feature of that study was that chord sequences comprised seven chords, and that they were composed in a way that Neapolitan sixth chords occurring at the fifth position were music-syntactically less irregular than Neapolitans at the third position (contrary to the sequences presented in Figure 9.3a). Consequently, the ERAN elicited at the fifth position was smaller than the ERAN elicited at the third position, and the ERAN was largest when elicited by Neapolitan chords at the seventh position (where they were syntactically most irregular).

Importantly, irregular chords at both the third and at the seventh position followed dominant-seventh chords. Therefore, the fact that the ERAN was larger at the seventh than at the third position cannot simply be explained by statistical probabilities for the local transition of chord functions. This suggests that the generation of the ERAN involves hierarchical processing, because within the hierarchical structure of the chord sequences, the degree of irregularity was higher at the seventh than at the third position of the chord sequences.

Further solutions As mentioned above, music-syntactic ir/regularity of chord-functions often confounds acoustic dis/similarity. In fact, tonal hierarchies, and music-syntactic regularities of major–minor tonal music are partly grounded on acoustic similarity (as described in Chapter 3 and Section 9.2). This poses considerable difficulty for the investigation of music-syntactic processing (particularly with regard to isolating neural correlates of *knowledge-free structuring*, *musical expectancy formation*, and syntactic *structure building*). However, several ERP studies have aimed at disentangling 'cognitive' mechanisms from 'sensory' mechanisms related to music-syntactic processing, and at constructing chord sequences in which acoustic irregularity does not confound music-syntactic irregularity.

The sequence shown in Figure 9.5A consists of regular chord functions (tonic, subdominant, supertonic, dominant, tonic). That is, according to the theory of harmony, the first four chords of the sequences were arranged in such a fashion

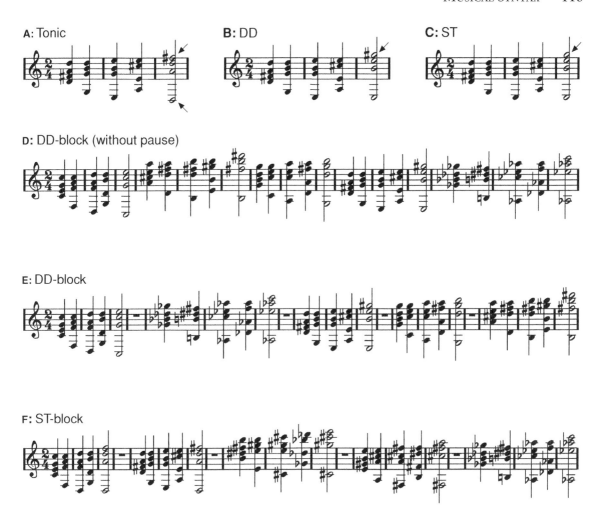

Figure 9.5 Examples of musical stimuli using double dominants (DDs) and supertonics (STs) as irregular chords. Top row: Chord sequences in (D major), ending either on a tonic chord (regular, A), on a double dominant (irregular, B), or on a supertonic (irregular, C). Arrows indicate pitches that were not contained in the preceding chords. Sequences from all 12 major keys were presented in pseudorandom order, and each sequence was presented in a tonal key that differed from the key of the preceding sequence, regular and irregular sequence endings occurred with equal probability ($p = 0.5$). Sequences were presented in direct succession (D), or separated by a pause (E, F), which allowed us to investigate late ERPs elicited by the final chords without overlap with ERPs elicited by the next chord. Pauses can also be used for ratings, or other tasks (Müller *et al.*, 2010). Reprinted with permission from Koelsch *et al.* (2007).

that a tonic at the fifth position was the most regular chord function. The first four chords of the sequence shown in Figure 9.5B are identical to those of the regular sequence (shown in Figure 9.5A), but the final chord of this sequence is a double dominant (DD).[27] Similarly, the first four chords of the sequence shown

---

[27] For explanation of the term *double dominant* see p. 17 and Figure 2.10.

Figure 9.6 Tree structures (according to the GSM)for the sequences shown in Figure 9.5, ending on a regular tonic (left), and on a DD (right). Dashed line: Expected structure (≠: The tonic chord is expected, but a DD is presented); dotted lines: A possible solution for the integration of the DD. $TP(\rightarrow DP)$ indicates that the DD can still be integrated, e.g., if the expected tonic phrase is re-structured into a dominant phrase.

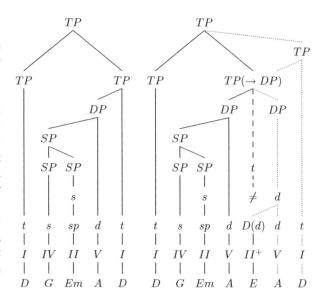

in Figure 9.5C are identical to those of the sequences shown in 9.5A and 9.5B, but the final chord of that sequence is a supertonic (ST).[28]

Figure 9.6 shows the hierarchical organization of these sequences according to Rohrmeier's GSM (see Section 9.1). The final DDs (or STs) follow the dominant instead of the tonic (which is the most regular chord at the final position), thus not fulfilling the expected tree-structure. Moreover, the tree is not revised (because the sequence ends on the DD or the ST, this is soon learnt by the participants during the course of an experiment).

Importantly, with respect to the first four chords, both DDs and final tonics contained new pitches, that is, pitches that were not contained in any of the previous chords: Tonic chords contained two new pitches (in both the top voice and the base voice; see the *f#* and the *d* indicated by the arrows in Figure 9.5A), and DDs contained one new pitch (in the top voice; see arrow in Figure 9.5B). In contrast to DDs, tonic chords did not introduce a new pitch class (i.e., the new pitches of tonic chords had been presented previously, namely in the first chord, either one octave lower or one octave higher). Thus, the new pitches of final tonics were perceptually more similar to pitches of the first chord than the new pitch of the DD was. However, because the octaves of the two new pitches of final tonics were only contained once in the very first chord of the sequence, it is reasonable to assume that these tones were masked by the second, third, and fourth chord. Therefore, the new pitch of the DD presumably did not represent a greater frequency deviant for the auditory sensory memory than the two new pitches of tonic chords (a potentially larger degree of sensory dissonance is dealt with further below).

---

[28]   For explanation of the term *supertonic* see Figure 2.6 on p. 16.

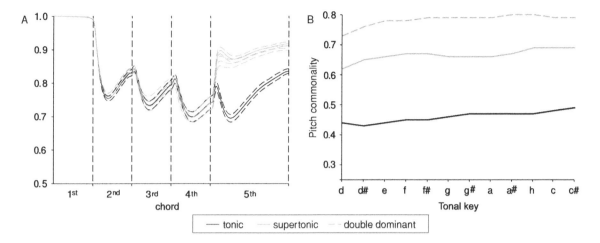

Figure 9.7   A: Correlation of local context (pitch image of the current chord) with global context (echoic memory representation as established by previously heard chords). The data show that music-syntactically irregular chord sequence endings (STs: Solid grey line, DDs: Dashed grey line) were more congruent with the preceding harmonic context than music-syntactically regular endings (tonic chords: Black line). For each sequence type, correlations were calculated for all 12 major keys (the line for each sequence type represents the mean correlation; the thin dotted lines indicate standard error of the mean). Auditory modelling was performed using the Contextuality Module of the IPEM Toolbox (Leman *et al.*, 2005). Length of local context integration window was 0.1 s, global context integration window was 1.5 s. The abscissa represents the time line (each of the first four chords had a duration of 600 ms, the fifth chord was presented for 1200 ms). The ordinate depicts correlation values. Note that these values indicate half decay values and that, particularly due to the use of the 1.5 s gliding window, information of all preceding four chords affects the correlations between the last chord and the preceding chords. **B**: Pitch commonality calculated for the different chord sequence endings (tonic chord, double dominant, supertonic) and the penultimate (dominant) chord. Values were computed separately for all 12 major keys according to Parncutt (1989) and connected with lines for better visualization. The graphs show that DDs and STs have an even higher pitch commonality with the directly preceding dominant than tonic chords. Pitch commonality values were calculated for the 12 keys to illustrate the effect of transposition on pitch commonality and to show that the pitch commonality ranges for the three analyzed chord types do not overlap. Modified with permission from Koelsch *et al.* (2007).

To test this assumption, it is possible to model the acoustic congruency of the final chords with the information contained in the auditory sensory memory traces established by the first four chords using the IPEM toolbox (Leman, 2000). This auditory modelling estimates pitch images of the echoic memory: as described in the section on the MMN (p. 45), acoustic information decays, but is kept in the echoic memory for a certain time. The IPEM toolbox can be used to model the correlation of the pitch image of a final chord with the pitch image of the first four chords stored in the echoic memory. Figure 9.7A shows the results of the modelling for the sequences shown in Figure 9.5, suggesting that the pitch images of the final DDs correlated even higher with the pitch images established by the first four chords than the pitch images of final tonic chords.

Another important factor to take into account when constructing chord sequences is the pitch commonality between a critical chord (including the overtones) and the preceding chord(s) or tones (including the overtones). Richard Parncutt (1989) suggested that the relations between successive sounds can also give rise to 'sensory dissonance.' That is, for example, two successive chords that are constituted by the same pitch classes (thus having a high pitch commonality) have a lower degree of sensory dissonance than two successive chords that do not share any pitch class (thus having a low pitch commonality). In ERP experiments on music-syntactic processing, it is thus important to make sure that early negativities elicited by irregular chords are not simply due to a higher degree of sensory dissonance related to the perception of the penultimate and the final chord.

For example, in both sequences of Figures 9.5A and B, the final and penultimate chord share one pitch class (*A* in Figure 9.5A, and *E* in Figure 9.5B). However, note that the final tonic only contains one tone of the shared pitch class (the *a'*), whereas the DD contains two tones of the shared pitch class (the *e* and the *e'*). Moreover, when counting the minor and major seconds (including the ninths) made up by penultimate and final chords, the dominant – tonic ending contains six major seconds (e.g., between *e'* and *f#'*) and two minor seconds (e.g., between *c#'* and *d'*), whereas the dominant – DD progression only contains two major seconds and one minor second. In this regard, the final DD should have a higher pitch commonality, and a lower sensory dissonance, with regard to the penultimate chord than the final tonic chord. This assumption can also be tested using a model provided in Parncutt (1989); the results of this modelling are shown in Figure 9.7B.[29]

Thus, with respect to both (a) the pitch commonality between final and penultimate chord and (b) the acoustic congruency between the final chord and the information of the first four chords stored in the echoic memory, it is reasonable to assume that any early negative ERP effects elicited by final DDs (compared to final tonics) are not simply due to a higher degree of sensory dissonance or a higher incongruency with the memory traces stored in auditory sensory memory. In fact, the models suggest that music-syntactically irregular DDs were acoustically even more similar to the previous chord(s) than music-syntactically regular endings.

To maintain this acoustic balance, it is important that the tonal key changes from sequence to sequence. That is, when using sequences such as those shown in Figure 9.5, each chord sequence should be presented in a tonal key that differs from the key of the preceding sequence (otherwise, for example in C major, the

---

[29] However, it should be noted that, despite the appeal of intricate computer-based models, simply counting all intervals between each of the tones of two chords is a valid method for the estimation of the sensory dissonance (in comparison to two other chords). Parncutt himself noted that, compared to modelling, the correlation between roughness and tension-values 'improved when we simply counted the number of semitones, tritones, major sevenths and their compounds in each sonority' (Parncutt 2006).

probability for a *C* major chord would be higher than the probability for a *D* major chord). Moreover, both sequence types should occur randomly with equal probability (to avoid that ERP effects are simply due to statistical probabilities within the experiment itself). Finally, also note that the direction of the voices from the penultimate to the final chord is identical for tonics and DDs, and that the superposition of intervals is identical for both final tonics and DDs. The latter point is important, because physically identical chords were music-syntactically regular in one sequence, but irregular in another (e.g., the final tonic chord of Figure 9.5A was a DD of sequences starting in *C* major, and the final DD of Figure 9.5B was a tonic in sequences starting in *E* major). Therefore, any effect elicited by a DD could not be due to the acoustical properties of the chord itself.

Figure 9.8 (left and right panel) shows that DDs elicited a clear ERAN, indicating that electrical reflections of music-syntactic processing can be measured in the absence of a physical irregularity. That is, in that study (Koelsch *et al.*, 2007) the ERAN was presumably due to (a) a disruption of musical *structure building*, (b) the violation of a local prediction based on *musical expectancy formation*, but not due to the processing of acoustic deviance. That is, it seems likely that processing of local dependencies, and processing of (hierarchically organized) long-distance dependencies, elicited early negative potentials, and that the observed ERAN-effect was a conglomerate of these potentials. The electrophysiological correlates of *musical expectancy formation* on the one hand, and hierarchical *structure building* on the other, have not been separated so far. It seems likely that both these processes elicit early anterior negativities, but future studies are required to specify such correlates (and thus specify possible sub-components of the ERAN).[30] Figure 9.8 also shows that the ERAN was followed by an N5 (the N5 was maximal over frontal leads bilaterally).

As already mentioned above, the high pitch commonality of the DDs with the previous chords (shown in Figure 9.5B) came at the price of introducing a new pitch class (which was not the case for tonic chords). In addition, the DDs did not belong to the tonal key established by the preceding chords (similar to the Neapolitans of previous studies). Figure 9.5C shows sequences ending on a supertonic (ST) instead of a DD. As the regular tonic chords, the STs did not introduce a new pitch class, and similar to DDs, the pitch commonality with the penultimate chord, and the correlation with the auditory image established by the preceding chords, was even higher than for regular tonic chords (see also Figure 9.7; because the final STs repeated the supertonic presented at the third position of the sequences, the final STs were acoustically even more similar to the preceding context than DDs). However, here the avoidance of a new pitch class comes at the price that final STs are minor chords, and because the first four chords contain only one minor chord (the supertonic at the third position),

---

[30]   In a auditory language experiment using an artificial finite-state grammar, syntactic violations elicited an early anterior negativity (Friederici *et al.*, 2002b), rendering it likely that similar potentials would also be elicited by similar grammar violations with musical sounds.

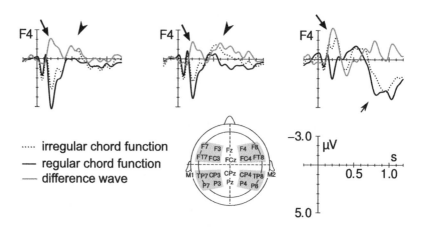

Figure 9.8 Grand-average ERPs elicited by DDs (top left) and STs (top middle), recorded from 24 participants (16 non-musicians, 8 amateur musicians). Chords were task-irrelevant (participants listened to the sequences under the instruction to detect timbre-deviants, and participants were not informed about the presence of irregular harmonies, nor about the true purpose of the experiment). The top right image shows grand-average ERPs of DDs from an experiment in which chords were task-relevant (participants were informed about the irregular chords and were asked to press one response button for the regular chords and another button for the irregular ones). Data were recorded from 20 non-musicians. Long arrows indicate the ERAN, arrowheads indicate the N5, and the short arrow indicates the P3. Note that the P3 potentials elicited when chords are task-relevant (short arrow) obscure N5 potentials (although the N5 can still be seen in the difference wave). Modified with permission from (Koelsch *et al*., 2007a).

minor chords occur with a probability of $p = 0.3$, whereas major chords occur with a probability of $p = 0.7$. Although it seems that this cannot account for an ERAN effect (because the supertonics at the third position do not seem to elicit an ERAN, for details see Koelsch & Jentschke, 2008), it cannot be excluded that this difference in probability partly contributes to the ERAN. Figure 9.8 (middle panel) shows that STs elicited a clear ERAN (followed by an N5), indicating that an ERAN can be elicited by in-key chords, and that the elicitation of the ERAN does not require out-of-key notes (for further experiments using STs see, e.g., Koelsch et al., 2007a; Koelsch & Jentschke, 2008). These findings also corroborate the notion that electrical reflections of music-syntactic processing can be measured in the absence of a physical irregularity.

9.3.2   Effects of task-relevance

The task that participants perform during an experiment has a strong influence on ERPs. The ERPs shown in the top left of Figure 9.8 were elicited when chords were task-irrelevant – every so often a chord was not played with a piano sound but with a deviant timbre (e.g., marimba, organ, or guitar), and participants listened to the sequences under the instruction to detect these timbre-deviants (participants were not informed about the presence of irregular harmonies, nor about the true purpose of the experiment). As stated earlier, the irregular chords (DDs) elicited an ERAN and an N5 under this condition (best to be seen in the difference wave of Figure 9.8).

The top right panel of Figure 9.8 shows ERPs elicited in an experiment in which chords were task-relevant – participants were informed about the regularity of the chords, presented with practice trials, and then responded with button presses as fast as possible after the final chords, pressing one button in response to the regular chords and another button in response to the irregular ones. Under this condition, the ERAN was overlapped by an N2b (see also Section 5.4), making it difficult to differentiate electric potentials due to music-syntactic processing from those related to the detectional and decisional processes reflected partly in the N2b (factor analysis is one feasible option due to the different scalp distributions of ERAN and N2b).[31]

Moreover, the task-relevant DDs elicited strong P3-potentials (presumably P3b, rather than P3a potentials, due to the probability of 0.5 for DDs; for a description of the P3 see also Section 5.4). Although the P3b was maximal over parietal leads (not shown), the P3 potentials clearly extended to frontal sites (see top left vs. top right of Figure 9.8). This led to an overlap, and perhaps to an interaction, with the N5, which was observable only in the difference waveform in that experiment.

The timbre-detection task described above has been used in numerous studies to avoid the overlap of ERPs reflecting music-syntactic processing with ERPs reflecting detectional and decisional processes (that is, to avoid an overlap of ERAN and N2b potentials, as well as of N5 and P3 potentials). At the same time, performance in the timbre-detection task informs the investigator whether participants listened to the musical stimulus.

9.3.3   Polyphonic stimuli

As mentioned above, the double dominants (DDs) of the homophonic sequences shown in Figure 9.5B introduced a new pitch class (the *g#'* in the top voice of the final chord). Thus, the DDs introduced a pitch that had not been presented either one octave lower or one octave higher in the previous harmonic context. Therefore, the ERP effects elicited by the DDs could still have been driven in part by the occurrence of a new pitch class which was perceptually less similar to the tones occurring in the previous harmonic context (compared to pitches of the final tonic). Figure 9.9 shows polyphonic sequences composed to deal with this problem. Here, all pitch classes of DDs were already presented in the previous harmonic context (moreover, the auxiliary notes and passing notes made the sequences

---

[31]   The ERAN, but not the N2b, inverts polarity at mastoid leads with nose- or common average reference.

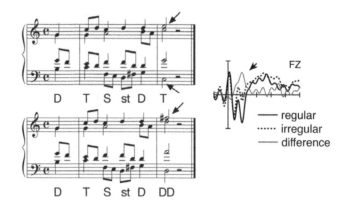

Figure 9.9 Left: Polyphonic sequences. As in the homophonic sequences shown in Figure 9.5A and B, chord sequences ended either on a regular tonic chord (T), or on an irregular double dominant (DD). Arrows indicate pitches that were not contained in the preceding chords. Note that sequences start with a dominant (in contrast to the stimuli presented in the previous figures, which started on a tonic). Therefore, effects of DDs could not be due to a template match/mismatch between the first and the last chord. The tonal key changed from sequence to sequence, and both regular and irregular sequence endings occurred randomly with equal probability ($p = 0.5$). Right: ERPs elicited by the final tonics and DDs, recorded from 12 musicians and 12 non-musicians under the task to detect timbre-deviants (that is, the harmonies were not task-relevant, and participants were not informed about the ir/regularity of chord functions). ERPs were referenced to the algebraic mean of both mastoid electrodes. Modified with permission from Koelsch (2005).

sound more natural, and more interesting, than the homophonic sequences). In addition, and in contrast to the sequences shown in Figure 9.5 (which began with a tonic chord), the polyphonic sequences began with a dominant, avoiding that final tonic chords sounded more regular simply because they repeated the first chord function of the sequence. That is, in the polyphonic sequences shown in Figure 9.9, all pitch classes of DDs occurred in the previous harmonic context (i.e., all notes of DDs occurred one or two octaves above or below in the previous context), and DDs repeated even more pitches of the preceding chords than tonic chords. In addition, the DDs had more pitches in common with the penultimate chord, thus the 'sensory dissonance' between final and penultimate chord (of which pitch commonality is the major component) was not greater for DDs than for final tonics. Hence, final DDs were acoustically even more similar to the preceding acoustic context than final tonic chords. Again, acoustic modelling confirmed the assumption that the pitch images of the final DDs correlated even higher than those of final tonics with the pitch images established by the previous chords (data are shown in Koelsch & Sammler, 2008). Also note that the superposition of intervals was again identical for both final tonics and DDs. Because sequences were presented in different keys during the experiment, physically identical chords were music-syntactically regular in one sequence, but irregular in another (for example, the final tonic chord in the top panel of Figure 9.5 was a DD of sequences starting in B-flat major, and the final DD in the bottom panel of Figure 9.5 was a tonic in sequences starting in D major). Therefore, any effect elicited by a DD could again not be due to the acoustical properties of the chord itself.

The ERPs shown in the right panel of Figure 9.9 reveal that the DDs of the polyphonic sequences also elicited an ERAN, demonstrating that the generation of the ERAN is not dependent on the occurrence of new pitch classes (i.e., new notes), and indicating that the ERAN effect elicited in that experiment was due to music-syntactic processing (and not due to pitch repetition effects). It is still true that the pitch class of the new note introduced by DDs occurred only once in the previous context, whereas the pitch class of the top voice of final tonic chords occurred twice, but it is unlikely that this accounts for the ERAN effect, particularly because the ERAN can even be elicited when irregular chords do not introduce any new pitch class (see ERPs elicited by STs, as described in the previous section). The ERAN amplitude was larger when elicited by the homophonic sequences shown in Figure 9.5 compared to when elicited by the polyphonic sequences, for which several reasons might account: (1) The occurrence of a new (out-of-key) note in the DDs of the homophonic sequences made DDs possibly slightly more unexpected than DDs of the polyphonic sequences; (2) The difference in sequence length (the homophonic sequences were five chords long, the polyphonic sequences consisted of six chords) might have led to an interaction between music-syntactic processing and working memory operations; (3) Polyphonic sequences did not begin with a tonic chord, perhaps making the extraction of the tonal centre more difficult than for homophonic sequences. These issues could be specified in future studies. As will be dealt with in more detail further below, the ERAN tended to be larger in musicians than in non-musicians.

Authentic musical stimuli   Both ERAN and N5 can also be elicited by syntactically irregular chords occurring in authentic musical stimuli (i.e., in music as it was composed by a composer). Figure 9.10 illustrates the design of one of these studies (Koelsch *et al.*, 2008). From an excerpt of an original composition containing a syntactically slightly irregular chord (see middle panel in the lower right of Fig. 9.10), two additional versions were created; in one version the original (slightly irregular) chord was rendered regular (top panel in the lower right of Fig. 9.10), and in another version it was rendered highly irregular (bottom panel in the lower right of Fig. 9.10). Duration of excerpts was around 10–12 sec. Pieces were performed by professional pianists and recorded using MIDI, and musical stimuli could thus be presented with musical expression (in contrast to the other studies that have been mentioned so far in this chapter, where stimuli were presented under computerized control without musical expression). Note that the critical chords were modified in the MIDI files; therefore, all three versions were identical with regard to the expressive variations in tempo and loudness. In addition, for

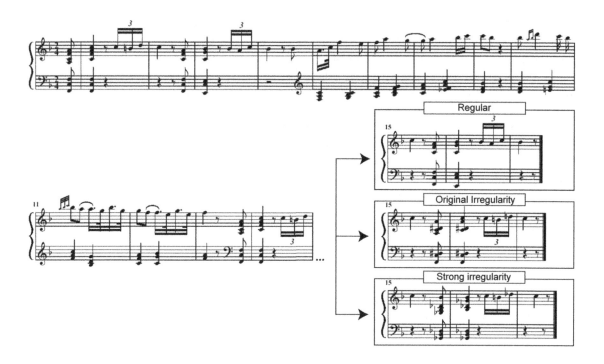

Figure 9.10 Beginning of Beethoven's piano sonata in *F* major (Op. 10, Nr.2), followed by three types of continuations. The continuation shown in the middle panel in the lower right is the original continuation as composed by Beethoven, containing an unexpected chord (introducing a modulation to *C* major). The top and bottom panels show expected and very unexpected continuations. Reprinted with permission from (Koelsch *et al*., 2008b).

each composition three versions without musical expression were created from the three versions with musical expression (to compare ERPs between expressive and non-expressive conditions).

ERPs obtained in that study (Koelsch *et al*., 2008b) are shown in Figure 9.11 (averaged across expressive and non-expressive conditions, because ERPs of these conditions did not interact with syntactic regularity). Compared to the expected chords, both slightly irregular chords (i.e., the chords composed by the composer), as well as the strongly irregular chords elicited an ERAN as well as an N5. Both ERAN and N5 had the typical latency, and a similar scalp distribution, compared to the previous studies that used shorter, more artificial (and non-expressive) musical stimuli. This shows that unexpected chords, as arranged by a composer, elicit an ERAN (and an N5). Therefore, the processes reflected in the ERAN and the N5, as described in the previous sections of this chapter, are not only experimental artefacts, but also occur when listening to 'real', naturalistic music.

The three different chord functions (regular, slightly irregular, strongly irregular) also elicited different emotional responses. This was indicated by behavioural ratings (regular endings were perceived as most pleasant, least

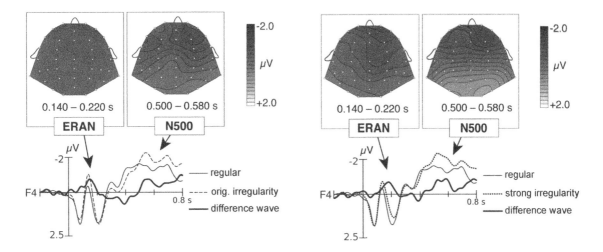

Figure 9.11   ERPs elicited by the chords shown in Figure 9.10, recorded from 20 non-musicians. The left panel shows ERPs elicited by regular and by slightly irregular chords (those originally composed by the composer). The right panel shows ERPs elicited by regular and by highly irregular chords. Modified with permission from (Koelsch *et al.*, 2008b).

arousing, and least surprising, whereas strongly irregular endings were perceived as least pleasant, most arousing, and most surprising), as well as by electrodermal responses (which were strongest for highly irregular chords and weakest for regular chords). Moreover, compared to non-expressive versions, expressive versions were perceived by the participants as more pleasant, more arousing, and more surprising. Correspondingly, chords played with expression elicited a stronger electrodermal response than chords presented without expression (this was probably in part due to chords with musical expression often being played with an accent). Nevertheless, the ERAN did not differ between expressive and non-expressive conditions, suggesting that the neural mechanisms of music-syntactic processing operate independently (a) of the emotional effects of music-syntactic (ir)regularity, and (b) of the emotional effects due to (non-)expressive musical performance. This indicates that the ERAN reflects cognitive, rather than emotional, processes. Note, however, that processing of a music-syntactically irregular chord can lead to the percept of unexpectedness, and thus affect emotional processes (as shown by the behavioural ratings, and by the electrodermal responses to unexpected chords, which were stronger than those to expected chords).

A similar study by Steinbeis *et al.* (2006) used chorales from J.S. Bach as stimuli. Each chorale was used in three different versions. One version was the original (the critical, slightly irregular chord was originally composed by Bach). From this version, a second version was created in which the critical chord was syntactically highly regular, and in a third version the critical chord was rendered highly irregular. In that study (Steinbeis *et al.*, 2006) stimuli were not played with musical expression. Similar to the study by Koelsch *et al.* (2008b), irregular (unexpected) chords elicited ERAN and N5 potentials, as well as stronger electrodermal responses (compared to regular chords). Other than in the above-mentioned studies, the ERAN had a relatively long peak latency (around 230 ms), and a more centro-temporal scalp distribution.

Three further studies also used more naturalistic musical stimuli, although in those studies irregular tones were introduced by the investigators (that is, irregular tones or chords were not originally composed that way); a study by Besson & Faita (1995) used opera melodies played by a computer, and a study by Besson *et al.* (1998) used opera melodies sung with musical expression. As described in Chapter 6, irregular tones (introduced by the investigators) elicited ERAN-like early negativities (as well as a P3-like *Late Positive Component*). Moreover, a study by Koelsch & Mulder (2002) used musical excerpts recorded from CDs in which the critical chord was regular, and this chord was transposed electronically to obtain stimuli with irregular chords. Again, irregular chords elicited ERAN and N5 potentials. Similar to the study by Steinbeis *et al.* (2006), the ERAN had a relatively long peak latency (around 250 ms), and a centro-temporal scalp distribution. Reasons for the longer peak latency and the centro-temporal scalp distribution will be discussed in Section 9.3.4.

Further paradigms   Regnault *et al.* (2001) and Poulin-Charronnat *et al.* (2006) used the sequences shown in Figure 3.8 (Chapter 3) in ERP studies. As illustrated in Chapter 3, these sequences avoid that psychoacoustical factors confound music-syntactic regularity. However, they have the disadvantage that the final tonic chords occur on a light beat (hampering syntactic integration). Therefore, the final tonic chords themselves are not regular, and are thus a suboptimal control condition for the final subdominant chords.

The study by Regnault *et al.* (2001, in which chords were task-relevant) reported P3 effects, and the study by Poulin-Charronnat *et al.* (2006, in which chords were task-irrelevant and participants watched a silent movie) reported an increased N5 for final subdominants compared to final tonics. No ERAN was elicited by the final subdominants, in either of the two studies, perhaps because the final tonic already represented an irregularity, and in comparison to that irregularity the subdominant at that position (which is a *regular* chord function at this cadential position, though not a regular chord function to terminate the sequence) did not elicit additional ERAN-potentials. That is, perhaps both the final tonics and the final subdominants elicited ERAN potentials which did not differ from each other.

Interestingly, the finding of an increased N5 in response to final subdominants reported by Poulin-Charronnat *et al.* (2006) suggests that these chords gave rise to the processing of (intra-musical) meaning (the N5 as an index of the processing of music-semantic information will be described in detail in Chapter 10). In this regard, the clear and stable behavioural effects reported in the studies using the chord sequences of Bigand *et al.* (2003)[32] presumably not only reflect music-syntactic processes, but also music-semantic processes.

9.3.4   Latency of the ERAN

The peak latency of the ERAN is often between 170 and 220 ms, with the exception of four studies: Koelsch & Mulder (2002) reported an ERAN with a latency of around 250 ms, Steinbeis *et al.* (2006) reported an ERAN with a latency of 230 ms (in the group of non-musicians), James *et al.* (2008) an ERAN with a latency of 230 ms in a group of musicians, and Patel *et al.* (1998) reported an ERAN-like response (the right anterior temporal negativity, RATN) with a peak latency of around 350 ms. The commonality of these four studies was that they used non-repetitive sequences, in which the position at which irregular chords could occur was unpredictable. It is also conceivable that the greater rhythmic complexity of the stimuli used in those studies had effects on the generation of the ERAN (leading to longer ERAN latencies), but possible effects of rhythmic structure on the processing of harmonic structure remain to be investigated. It is also interesting that all four of these studies (Patel *et al.*, 1998; Koelsch & Mulder, 2002; Steinbeis *et al.*, 2006; James *et al.*, 2008) reported a more centro-temporal than frontal scalp distribution of the ERAN. However, in another study that used non-repetitive sequences in which the occurrence of an irregular chord was unpredictable (Koelsch *et al.*, 2008), the peak amplitude of the ERAN was between 158 and 180 ms, and the maximum of the ERAN was over frontal electrodes. The specific effects on latency and scalp distribution of the ERAN are, thus, not yet understood. In the next section, matters will be complicated further by differentiating between ERAN potentials elicited by melodies and by chords.

9.3.5   Melodies

The ERAN can not only be elicited by chords. As reviewed in Chapter 6 (p. 57 ff.), studies investigating the processing of unexpected tones in melodies also report frontal negative ERPs emerging at around 100 ms after stimulus onset, peaking at around 120 to 180 ms after the onset of the unexpected tone (Besson & Macar, 1986; Verleger, 1990; Paller *et al.*, 1992; Besson & Faita, 1995; Besson *et al.*, 1998; Miranda & Ullman, 2007; Brattico *et al.*, 2006; Peretz *et al.*, 2009; Koelsch & Jentschke, 2010). This ERP effect resembles the ERAN, although it appears to have a shorter peak latency, and a smaller amplitude compared to the ERAN elicited by chords.

A study by Koelsch & Jentschke (2010) directly compared the processing of (ir)regular tones in melodies with the processing of (ir)regular chord functions.

---

[32]   For reviews see, e.g., Tillmann (2009) and Bigand & Charronnat (2006).

That study used the chord sequences shown in Figure 9.5 (presented randomly in all twelve major keys), as well as only the top voice of these sequences as melodic stimuli. Both irregular melody endings and irregular final chord functions elicited an early effect around 100 to 150 ms after the onset of the tones (peaking at around 125 ms, referred to in the following as *N125*). In addition to the N125, irregular chords (but not irregular melody endings) also elicited an effect around 160 to 210 ms (peaking at around 180 ms, referred to in the following as *N180*). The N125 had a different scalp distribution than the N180: The N125 was more frontally, and the N180 more broadly, distributed over the scalp. These findings suggest that melodic information (which is not only present in melodies, but also in the top voice of chords) is processed earlier, and with partly different neural mechanisms, than harmonic information specific for chords.[33] The observation of the N125 in response to irregular tones of melodies is consistent with the results of the above-mentioned studies investigating processing of familiar melodies with ERPs. Similar responses with an onset around the onset of the N1, but with longer peak latencies, were reported in studies by Miranda & Ullman (2007) and Brattico *et al.* (2006). In the latter two studies, the latency of effects was perhaps longer because incongruous tones occurred in the middle of melodies (where expectancies for tones were presumably less specific compared to phrase endings).

The assumption that irregular chord functions elicit N180 potentials does not rule out the possibility that this effect can also be elicited by melodies: The data from Miranda & Ullman (2007) indicate that irregular tones of melodies that establish a tonal key and a harmonic context more clearly than the melodies used in the study by Koelsch & Jentschke (2010), also elicit N180 potentials (see also Brattico *et al.*, 2006; Peretz *et al.*, 2009). This is perhaps due to the phenomenon that out-of-key tones of melodies automatically engage processes related to harmonic processing. It is an open question how different degrees of the establishment of tonal key and harmonic context, as well as different degrees of rhythmic and spectral complexity, influence N125 and N180 potentials.

The notion that N125 and N180 reflect different cognitive processes is supported by the observation that oscillatory activities elicited by regular and irregular chord functions differ between these time windows: Using the same chord sequences as those shown in Figure 9.5A & C, a study by Herrojo-Ruiz *et al.*

---

[33] Note that, in the melody condition of the study by Koelsch & Jentschke (2010), the amount of pitch distance between the last regular tone and the two preceding tones was ∼10% (two semitones), whereas it was ∼16% (three semitones) or ∼20% (four semitones) for the irregular endings. Thus, the strength of refractoriness effects could have been slightly larger for the irregular than for the regular tones, and thus also account for, or at least contribute to, the N125 effect. Nevertheless, because tones were spectrally rich piano tones, and because irregular chords had even more pitches in common with the preceding two chords, it is unlikely that the N125 was only due to refractoriness effects. To rule out such a possibility, future studies could devise stimuli in which possible refractoriness effects would be equal for regular and irregular events.

(2009) reported a difference of spectral power between supertonic and tonic chords in the delta band (<4 Hz) at right anterior and right central electrode sites between 100 and 150 ms after chord onset (corresponding to the peak latency of the N125); similar differences were observed for the spectral power of phase-locked theta activity (4–7 Hz) in the time range from 100 to 160 ms at left- and right-anterior electrode sites.[34] By contrast, in the time span from 175 to 250 ms (corresponding to the peak latency of the N180), the processing of irregular chords (compared with regular chords) was associated with a decrease in the degree of global phase synchrony in the lower alpha band (8–10 Hz; this effect was maximal at FC4, and presumably reflected the decoupling between the oscillatory activities at right fronto-central and left temporo-parietal regions mediated by long-range alpha phase synchrony).[35]

The findings that electrophysiological effects elicited by melodic and harmonic information differ with regard to latency, scalp distribution, and probably oscillatory correlates, suggest that melodic and harmonic information contributes differently to the ERAN: Whereas processing of melodic information contributes earlier to the ERAN (as reflected in the N125), processing of harmonic information contributes in later time windows to the ERAN (as reflected in the N180).[36] This interpretation is consistent with the notion that the hierarchy of stability of chords is more complex than the hierarchy of stability of tones (see Chapter 3), and consistent with the assumption that melodic processing develops earlier during childhood than the processing of chord functions (e.g., Trainor & Trehub, 1994; see also Section 9.9 below). Note that melodic information is also more concrete than the more abstract functional-harmonic information of a chord, because melodies, but not chords, can be sung (by a single individual).

9.3.6   Lateralization of the ERAN

Although the ERAN was significantly lateralized in some of the reported studies (e.g., Koelsch *et al.*, 2000, 2007a; Koelsch & Sammler, 2008; Koelsch & Jentschke, 2008; Müller *et al.*, 2010), it was not right-lateralized in a few studies also using chord-sequence paradigms (Steinbeis *et al.*, 2006; Loui *et al.*, 2005; Leino *et al.*, 2007), or melodies (Miranda & Ullman, 2007). One difference between those studies is that the studies in which the ERAN was right-lateralized had

---

[34]   The spectral energy (total and evoked) of both delta and theta oscillations increased in response to final chords, and this increase was stronger for the regular than for the irregular chords. The difference in oscillatory activity between regular and irregular chords was presumably due to phase-resetting of oscillations in the theta band, as well as due to both phase-resetting and increase in amplitude of oscillations in the delta band.
[35]   Interestingly, in the ERAN-time window no effects of irregular chords were found in the gamma band. Differences in the gamma band were found, however, in the N5 time window, around 500–550 ms after stimulus onset.
[36]   Note that the degree of consonance / dissonance of a chord (in terms of the vertical dimension of harmony; Tramo *et al.*, 2001) is presumably processed earlier than functional-harmonic information.

relatively large numbers of participants (twenty or more; Koelsch *et al.*, 2007a; Koelsch & Sammler, 2008; Koelsch & Jentschke, 2008; Müller *et al.*, 2010), whereas studies in which no lateralization of the ERAN was reported have mostly measured less than twenty subjects, namely eighteen in the study by Loui *et al.* (2005), or ten in the study by Leino *et al.* (2007). This difference is important because it seems that the ERAN is lateralized more strongly in males than in females (females often showing a rather bilateral ERAN; Koelsch *et al.*, 2003). Thus, a relatively large number of subjects is required until the lateralization of the ERAN reaches statistical significance. Additional factors that modulate the lateralization of the ERAN might include the salience of irregular chords, attentional factors, and the signal-to-noise ratio of ERP data.

In some earlier studies, I have also referred to the ERAN-effect as *music-syntactic MMN* (Koelsch *et al.*, 2002b, 2003a,b,c), not only due to the resemblance with the MMN, but also because the term 'early right anterior negativity' falls short when the effect elicited by irregular chords is not significantly lateralized (which seemed to be the case in females compared to males, Koelsch *et al.*, 2002b). Lack of lateralization also led authors to label effects elicited by music-syntactically irregular events as *early negativity* (Steinbeis *et al.*, 2006), or *early anterior negativity* (Loui *et al.*, 2005).[37] However, other studies used the term ERAN even when the effect was not significantly right-lateralized over the scalp for two reasons, as follows. (1) Functional neuroimaging studies consistently showed right-hemispheric predominance for the activation of the structures that are assumed to be involved in the generation of the ERAN (see Section 9.4). That is, although the ERAN was sometimes not significantly lateralized in the scalp topographies of electroencephalographic data, it is likely that the brain activity underlying the generation of the ERAN was nevertheless stronger in the right than in the left hemisphere. (2) The term ERAN was also established with regard to the functional significance of this ERP component, rather than only for its scalp distribution (Koelsch *et al.*, 2007a; Maess *et al.*, 2001; Miranda & Ullman, 2007; Maidhof & Koelsch, 2011). Note that similar conflicts exist for most (if not all) endogenous ERP components, for example: The P300 is often not maximal around 300 ms (e.g., McCarthy & Donchin, 1981), the N400 elicited by violations in high-cloze probability sentences typically starts around the P2 latency range (Gunter *et al.*, 2000; Van Den Brink *et al.*, 2001), and the MMN sometimes has positive polarity in infants (e.g., Winkler *et al.*, 2003; Friederici *et al.*, 2002a). Thus, the ERAN-label is appropriate for (relatively early) ERP effects with frontal to centro-temporal scalp distribution that are elicited by music-syntactic irregularities.

---

[37] In another study by Loui *et al.* (2009), however, the term *early anterior negativity* was also used to label effects elicited by acoustical pattern deviants occurring in an auditory oddball paradigm. Such acoustic pattern deviants elicit an MMN (Tervaniemi *et al.*, 2001). Therefore, the label 'early anterior negativity' is rather unspecific with regard to the functional significance of underlying neural operations.

9.4   Neuroanatomical correlates

Music-syntactic processing appears to rely on neural sources located (1) in the inferior pars opercularis of the inferior fronto-lateral cortex (corresponding to inferior BA 44, or BA 44v according to the anatomical labelling of Broca's area proposed by Amunts *et al.*, 2010), presumably with (2) additional contributions from the ventrolateral premotor cortex and (3) the anterior superior temporal gyrus (planum polare; Koelsch, 2006). A study with magnetoencephalography (Maess *et al.*, 2001) using a chord sequence paradigm with the stimuli depicted in Figure 9.3a & b, reported a dipole solution of the ERAN with a two-dipole model, the dipoles being located bilaterally in inferior BA 44 (the dipole strength was nominally stronger in the right hemisphere, but this hemispheric difference was statistically not significant). The results of that MEG study (Maess *et al.*, 2001) were supported by a source analysis of the ERAN with EEG data (Villarreal *et al.*, 2011). In that study, two different analyses were conducted: (1) regional (discrete) sources were localized,[38] and (2) neural activity was localized using Multiple Sparse Priors.[39] Main sources were located within and around BA 44.[40] The source localization results of these two studies (Maess *et al.*, 2001; Villarreal *et al.*, 2011) are corroborated by results of functional neuroimaging studies that used chord sequence paradigms (Koelsch *et al.*, 2002a, 2005b; Tillmann *et al.*, 2006), polyphonic music (Janata *et al.*, 2002b), and melodies (Janata *et al.*, 2002a). These studies showed activations of inferior fronto-lateral cortex at coordinates close to those reported in the MEG and the EEG study (Figure 9.12).[41] In particular, the fMRI study by Koelsch *et al.* (2005) supported the assumption that neural generators of the ERAN are located in inferior BA 44: As will be reported in more detail below, the ERAN has been shown to be larger in musicians than in non-musicians (Koelsch *et al.*, 2002a), and in the fMRI study by Koelsch *et al.* (2005b) effects of musical training were related to activations of inferior BA 44, in both adults and children. Data obtained from patients with lesions of the left inferior frontal cortex (with the strongest overlap of lesions in the left pars opercularis) showed that the scalp distribution of the ERAN differs from that of healthy controls (Sammler *et al.*, 2011), supporting the assumption that neural sources located in the pars opercularis are involved in the generation of the ERAN. Moreover, data recorded from intracranial grid-electrodes from patients with epilepsy identified two ERAN sources, one in the inferior fronto-lateral

---

[38]   Using *Brain Electrical Source Analysis* (BESA).

[39]   An L2-norm-like approach, computed with SPM 8.

[40]   A nice additional feature of that study was that it also performed a source analysis of MMN potentials elicited by mistuned chords in the same subjects. In contrast to the ERAN, main sources of MMN potentials were located within, and around, BA 41.

[41]   An EEG study by James *et al.* (2008) reported a localization of the main generators of an ERAN potential in medial temporal areas (supposedly hippocampus and amygdala) and the insula, but no control ERP component (such as P1 or N1) was localized in that study.

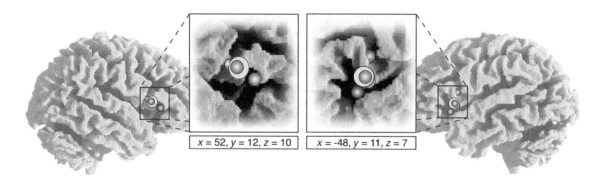

Figure 9.12   Activation foci (small spheres) reported by functional imaging studies on music-syntactic processing using chord sequence paradigms (Koelsch *et al.*, 2005b; Maess *et al.*, 2001; Tillmann *et al.*, 2003; Koelsch *et al.*, 2002a) and melodies (Janata *et al.*, 2002a). The two larger spheres show the mean coordinates of foci (averaged for each hemisphere across studies, coordinates refer to standard stereotaxic space). Modified with permission from Koelsch & Siebel (2005).

cortex, and one in the superior temporal gyrus (Sammler, 2008, the latter one was inconsistently located in anterior, middle, and posterior superior temporal gyrus).

Note that the main frontal contribution to the ERAN stays in contrast to the phMMN which receives its main contributions from neural sources located within, and in the vicinity of, the primary auditory cortex (for a direct comparison see Villarreal *et al.*, 2011).[42] Likewise, the main generators of the afMMN appear to be located in the temporal lobe (Korzyukov *et al.*, 2003). That is, whereas the phMMN (as well as the afMMN) receives main contributions from temporal areas, the ERAN appears to receive its main contributions from frontal areas. This indicates (a) that the ERAN should be differentiated from the phMMN and the afMMN, and (b) that the ERAN reflects processes of syntactic analysis (and not merely acoustical comparison processes; this issue is discussed in more detail in Section 9.5). Corroborating this notion, EEG studies investigating the ERAN and the phMMN under propofol sedation showed that the phMMN is strongly reduced, but still significant, under deep propofol sedation (Modified Observer's Assessment of Alertness and Sedation Scale level 2–3, mean Bispectral Index = 68),[43] whereas the ERAN is abolished during this level of sedation (Koelsch *et al.*, 2006b). This highlights the importance of the frontal cortex for

---

[42]   The MMN often receives additional (but smaller) contributions from frontal cortical areas, for a review see Deouell (2007).

[43]   The Modified Observer's Assessment of Alertness and Sedation Scale (MOAAS) is a clinical score used to assess the level of sedation; e.g., awake state corresponds to MOAAS level 5 ('Responds readily to name spoken in normal tone'), level 3 corresponds to 'Responds only after name is called loudly and/or repeatedly,' and level 2 to 'Responds only after mild prodding or shaking.' The Bispectral Index (BIS) is an electrophysiological indicator of the depth of sedation. Both MOAAS and BIS are well established techniques to adjust anaesthesia at a certain depth.

music-syntactic processing, because propofol sedation appears to affect hetero-modal frontal cortices earlier, and more strongly than unimodal sensory cortices (Heinke *et al.*, 2004; Heinke & Koelsch, 2005).

It is important to note that activity in BA 44 (which in the left hemisphere is often referred to as part of Broca's area) appears to be required for the processing of phrase structure grammar during language perception (e.g., Friederici *et al.*, 2006), as well as for the hierarchical processing of action sequences (e.g., Koechlin & Jubault, 2006; Fazio *et al.*, 2009), and possibly also for the processing of hier-archically organized mathematical formulae and termini (Friedrich & Friederici, 2009). Thus, with regard to language-syntactic processing, the neural resources for the processing of musical and linguistic syntax appear to overlap strongly, and this view is particularly supported by studies showing interactions between music-syntactic and language-syntactic processing (Koelsch *et al.*, 2005b; Stein-beis & Koelsch, 2008b; Slevc *et al.*, 2009, for details see Section 9.6). On a more abstract level, it is likely that Broca's area is involved in the processing of hierarchi-cally organized sequences in general, be they musical, linguistic, action-related, or mathematical. This notion has been discussed previously by several authors, and will be formally stated in Section 9.6.1 as the *Syntactic Equivalence Hypoth-esis*.[44] However, only studies investigating interactions between music-syntactic and language-syntactic processing are in the position to test the hypothesis that at least partly identical cognitive operations (and neural populations underlying these operations) are involved in the syntactic processing across different domains (details are discussed in Section 9.6).[45]

9.5   Processing of acoustic vs. music-syntactic irregularities

There is a crucial difference between the neural mechanisms underlying music-syntactic processing (as reflected in the ERAN) on the one side, and the processing of acoustic irregularities (as reflected in phMMN and afMMN) on the other. The generation of both phMMN and afMMN is based on an on-line establishment

[44]   See also, e.g., the special issue of *Cortex* on *Integrative Models of Broca's Area and the Ventral Premotor Cortex* (Fiebach & Schubotz, 2006); see also Tettamanti & Weniger (2006), and Bahlmann *et al.* (2009); for example, Tettamanti & Weniger (2006) discussed Broca's area as 'a supramodal hierarchical processor,' and similarly, Bahlmann *et al.* (2009) proposed that Broca's area reflects the 'least common denominator for hierarchical processing' in different cognitive domains (such as language, music, and action).

[45]   Logically, it is not possible to prove a distinction between processes underlying 'music' and 'language,' because it is always possible that a musical stimulus was chosen in a way that it did not tap into processing resources required for a given linguistic stimulus. Therefore, studies such as the one published recently in the *Proceedings of the National Academy of Sciences* by Fedorenko, Behr, and Kanwisher make their conclusions of 'functional specificity' (and thus of a distinction of neural resources underlying processing of music and language) based on a logical fallacy. Without logic, however, we do not deal with science, but with hocus-pocus.

of regularities – that is, based on representations of regularities that are extracted on-line from the acoustic environment. By contrast, music-syntactic processing (as reflected in the ERAN) relies on representations of music-syntactic regularities that already exist in a long-term memory format (although music-syntactic processing can modify such representations). That is, the statistical probabilities that make up music-syntactic regularities are not learned within a few moments, and the representations of such regularities are stored in a long-term memory format (as described in Section 9.2).[46]

With regard to the MMN, it is important to not confuse the on-line establishment of regularities with long-term experience or long-term representations that might influence the generation of the MMN. For example, pitch information can be decoded with higher resolution by some musical experts (leading to a phMMN to frequency deviants that are not discriminable for most non-experts; Koelsch et al., 1999). Another example is that the detection of a phoneme is facilitated when that phoneme is a prototype of one's language (leading to a phMMN that has a larger amplitude in individuals with a long-term representation of a certain phoneme compared to individuals who do not have such a representation; Näätänen et al., 1997; Winkler et al., 1999; Ylinen et al., 2006). In this regard, long-term experience has clear effects on the processing of physical oddballs (as reflected in the MMN).

However, in all of these MMN-studies (Koelsch et al., 1999; Näätänen et al., 1997; Winkler et al., 1999; Ylinen et al., 2006), the generation of the MMN was dependent on representations of regularities that were extracted on-line from the acoustic environment; for example, in a seminal study by Näätänen et al. (1997), the standard stimulus was the phoneme /e/, and one of the deviant stimuli was the phoneme /õ/, which is a prototype in Estonian (but not in Finnish). This deviant elicited a larger phMMN in Estonian compared to Finnish subjects, reflecting that Estonians have a long-term representation of the phoneme /õ/ (and that Estonians were, thus, more sensitive to detect this phoneme). However, the regularity for this experimental condition ('/e/ is the standard and /õ/ is a deviant') was independent of the long-term representation of the phonemes, and this regularity was established on-line by the Estonian subjects during the experiment (and could have been changed easily into '/õ/ is the standard and /e/ is the deviant'). That is, the statistical probabilities that make up the regularities in such an experimental condition are learned within a few moments, and the representations of such regularities are not stored in a long-term memory format.

---

[46]   This important distinction is often not made – for example, in my own studies using Neapolitan chords (e.g., Koelsch et al., 2000), effects of the processing of acoustic deviance contributed to the ERAN effect (see Section 9.3.1), Fujioka et al. (2005) used the term MMN for effects elicited by out-of-key notes in unfamiliar melodies (although that effect was based on music-syntactic, not acoustic irregularities), Herholz et al. (2008) also used the term MMN for effects elicited by wrong tones of familiar melodies (in the absence of acoustic irregularity), and Loui et al. (2009) labelled effects elicited by physical pattern deviants occurring in an auditory oddball paradigm as early anterior negativity (instead of pattern-MMN).

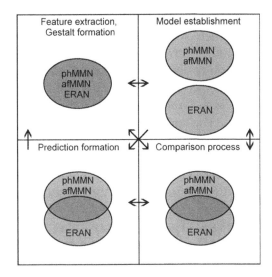

Figure 9.13 Illustration of overlaps and differences with regard to processes required to elicit MMN and ERAN (see text for details). The elicitation of both MMN, as well as ERAN, requires extraction of acoustic features, and the formation of auditory Gestalten (top left quadrant). The establishment of a model of intersound-relationships differs between MMN and ERAN (top right quadrant); in the case of the MMN, a model of regularities is based on inter-sound relationships that are extracted on-line from the acoustic environment. These processes are linked to the establishment and maintenance of representations of the acoustic environment, and thus to processes of auditory scene analysis. In the case of the ERAN, a model of inter-sound relationships is built based on representations of music-syntactic regularities that already exist in a long-term memory format. The bottom quadrants illustrate that the processes underlying the prediction of subsequent acoustic events, and the comparison of new acoustic information with the predicted sound, presumably overlap in part for MMN and ERAN. Modified with permission from Koelsch (2009).

With regards to the phMMN and the afMMN, Erich Schröger (2007) described four processes that are required for the elicitation of an MMN, which are related below to the processes underlying the generation of the ERAN (see also Figure 9.13).

(1) *Feature extraction* and *Gestalt formation* (see also Table 9.1), including the extraction of sound features, the separation of sound sources, and the establishment of representations of auditory objects are required for both the elicitation of an MMN, and the elicitation of an ERAN (see also top left of Figure 9.13).[47] Note that the establishment of a representation of an auditory object is based on operations of the echoic (or auditory sensory) memory, but not sufficient to evoke an MMN or an ERAN (a tone or chord perceived without any context elicits neither MMN nor ERAN).

---

[47]   For exceptions see Widmann *et al.* (2004) and Schon & Besson (2005).

(2) With regard to the MMN, regularities inherent in the sequential presentation of discrete events are detected and integrated into a model of the acoustic environment. Winkler (2007) stated that an MMN can only be elicited when sounds violate some previously detected *inter-sound relationship*. This statement illustrates a crucial difference between the cognitive operations underlying music-syntactic processing (as reflected in the ERAN) and processing of acoustic oddballs (as reflected in the MMN): The regularity underlying the elicitation of the MMN can be elicited without long-term knowledge (see also *knowledge-free structuring* in Table 9.1). By contrast, the music-syntactic processes underlying the generation of an ERAN involve representations of regularities that already exist in a long-term memory format (analogous to the representations of syntactic regularities of language). These representations automatically determine the processing of music-syntactic information. That is, the regularities themselves do not have to be detected, and it is not the regularity that is integrated into a model of the acoustic environment, but it is the actual sound (or chord) that is integrated into a cognitive (structural) model according to long-term representations of regularities. That is, the representations of (music-) syntactic regularities are usually not established on-line, and they are, moreover, not necessarily based on the inter-sound relationships of the acoustic input (see top right of Figure 9.13). Note that, due to its relation to representations that are stored in a long-term format, music-syntactic processing is intrinsically connected to learning and memory. Unless an individual is formally trained in music theory, these memory representations are implicit and cannot be described with words, or concepts.

In addition to *musical expectancy formation*, the establishment of a music-structural model also involves *structure building* (if the music is hierarchically organized). That is, as described before, syntactic processing of tonal music usually requires processing of long-distance dependencies at a phrase structure level.[48] By contrast, processing of acoustic oddballs involves sequential processing guided by local organizational principles with regularities usually limited to neighbouring units, but not by hierarchical processing. As already mentioned in Chapter 8, the ability to process phrase structure grammar is available to all humans, whereas non-human primates are apparently not able to master such grammars (Fitch & Hauser, 2004). Thus, it is likely that only humans can adequately process music-syntactic information at the phrase structure level. Auditory oddballs, by contrast, can be detected by non-human mammals such as cats and macaques (and auditory oddballs elicit MMN-like responses in these animals; for a brief overview see Näätänen *et al.*, 2005, p. 26).

(3) Predictions about forthcoming auditory events are derived from the model.[49] This is the case for both auditory sensory memory representations of

---

[48]   For studies comparing neural correlates of phrase structure grammar and finite state grammar see, e.g., Friederici *et al.* (2006) and Opitz & Friederici (2007).

[49]   For elaborations with regard to the generation of the MMN see Winkler (2007) and Garrido *et al.* (2009).

acoustic regularity, and musical expectancy.[50] As mentioned in (2), however, in the case of the generation of an MMN these predictions are derived from regularities established on-line based on the inter-sound relationships of the acoustic input, whereas in the case of the generation of an ERAN the predictions are based on representations of music-syntactic regularities that already exist in a long-term memory format. In addition, in the case of the processing of auditory oddballs, the predictions are established based on local organizational principles, whereas in the case of music-syntactic processing, the predictions can also be based in part on the processing of phrase structure grammar involving probabilities for the transition of chord functions within hierarchical structures.[51]

To date it is not known to which degree the predictions preceding the generation of the MMN and the ERAN are established in the same brain areas. Because the premotor cortex (PMC, corresponding to lateral BA 6) has been implicated in serial prediction (Schubotz, 2007), it is likely that PMC serves the formation of both (i) predictions preceding the detection of auditory oddballs and (ii) predictions preceding the detection of music-syntactic irregularities. However, it is also likely that, in addition to such overlap, predictions based on auditory sensory memory operations are generated in sensory-related areas (i.e., in the auditory cortex), and predictions based on music-syntactic knowledge in hetero-modal areas such as Broca's area (BA 44/45; see bottom left of Figure 9.13; for neural generators of the ERAN see Section 9.4).

(4) If a new sound enters the system, the representation of the incoming sound is compared with the sound predicted by the model. Such comparison processes presumably overlap at least partly with regard to the generation of the ERAN on the one side, and the generation of the MMN on the other (see bottom right of Figure 9.13). However, similarly to (3) it is unknown whether such processing

---

[50] The syntactic expectancy of listeners (particularly of those not formally trained) is usually an *unaware expectancy* rather than an *aware expectancy*, meaning that individuals do not consciously (or effortfully) anticipate the sound of a critical chord (or tone). Moreover, the musical sound expectancy is often a probabilistic one, rather than a very specific one. Finally, non-musicians expect a sound, rather than a chord-function. For these reasons I usually use the term *musical sound expectancy* to refer to music-syntactic expectancies of listeners. By contrast, I prefer the term *prediction* for more specific (and high-probability) expectations and forecasts that involve conscious awareness. However, I presume that musical expectancy and musical prediction largely overlap, and perhaps represent a continuum, rather than alternative processes. For example, during the perception of a chord sequence in the background, listeners build up music-syntactic expectancies, and if one pays conscious attention to a sequence moving towards its end, one might form the prediction for a final tonic chord. For reasons of better readability I mainly use the term 'prediction' in this section, although 'prediction' would be more suitable for the high-probability predictions underlying the generation of the MMN, and 'expectancy' for the more passive predictive processes preceding the generation of the ERAN.
[51] Although hierarchically organized structures are in principle not predictable, several constraints of major–minor tonal music enable an individual to make predictions based on hierarchical organization. For example, the metrical grid, and the usually symmetric organization of phrases, in addition to the Gestalt of a phrase, make it possible to realize, for example, that a phrase comes to an end, and to establish representations of probable chord functions appropriate to mark the closure of a cadence.

involves only auditory areas in the case of generation of an MMN (in which the sound representation might be more concrete, or 'sensory', due to directly preceding stimuli that established the regularities), and primarily frontal areas in the case of generation of an ERAN.

Notably, Winkler (2007) stated that the primary function of the MMN-generating process is to maintain neuronal models underlying the detection and separation of auditory objects. This also differentiates the detection of acoustic irregularities from music-syntactic processing, because syntactic processing serves the computation of a string of auditory structural elements that – in their whole – represent a form that conveys meaning which can be understood by a listener familiar with the syntactic regularities (Koelsch & Siebel, 2005; Steinbeis & Koelsch, 2008b).

9.6   Interactions between music- and language-syntactic processing

Section 9.4 described a topological overlap of neural resources underlying syntactic processing of language and music. However, it is theoretically possible that entirely different neurons (located, e.g., in BA 44) are involved in music- and language-syntactic processing. This would still lead to identical EEG topographies (as reported, e.g., by Patel *et al.*, 1998), and identical activation patterns in functional neuroimaging studies (which has so far not been shown, but implicitly assumed, e.g., by Maess *et al.*, 2001; Koelsch & Siebel, 2005). Therefore, the strongest evidence for cognitive operations, and for neural resources mediating these operations, are (a) interactions between language- and music-syntactic processing, and (b) patient studies showing that lesions result in selective syntactic processing deficits in both language and music.

Four studies have so far revealed interactions between music-syntactic and language-syntactic processing (Koelsch *et al.*, 2005; Steinbeis & Koelsch, 2008b; Slevc *et al.*, 2009; Fedorenko *et al.*, 2009). The first of these studies (Koelsch *et al.*, 2005b) investigated whether processing of syntax in music interacts with the processing of syntax and semantics in language during the simultaneous processing of music (chords) and language (words). As illustrated in Figure 9.14, five-word sentences were presented visually, simultaneously with auditorily presented sequences of five chords (i.e., each word was presented with the onset of a chord). Three different sentence types were used: The first type was a syntactically correct sentence in which the final noun had a high semantic cloze probability. The other two types were modified versions of the first sentence type: Firstly, a sentence with a gender disagreement between the last word (noun) on the one hand, and the prenominal adjective as well as the definite article that preceded the adjective on the other; such gender disagreements elicit a left anterior negativity (LAN; see Section 5.5.2). Secondly, a sentence in which the final noun

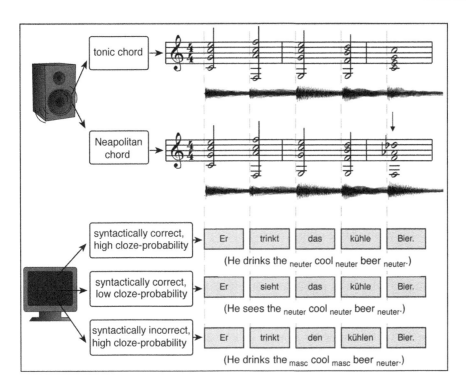

Figure 9.14 Examples of experimental stimuli used in the studies by Koelsch *et al.* (2005b) and Steinbeis & Koelsch (2008b). Top: Examples of two chord sequences in C major, ending on a regular (upper row) and an irregular chord (lower row, the irregular chord is indicated by the arrow). Bottom: Examples of the three different sentence types. Onsets of chords (presented auditorily) and words (presented visually) were synchronous. Reprinted with permission from Steinbeis & Koelsch (2008b).

was syntactically correct, but had a low semantic cloze probability (such low-cloze probability words elicit an N400; see Section 5.5.1).[52]

Half of the musical sequences ended on a regular tonic chord, the other half ended on a music-syntactically irregular chord function (Neapolitan chords; the same chord sequences had also been used in the studies described in Section 9.3). Sentences and chord sequences were combined in a 3 × 2 design (3 sentence types, 2 chord types) so that six different experimental conditions could be investigated: Final nouns of sentences that were syntactically correct and had a high-cloze probability were presented simultaneously with either a regular or an irregular final chord function. Analogously, syntactically correct, but semantically unexpected (low-cloze probability) final words were presented with either a regular or an

---

[52]   Sentence material was taken from a study by Gunter *et al.* (2000) and slightly modified: To all sentences, an adjective was added after the third word, so that both sentences and chord sequences had the same number of elements.

irregular chord function. Finally, final words of sentences with a syntactic gender disagreement (and high semantic cloze probability) were presented with either a regular or an irregular chord function (Figure 9.14). Participants were asked to ignore the musical stimulus, to concentrate on the words, and to answer in 10% of the trials whether the last sentence was correct or (syntactically or semantically) incorrect.

The rationale for this experimental design was that if language processing operates independently of music processing, neither LAN nor N400 should be influenced by the syntactic irregularities in the music (and vice versa). Because of the mentioned overlap of cerebral structures and neuronal processes involved in the syntactic analysis of music and of language, it was hypothesized that the presentation of a music-syntactic irregularity would influence the processing of syntactic violations within the sentences.

Syntactically irregular words elicited an LAN (Figure 9.15A) which was reduced when the irregular word was presented simultaneously with an irregular chord (compared to when the irregular word was presented with a regular chord; see Figure 9.15B). The N400 peak amplitude (elicited by the low-cloze probability words) was not affected by the regularity of chords (Figure 9.15D).[53] This finding is consistent with studies by Bonnel *et al.* (2001) and Besson *et al.* (1998), in which the occurrence of harmonically regular and irregular notes at the end of a melody did not have an effect on the processing of semantically congruous and incongruous words that were sung on these notes (in the study by Besson *et al.*, 1998, the N400 was used as an electrophysiological index of semantic processing; irregular notes elicited a late positive component).

The reduced LAN elicited by syntactically irregular words (compared to regular words) on music-syntactically irregular chords (compared to regular chords) indicates an interaction between the processing of the syntactic properties of the words (that is, morpho-syntactic processing) and the processing of the music-syntactic information. This interaction indicates an intersection of cognitive operations of (a) music-syntactic and (b) language-syntactic processing (and of neural populations mediating music-syntactic and language-syntactic processing). That is, this interaction shows the existence of a cognitive operation (and a neural population mediating this operation) x, that is part of music-syntactic processing (M_{syn}) as well as of language-syntactic processing (L_{syn}), formally stated as follows:

$$\exists\{x|(x \in M_{syn}) \wedge (x \in L_{syn})\}$$

M_{syn}:   Music-syntactic processing
L_{syn}:   Language-syntactic processing
x: A cognitive operation (or a neural population mediating this operation)

---

[53]   However, the ERP traces (Figure 6 of Koelsch *et al.*, 2005b) suggest an interaction between the semantic cloze probability and the syntactic irregularity of the chords in a later time window (around 500 to 900 ms). Unfortunately, this potential effect was not evaluated statistically.

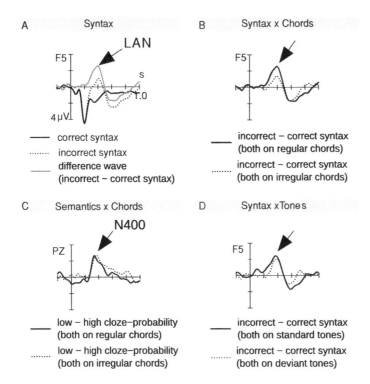

Figure 9.15 Grand-average ERPs elicited by the stimuli shown in Figure 9.14; participants ignored the musical stimulus, concentrated on the words, and answered in 10% of the trials whether the last sentence was correct or (syntactically or semantically) incorrect. **A**: Compared to regular words, morpho-syntactically irregular words elicited a LAN, best to be seen in the difference wave (grey line, indicated by the arrow). All ERPs were elicited on regular chords. **B** shows the LAN effects (difference waves), separately for the condition in which words were presented on regular chords (solid line, this line is identical with the grey difference wave in **A**), and in which words were presented on irregular chords (dotted line). The data indicate that the morpho-syntactic processing (as reflected in the LAN) is reduced when words have to be processed simultaneously with a syntactically irregular chord. **C** shows the ERPs elicited by words with high semantic cloze probability subtracted from ERPs elicited by words with low semantic cloze probability. The solid line represents the condition in which words were presented on regular chords, the dotted line represents the condition in which words were presented on irregular chords. In both conditions, semantically irregular (low-cloze probability) words elicited an N400 effect. Importantly, the N400 was not influenced by the syntactic irregularity of chords (the same N400 effect can be seen in both difference waves). **D** shows waves analogous to those shown in **B**, except that tones were presented (instead of chords) in an auditory oddball paradigm (tones presented at positions 1 to 4 were standard tones, and the tone at the fifth position was either a standard tone as well, or a deviant tone, analogous to the chord sequences). As in the chord condition, morphosyntactically irregular words elicited a clear LAN effect (solid difference wave). In contrast to the chord condition, virtually the same LAN effect was elicited when words were presented on deviant tones. This indicates that morpho-syntactic processing (as reflected in the LAN) is not influenced when words have to be processed simultaneously with an acoustically deviant tone. Therefore, the interaction between language- and music-syntactic processing shown in **B** was not due to any acoustic irregularity, but specifically due to syntactic irregularity. Scale in **B – D** is identical to scale in **A**. Modified with permission from Koelsch *et al.* (2005b).

The ERAN amplitude was not affected by the language-syntactic regularity in this study, but in a study by Steinbeis & Koelsch (2008b) which will be reported further below. Importantly, neither of the two studies (Koelsch *et al.*, 2005b; Steinbeis & Koelsch, 2008b) showed an influence of the semantic cloze probability of the words on the ERAN, suggesting that the cognitive operations underlying music-syntactic processing (and the neural population mediating these operations) do not intersect with those underlying language-semantic processing (L_{sem}):

$$\exists\{x|(x \in M_{syn}) \wedge (x \in L_{syn}) \wedge (x \notin L_{sem})\}$$

M_{syn}:   Music-syntactic processing
L_{syn}:   Language-syntactic processing
L_{sem}:   Semantic processing

In the study by Koelsch *et al.* (2005b) a control experiment was conducted in which the same sentences were presented simultaneously with sequences of single tones. The tone sequences ended either on a standard tone or on a frequency deviant (similar to the sequences shown in Figure 5.2 on p. 47). The phMMN elicited by the frequency deviants did not interact with the LAN (in contrast to the ERAN; see Figure 9.15B vs. D), indicating that the processing of auditory oddballs (as reflected in the phMMN) does not consume resources related to syntactic processing. It also shows that the interaction between the processes underlying the generation of the LAN and those underlying the processing of music-syntactic information is not simply due to any type of acoustic irregularity (and that the LAN is not influenced by the generation of any negative potential such as the phMMN), but that the interaction between LAN and music-syntactic information is quite specific (whether the afMMN interacts with music-syntactic or language-syntactic processing is not known):

$$\exists\{x|(x \in M_{syn}) \wedge (x \in L_{syn}) \wedge (x \notin A_{dev}) \wedge (x \notin L_{sem})\}$$

M_{syn}:   Music-syntactic processing
L_{syn}:   Language-syntactic processing
A_{dev}:   Acoustic deviance processing
L_{sem}:   Semantic processing

Interestingly, no N5 was elicited in that study. This absence of the N5 cannot simply be due to the fact that subjects ignored the musical stimulus, or that they were reading words (because other studies showed that under a condition in which individuals read a self-selected book and ignore the music, an N5 can still be observed; see Section 9.7 for details). Therefore, the absence of the N5 can only be due to the fact that participants were forced to process words and chords *synchronously* under the task to ignore the music and to analyze the syntax and the semantics of the sentences. Notably, the fact that an ERAN (but no N5) was observed, in addition to the fact that there was an interaction between the LAN and the regularity of chords, is strong evidence for the assumption that the ERAN reflects (music-)syntactic processing.

A similar study by Steinbeis & Koelsch (2008b) used the same stimulus material, but a different task than the study by Koelsch *et al.* (2005b). Here, participants were instructed to concentrate on the words (and to answer in 10% of the trials whether the last sentence was syntactically or semantically correct or incorrect, as in the previous study), and in addition to monitor the timbre of the chord sequences and to detect timbre deviants that were presented every so often. This new task led to two changes in the music-related ERPs: The ERAN amplitude was (slightly) larger, and a clear N5 was elicited. That latter phenomenon was important to investigate possible interactions between the N5 and the N400.

As in the previous study (Koelsch *et al.*, 2005), results showed a reduction of the LAN when words were presented on irregular chords (compared to when words were presented on regular chords). In addition, the amplitude of the ERAN was reduced when chords were presented on syntactically (but not semantically) irregular words (Figure 9.16A), lending strong support for the idea that the ERAN reflects syntactic processing. Such an interaction between the ERAN and the (morpho-)syntactic processing of words was not found in the previous study, probably due to the different task. Finally, results of that study (Steinbeis & Koelsch, 2008b) also showed an interaction between the N5 and the semantic cloze probability of words (in the absence of an interaction between the N5 and the syntactic regularity of words; Figure 9.16B,C).[54] This is one indication for the N5 reflecting music-semantic processing (this will be dealt with in more detail in Chapter 10).

Results of these ERP studies (Koelsch *et al.*, 2005b; Steinbeis & Koelsch, 2008b) were corroborated by behavioural studies. In a study by Slevc *et al.* (2009) participants performed a self-paced reading of 'garden-path' sentences (that is, of sentences which have a different syntactic structure than initially expected). Words (presented visually) occurred simultaneously with chords (presented auditorily). When a syntactically unexpected word occurred together with a music-syntactically irregular (out-of-key) chord, participants needed more time to read the word (that is, participants showed stronger garden-path effects). No such interaction between language-syntactic and music-syntactic processing was observed when words were semantically unexpected, nor when the chord presented with the unexpected word had an unexpected timbre (but was harmonically correct).

Similar results were reported by Fedorenko *et al.* (2009) in a study in which sentences were sung. Sentences were either subject-extracted or object-extracted relative clauses, and the note sung on the critical word of a sentence was either in-key or out-of-key. Participants were less accurate in their understanding of object-related extractions compared to subject-extracted extractions (as expected).

---

[54]   This could not be observed in the previous study by Koelsch *et al.* (2005b) because no N5 was elicited, due to the task.

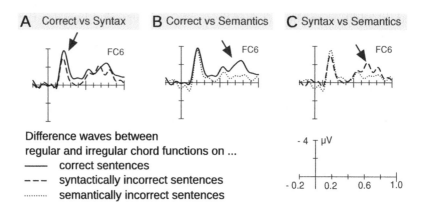

Figure 9.16 Grand-average ERPs elicited by the stimuli shown in Figure 9.14; participants concentrated on the words, and answered in 10% of the trials whether the last sentence was correct or (syntactically or semantically) incorrect; in addition, they monitored the timbre of the chord sequences and detected infrequently occurring timbre deviants. Other than in Figure 9.15A-C, where ERPs of words are shown for different chord conditions, the ERPs presented here were elicited on chords, and are shown for the different word conditions (all waves are difference waves). **A**: The solid difference wave shows the ERAN (indicated by the arrow) and the N5 elicited on syntactically and semantically correct words. The dashed difference wave shows ERAN and N5, elicited when chords were presented on morpho-syntactically incorrect (but semantically correct) words. Under the latter condition, the ERAN (but not the N5) was reduced. **B**: The solid difference wave is identical to the solid difference wave of **A**, showing the ERAN and the N5 elicited on syntactically and semantically correct words (the N5 is indicated by the arrow). The dotted difference wave shows ERAN and N5, elicited when chords were presented on semantically incorrect (but morpho-syntactically correct words). Under the latter condition, the N5 (but not the ERAN) was reduced. **C** shows the direct comparison of the difference waves in which words were syntactically incorrect (dashed line) or semantically incorrect (dotted line). These ERPs show that the ERAN was influenced by the morpho-syntactic processing of words, but not by the semantic processing of words. By contrast, the N5 was influenced by the semantic processing of words, but not by the morpho-syntactic processing of words. Modified with permission from Steinbeis & Koelsch (2008b).

Importantly, the difference between the comprehension accuracies of these two sentence types was larger when the critical word (the last word of a relative clause) was sung on an out-of-key note. No such interaction was observed when the critical word was sung with greater loudness. Thus, both of these studies (Slevc *et al.*, 2009; Fedorenko *et al.*, 2009) showed that music- and language-syntactic processes specifically interact with each other, presumably because they both rely on common processing resources.

   The findings of the mentioned EEG and behavioural studies showing interactions between language- and music-syntactic processing were also corroborated by a study investigating patients with Broca's aphasia (Patel *et al.*, 2008). That study showed that Broca-aphasics also show impaired music-syntactic processing in response to out-of-key chords occurring in harmonic progressions. Note that the term *Broca's aphasia* was used in that study to refer to the phenomenology of this type of aphasia, not to the lesion site (thus, it is likely that not all of the patients had a lesion that included Broca's area).

9.6.1   The Syntactic Equivalence Hypothesis

The combined results show that there is at least an intersection of cognitive operations of (a) music-syntactic and (b) language-syntactic processing (and of neural populations mediating music-syntactic and language-syntactic processing) that are different from cognitive operations (and neural populations mediating these operations) of acoustic deviance processing and of semantic information processing. In other words, there exist cognitive operations (and neural populations mediating these operations) that are required for music-syntactic and for language-syntactic processing, but that are not involved in the processing of acoustic deviance or language-semantic processing. Because activity in Broca's area appears to be a necessary condition for the hierarchical processing of action sequences (e.g., Koechlin & Jubault, 2006; Fazio *et al.*, 2009), and because Broca's area is also engaged for the processing of hierarchically organized mathematical formulae and termini (Friedrich & Friederici, 2009), such syntactic operations might, beyond music- and language-syntactic processing, intersect with the syntactic processing of actions, mathematical formulae, and other hierarchical structures involving long-distance dependencies on a phrase-structure level. That is, I presume that there exist cognitive operations (and neural populations mediating these operations) that are required for music-syntactic, language-syntactic, action-syntactic, and mathematical-syntactic processing, that are not involved in the processing of acoustic deviance or language-semantic processing. I refer to this statement as the *Syntactic Equivalence Hypothesis*,[55] formally stated as:

$$\exists \{x | (x \in M_{syn}) \wedge (x \in L_{syn}) \wedge (x \in Act_{syn}) \wedge (x \in Math_{syn}) \wedge (x \notin A_{dev}) \wedge (x \notin L_{sem})\}$$

M_{syn}:	Music-syntactic processing
L_{syn}:	Language-syntactic processing
Act_{syn}:	Action-syntactic processing
$Math_{syn}$:	Mathematical-syntactic processing
A_{dev}:	Acoustic deviance processing
L_{sem}:	Semantic processing

While music-syntactic processing clearly interacts with (morpho-)syntactic processing (Koelsch *et al.*, 2005b; Steinbeis & Koelsch, 2008b) and with hierarchical

---

I thank Mark Steedman for suggesting this term to me.

processing of language (garden-path effects and structural complexity; Slevc *et al.*, 2009; Fedorenko *et al.*, 2009), the interaction between music-syntactic processing of chord functions and word category violations seems to be considerably weaker. Maidhof & Koelsch (2011) presented music-syntactically irregular chord functions synchronously together with words that were either correct, or represented a word category violation.[56] Such violations elicit an early left anterior negativity (ELAN; see also Section 5.5.2), taken to reflect initial syntactic structure building based on word category information (Friederici, 2002). When the music was ignored, the ERAN (elicited by incorrect chords) was slightly reduced when presented on syntactically incorrect words (compared to when presented on correct words).[57] When the music was attended (and the speech stimulus ignored), no such interaction was observed (and there was no effect of the regularity of chords on the ELAN amplitude). Therefore, these findings provide a hint, though no clear support, for an interaction of neural resources for syntactic processing already at these early stages. This is consistent with data of functional neuroimaging studies using MEG (Friederici *et al.*, 2000) or fMRI (Rüschemeyer *et al.*, 2005) and the same sentence material as that used in the study by Maidhof & Koelsch (2011). Both MEG and fMRI data indicate that the word category violations of these sentences activate anterior-superior temporal cortical areas, rather than inferior frontolateral cortex. Using different sentence material, an MEG study by Herrmann *et al.* (2009) also showed activation of supratemporal cortex in response to word category violations.

By contrast, a study by Heim *et al.* (2006) showed that morphosyntactic violations within sentences (electrically reflected in an LAN) activate Broca's area (BA 44). Because the ERAN elicited by music-syntactically irregular chord functions interacts with the LAN, and because the ERAN receives contributions from BA 44, music-syntactic processes reflected in the ERAN appear to bear analogies, at least in part, with the morphosyntactic processing of words. This is consistent with the notion that a *chord function* parallels a *lexeme* (and that part of the ERAN is associated with the music-syntactic process of *musical expectancy formation* described in Section 9.2).

On the other hand, the late processes of music-syntactic (re-)analysis and revision (reflected electrically in a P600, Patel *et al.*, 1998) appear to bear analogies with processes of syntactic (re-)analysis during language perception. The intersection between the latter processes is referred to as the *Shared Syntactic Integration Resource Hypothesis* (SSIRH), stating that music and language rely on shared, limited processing resources that activate separable syntactic representations

---

[56] Syntactically correct sentences consisted of a noun phrase, an auxiliary, and a past participle: 'Das Baby wurde gefüttert.' / 'The baby was fed.'. Syntactically incorrect sentences contained a word category violation: In those sentences, a noun phrase and an auxiliary was directly followed by a preposition and a past participle (e.g., 'Die Gans wurde im gefüttert.' / 'The goose was in-the fed'), which is a syntactically illegal phrase structure in German.

[57] This interaction between music-syntactic and language-syntactic processing only approached statistical significance.

(Patel, 2003). The SSIRH also states that, whereas the linguistic and musical knowledge systems may be independent, the system used for on-line structural integration may be shared between language and music (see also Fedorenko *et al.*, 2009). This system was argued to be involved in integrating incoming elements (words in language, tones/chords in music) into evolving structures (sentences in language, harmonic sequences in music). Beyond the *SSIRH*, which states that the (late) processes of music-syntactic integration and repair intersect with the (late) processes of language-syntactic integration, the *Syntactic Equivalence Hypothesis* advocated here also includes the studies showing early interactions (as reflected in the ERAN and LAN; Koelsch *et al.*, 2005b; Steinbeis & Koelsch, 2008b), indicating that processing of music- and language-syntactic information intersects also at levels of morpho-syntactic processing, phrase-structure processing, and possibly word-category information. Thus, the phenomena referred to by the *SSIRH* represent a subset of those referred to by the *Syntactic Equivalence Hypothesis*.

9.7   Attention and automaticity

How independent of attention is music-syntactic processing? If individuals have a conversation while music is playing in the background – to what extent is the music still processed? And to what extent is language processed even when it is ignored (because individuals attend to the music, and not to the conversation)? Early psychological theories of attention proposed a filter model with a structural limitation (the 'attentional bottleneck'), and subsequent theories proposed that stimuli are selected for further processing at an early (Broadbent, 1957a, 1958), or late stage (Deutsch & Deutsch, 1963; Duncan, 1980). An intermediate theory (Treisman, 1964) proposed that filtering attenuates processing of unattended stimuli, rather than completely preventing it. In contrast to filter models, capacity models of attention (e.g., Kahneman, 1973) assume a general limit of cognitive operations, and a flexible allocation of the processing capacity to any stage in the processing chain. Early stages are thought to require no attention at all and are not under the strategic control of participants (*preattentive* or *automatic* processing; but note also that Logan, 1992, for example, differentiates between automatic and preattentive processes), whereas later stages require increasing amounts of capacity and can be controlled by individuals (e.g., Schneider & Shiffrin, 1977; Shiffrin & Schneider, 1977). However, the all-or-none concept of automaticity, i.e., the assumption of fully automatic processes that are independent of attention (and do not use limited capacity resources), was extended: Subsequent views (e.g., Hackley, 1993) hold that processes can vary in their degree of automaticity, and these views distinguish between strongly automatic (obligatory and not modifiable by attention), partially automatic (obligatory but modifiable by attention) and controlled processes (nonobligatory and requiring attentional resources).[58]

---

[58]   The concept of automaticity is reviewed in Moors & De Houwer (2006).

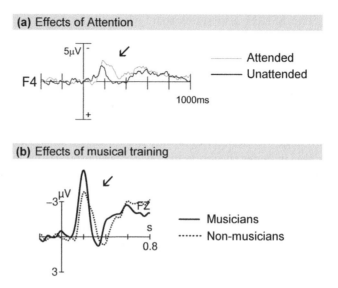

Figure 9.17 (a) shows difference ERPs (tonic subtracted from Neapolitan sixth chords) elicited when attention was focussed on the musical stimulus (grey line), and when attention was focussed on a reading comprehension task (black line). The ERAN (indicated by the arrow) clearly differed between conditions, being smaller in the ignore-condition (adapted with permission from Loui *et al.*, 2005). (b) shows difference ERPs (tonic subtracted from Neapolitan sixth chords) elicited in musicians (solid line) and non-musicians (dotted line). The ERAN (arrow) clearly differed between groups, being smaller in the group of non-musicians (adapted from Koelsch *et al.*, 2002d).

ERP studies investigating the automaticity of music-syntactic processing showed that the ERAN can be elicited while participants play a video game (Koelsch *et al.*, 2001), read a self-selected book (Koelsch *et al.*, 2002e), or perform a highly attention-demanding reading comprehension task (Loui *et al.*, 2005). In the latter study, participants performed the reading task while ignoring all chord sequences, or they attended to the chord sequences and detected chords which deviated in their sound intensity from standard chords. These conditions enabled a comparison of processing task-irrelevant irregular chords under an attend condition (intensity detection task) and an ignore condition (reading comprehension task). Results showed that an ERAN was elicited in both conditions and that the amplitude of the ERAN was reduced (but still significant) when the musical stimulus was ignored (Figure 9.17a; because the ERAN was not significantly lateralized, it was denoted as 'early anterior negativity' by the authors).

The neural mechanisms underlying the processing of music-syntactic information (as reflected in the ERAN) are active even when participants selectively attend to a speech stimulus (Maidhof & Koelsch, 2011). In that study, speech and music stimuli were presented simultaneously from different locations (20° and 340° in the azimuthal plane). The ERAN was elicited even when participants selectively attended to the speech stimulus, but its amplitude was significantly decreased compared to the condition in which participants listened to music only. The findings of the latter two studies (Loui *et al.*, 2005; Maidhof & Koelsch, 2011) indicate that the neural mechanisms underlying the processing of harmonic structure operate in the absence of attention, but that they can be clearly modulated by different attentional demands. In this regard, music-syntactic processing operates partially automatically. The study by Maidhof & Koelsch (2011) also showed that the ELAN (elicited by word-category violations in the speech stimulus) was nominally (but not significantly) smaller when the speech stimulus was to be ignored (and the music stimulus to be attended) compared to when the speech stimulus was to be attended (and the music stimulus was to be ignored). Moreover, attentional demands influenced the ERAN significantly more strongly than the ELAN.

That is, in a conscious individual, musical syntax is processed even when a listener is not paying attention to the music (and the brains of listeners register music-syntactic irregularities even when listeners are not aware of these irregularities). In unconscious individuals, music-syntactic processing is abolished. Under deep propofol sedation (where individuals are in a state similar to natural sleep), no ERAN is observable (Koelsch *et al.*, 2006; that study used the sequences shown in Figure 9.5). This stays in contrast to the phMMN, which is strongly reduced, but still significant under deep sedation Koelsch *et al.* (2006). This suggests that the elicitation of the ERAN requires a different state of consciousness on the part of the listener than the phMMN.[59]

With regard to the MMN, several studies have shown that the MMN amplitude can be reduced in some cases by attentional factors (for a review see Sussman, 2007). However, it has been argued that such modulations can be attributed to effects of attention on the formation of representations for standard stimuli, rather than to deviance detection processes (Sussman, 2007), and that the processes underlying the generation of the MMN are largely unaffected by attentional modulations (Grimm & Schroger, 2005; Sussman *et al.*, 2004; Gomes *et al.*, 2000). That is, the MMN seems to be considerably more resistant against attentional modulations than the ERAN.

9.8   Effects of musical training

Both long-term and short-term training modulate music-syntactic processing, as shown by effects of musical training on the ERAN, the LPC/P600, and the P3.[60] With regard to the ERAN, this issue has been investigated so far by five studies

---

[59]   See also Heinke *et al.* (2004) and Heinke & Koelsch (2005).
[60]   For training effects on LPC/P600 and P3 see, e.g., Besson & Faita (1995) and Regnault *et al.* (2001).

(Koelsch *et al.*, 2002d, 2007a; Koelsch & Sammler, 2008; Koelsch & Jentschke, 2008; Müller *et al.*, 2010). These studies showed that the ERAN is larger in musicians (Figure 9.17b; Koelsch *et al.*, 2002d), and in amateur musicians (Koelsch *et al.*, 2007a) compared to non-musicians. In addition, an MEG study by Fujioka *et al.* (2005) reported stronger event-related magnetic fields in musicians (compared to non-musicians) in response to out-of-key tones occurring at the end of two-part melodies. In one of the mentioned studies (Koelsch *et al.*, 2007a), the difference between groups was just above the threshold of statistical significance, and three recent studies reported nominally larger ERAN amplitude values for musicians compared to non-musicians (Koelsch & Sammler, 2008), musicology students (Müller *et al.*, 2010) and amateur musicians (Koelsch & Jentschke, 2008), although the group differences did not reach statistical significance in these studies. Despite this lack of statistical significance it is remarkable that all of the studies that have so far investigated training effects on the ERAN observed larger ERAN amplitudes for musically trained individuals compared to non-musicians. Corroborating these ERP data, significant effects of musical training on the processing of music-syntactic irregularities have also been shown in fMRI experiments for both adults and eleven-year-old children (Koelsch *et al.*, 2005a).

The evidence from the mentioned studies indicates that the effects of musical long-term training on the ERAN are often small, but reliable and consistent across studies. This is in line with behavioural studies showing that musicians respond faster, and more accurately to music-structural irregularities than non-musicians (e.g., Bigand *et al.*, 1999), and with ERP studies on the processing of musical structure showing effects of musical training on the generation of the P3 using chords (Regnault *et al.*, 2001), or the elicitation of a late positive component (LPC) using melodies (Besson & Faita, 1995).[61] The ERAN is presumably larger in musicians because musicians have (as an effect of the musical training) more specific representations of music-syntactic regularities and are, therefore, more sensitive to the violation of these regularities (for effects of musical training in children see next section).

To investigate short-term effects on music-syntactic processing, Koelsch & Jentschke (2008) presented the sequence types shown in Figure 9.5 for approximately two hours to participants (participants were watching a silent movie with subtitles). Results showed that music-syntactically irregular chords elicited an ERAN, and that the amplitude of the ERAN decreased over the course of the experimental session. These results thus indicated that neural mechanisms underlying the processing of music-syntactic information can be modified by short-term musical experience. Interestingly, although the ERAN amplitude was significantly reduced, it was still present at the end of the experiment, suggesting that cognitive representations of basic music-syntactic regularities are remarkably stable, and cannot easily be modified. This notion is consistent with results of a study by

---

[61] See also Schon *et al.* (2004), Magne *et al.* (2006), and Moreno & Besson (2006).

Carrión & Bly (2008) which reported that the amplitude of the ERAN elicited by irregular chords did not increase when participants underwent a training session in which they were presented with eighty-four training sequences that ended on syntactically correct chords. Notably, in contrast to the ERAN, the P3b amplitude increased in that study as an effect of training.

9.9   Development

The youngest individuals in whom music-syntactic processing has been investigated so far with ERPs were, to my knowledge, 4 month-old babies.[62] These babies did not to show an ERAN (unpublished data from our group, irregular chords were Neapolitan sixth chords; some of the babies were sleeping during the experiment). In 2.5 year-old children (30 months) we observed an ERAN in response to supertonics as well as to Neapolitan chords (unpublished data from our group). In this age group, the ERAN was quite small, suggesting that the development of the neural mechanisms underlying the generation of the ERAN commence around, or not long before, this age.[63]

Children at the age of 5 years show a clear ERAN (Figure 9.18), but with longer latency than adults (around 230–240 ms; Jentschke et al., 2008, in that study the ERAN was elicited by supertonics). Similar results were obtained in another study using Neapolitans as irregular chords (Koelsch et al., 2003). It is not known whether the longer latency in 5-year-olds (compared to adults) is due to neuroanatomical differences (such as fewer myelinated axons), or due to less specific representations of music-syntactic regularities, or both. The study by Jentschke et al. (2008) also investigated music processing in a group of 5-year-olds with specific language impairment (SLI, which is characterized by deficient processing of linguistic syntax).[64] In contrast to children with typical language development, children with SLI showed neither an ERAN, nor an N5. That is, children with

---

[62]   For an fMRI study with newborn infants see below and Perani et al. (2010).

[63]   For behavioural studies on the development of music-syntactic processing see, e.g., Schellenberg et al. (2005).

[64]   A main characteristic of SLI children is that they have severe difficulties with grammar. They perform worse on many measures of syntactic comprehension, especially those concerning syntactic complexity. In general, it seems that lexical and pragmatic skills are relatively intact, with phonology and argument structure abilities being slightly worse, and morphosyntactic skills (particularly processing of grammatical morphemes) being most impaired. In the study by Jentschke et al. (2008), linguistic skills were evaluated with the standardized language development test (SETK 3-5) which consisted of four subtests evaluating different aspects of language processing: (1) 'Sentence comprehension' reflecting the complex interplay of phonologic, lexical-semantic, and morphologic-syntactic processing steps. (2) 'Generation of plurals' being related to syntactic processing, especially knowledge of morphological rules. Children with SLI are hampered in their extraction of such rule-based patterns from spoken language. (3) 'Non-word repetition' being a measure of the ability to process and to store unknown phoneme patterns in short-term memory. Difficulties in this subtest are considered a classical marker of SLI. (4) 'Repetition of sentences' reflecting grammatical knowledge and

SLI showed deficient syntactic processing of language, as well as of music. It is tempting to speculate that engagement of the processes underlying the generation of ERAN and N5 leads to training effects of music(-syntactic) processing which transfer to language processing, and that, therefore, musical training can aid the therapy of children with SLI (and perhaps even prevent SLI from developing in children at risk for SLI). Notably, as will be reported further below, results from children with musical training indicate such transfer-effects.

At the age of nine, the ERAN appears to be very similar to the ERAN of adults. In a recent study, 9-year-olds with musical training showed a larger ERAN than children without musical training (Jentschke, 2007), and the latency of the ERAN was around 200 ms, in both children with and without musical training, thus still being longer than in older children and adults. With fMRI, it was observed that children at the age of 10 show an activation pattern in the right hemisphere that is strongly reminiscent to that of adults (with clear activations of inferior frontolateral cortex elicited by Neapolitan chords; Koelsch *et al.* 2005a). In that study, children also showed an effect of musical training, namely a stronger activation of the right pars opercularis in musically trained children (as observed in adults; see also Section 9.8 on musical training).

In 11-year-olds, the ERAN has a latency of around 180 ms, regardless of musical training, and is practically indistinguishable from the ERAN observed in adults (Jentschke & Koelsch, 2009). As in 9-year-olds, 11-year-old children with musical training show a larger ERAN than children without musical training (Jentschke & Koelsch, 2009). In that study, children with and without musical training participated in a music experiment suited to elicit an ERAN (using sequences such as those shown in Figure 9.5, ending on a regular tonic, or on an irregular supertonic), as well as in a language experiment suited to elicit an ELAN (see Section 5.5.2). In the group of children with musical training, each child had received musical training for several years.[65] The rationale of the study was that children with musical training would transfer their (music-)syntactic processing skills to the language domain, and thus show ELAN potentials that are more strongly developed than those of children without musical training.

The upper panel of Figure 9.19 shows that, in the music experiment, the ERAN (but not the N5) was larger in the group of children with musical training

---

working memory functions, i.e., the ability to employ knowledge of grammatical structures in order to process sentences and to store them in memory in a compact form. Scores of all four subtests of this language development test significantly correlated with the ERAN, but not with the N5 amplitude. Non-verbal intelligence in the children with SLI was lower than in the children with typical language development, unfortunately, and similar differences were present for the parents' duration of education (which was shorter for the parents of children with SLI), and the socioeconomic status of their occupation (which was also lower in this group). However, it is worth noting that neither the education, nor socioeconomic status of the parents were correlated with the ERAN, nor with the N5.

[65]   Most of the boys with musical training were boys of the Leipzig St. Thomas Boys Choir, the other boys as well as the girls were students of a public music school in Leipzig.

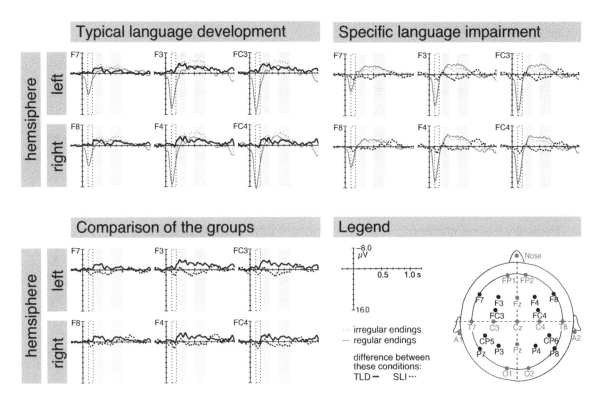

Figure 9.18 Top panel: Grand-average ERPs, separately for children with typical language development (top-left panel) and for children with specific language impairment (top-right panel). ERPs elicited by regular chord sequence endings (tonic chords) are indicated by the thin solid line, ERPs elicited by irregular endings (supertonics) are indicated by the thin dotted lines. The thick lines show the difference ERPs (regular subtracted from irregular), separately for children with typical language development (TLD, thick solid line) and specific language impairment (SLI, thick dotted line). The grey-shaded areas indicate the ERAN (230–350 ms) and N5 (500–700 ms) time windows. The direct comparison of the difference waveforms for both groups of children in the bottom panel shows differences between groups in both the ERAN and the N5 time window. The dotted rectangles show that ERPs reflecting early acoustic processing (in a time window from 100 to 180 ms) did not differ between groups. Adapted and reprinted with permission from Jentschke *et al.* (2008).

(consistent with the studies showing that musical training is reflected in increased ERAN amplitudes; see Section 9.8). The more important result, however, was that the amplitude of the ELAN elicited by syntactically incorrect words was about five times larger in children with musical training than in children without musical training (see lower panel of Figure 9.19). In addition, a later sustained negativity was observed bilaterally in both groups in response to the syntactically incorrect words, which was also considerably larger in the group of children with musical training. That is, neurophysiological correlates of syntactic language processing were more strongly developed in children with, compared to children without, musical training. Because syntactic processing is important

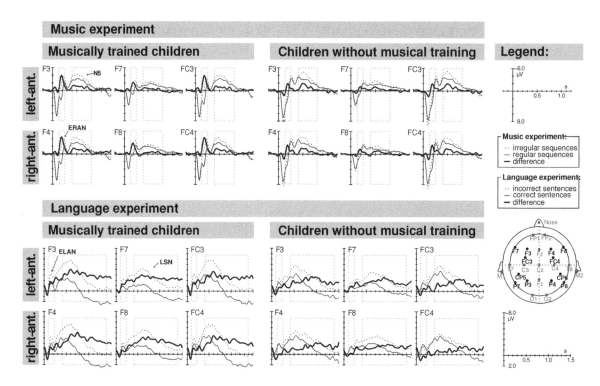

Figure 9.19 Grand-average ERPs from the music (upper panel) and the language experiment (bottom panel), separately for the group of musically trained children (left panels) and the group of children without musical training (right panels). Thin grey dotted lines represent ERP responses to irregular chords (supertonics) or to syntactically incorrect sentences, thin grey solid lines indicate ERP responses to regular (tonic) chords or to syntactically correct sentences. The thick black solid lines indicate the difference waves (incorrect–correct). In musically trained children, the ERAN was larger than in children without musical training (see slim dotted rectangles in the upper panels). The N5 did not differ between groups (wide dotted rectangles in the upper panels). Likewise, the ELAN was larger in musically trained children than in children without musical training (see slim dotted rectangles in the lower panels). The ELAN was followed by a later sustained negativity which also differed between groups (wide dotted rectangles in the lower panels). Adapted and reprinted with permission from Jentschke & Koelsch (2009).

for understanding complex hierarchical structures, and because a fast processing of such structures also serves fast thinking, these results indicate that musical training during childhood has beneficial effects on the intellectual abilities of children. With regard to semantic processing, the amplitude of the N400 (elicited by the semantically correct first word of each sentence) did not differ between groups, consistent with verbal IQ scores that did not differ between groups.[66]

Significance of acoustic change detection   Notably, MMN-like responses can be recorded even in the foetus (for details see Section 5.3), supporting the notion that

---

[66]   No semantic violations were presented in that study.

the generation of such discriminative responses is based on the (innate) capability to establish representations of inter-sound regularities that are extracted on-line from the acoustic environment (and the innate capability to perform auditory scene analysis), whereas the generation of the ERAN requires representations of musical regularities that have to be learned through listening experience, involving the detection of regularities (i.e., statistical probabilities) underlying, e.g., the succession of harmonic functions.

Notably, it is likely that, particularly during early childhood, the MMN system is of fundamental importance for music-syntactic processing: As described in Section 9.2, this system enables an individual during the perception of tonal music (a) to establish a representation of the metre of a piece, (b) to establish a representation of the tones of a tonal key, (c) to detect tones that do not belong to that key, and (d) to establish a representation of a harmonic hierarchy of stability. Because music-syntactic irregularity (and harmonic distance) is usually related to acoustic deviance, the acoustic deviance detection mechanism proliferates information about the irregularity of tones and chords. Such information aids the build-up of a music-syntactic model, the detection of music-syntactic (ir)regularities, and the memorization of statistical probabilities for chord- or sound-transitions. Notably, acoustically deviant music-syntactic irregularities automatically draw attention to the music (initiated by the MMN mechanism, and reflected in the P3a). This attention-catching mechanism presumably contributes to musical learning, and to the surprisingly sophisticated implicit musical knowledge that children already have at a remarkably young age.

The contributions of the MMN system to music-syntactic processing might be reflected in the activation of inferior fronto-lateral cortex in newborn infants in response to music-syntactic irregularities (Perani *et al.*, 2010). In that study, newborn infants (measured within one to three days after birth) were presented with classical piano pieces into which the investigators had inserted key shifts at the end of cadences (pieces were played without musical expression). Such key shifts evoked a change of BOLD activity in the auditory cortex as well as in the (left) inferior fronto-lateral cortex. However, because the key shifts also introduced a sensory dissonance and physical auditory oddballs, it is possible that this activation was due to (a) the processing of acoustic irregularity, (b) the processing of harmonic structure (in terms of establishing an implicit acoustic representation of key membership, and determining harmonic distance between harmonies based on acoustic similarity), (c) both the processing of acoustic irregularity and of harmonic structure. Because the newborns had only minimal exposure to music (prior to birth), it is unlikely that the newborns already had extensive knowledge of music-syntactic regularities. Thus, the data from Perani *et al.* (2010) illustrate the close interplay between acoustical and music-syntactic processing, and its relevance for the acquisition of music-syntactic knowledge.

10

Musical Semantics

10.1 What is musical semantics?

To communicate, an individual has to utter information that can be interpreted and understood by another individual. This chapter deals with neural correlates of the processing of meaning emerging from the interpretation of musical information by an individual. Theoretically, it is important to differentiate the 'meaning *of* music' (in the sense of 'how does a musical system work with regard to its capability of conveying meaning information'), and the processes underlying the emergence of meaning due to the interpretation of musical information by an individual. The former deals with music and its meaning as an object (and with questions such as 'does music have meaning?,' 'which types of meaning can music convey?,' 'what are the differences between music and language with regard to meaning?'), the latter deals with the subjective processes related to the interpretation of musical information (and the interpretation of psychological and physiological effects of music perception) that give rise to meaning. Such a distinction is important because on the one hand a musical system can be used to convey specific meaning information, and on the other musical information may evoke meaningful associations in a receiver although this specific meaning was not intended to be conveyed by the producer. One example for the latter phenomenon is a specific personal memory brought to mind in an individual by specific musical information. This chapter proposes a new neurobiological theory of musical meaning that considers both sides: Music as a system to convey meaning information, and the psychological reality that musical information means something for an individual.

Brain and Music, First Edition. Stefan Koelsch.
© 2013 John Wiley & Sons, Ltd. Published 2013 by John Wiley & Sons, Ltd.

Table 10.1 Synopsis of the dimensions of musical meaning.

extra-musical			intra-musical	musicogenic		
iconic	indexical	symbolic		physical	emotional	personal

Note that the term *meaning* is not used here to refer only to (directional) relations between two things (a sign and a referent), and – even when dealing with the reference of a sign to a referent – the term meaning is not used here to refer only to the conscious intentional use of a sign to refer to a referent. For example, when listening today to music from J.S. Bach, the concept 'baroque' might be evoked in a listener, although Bach himself could not have intended to convey such a concept (because it was not used in his time for his music).

In the following sections, seven dimensions of musical meaning will be described, divided into three classes of musical meaning: Extra-musical, intra-musical, and musicogenic meaning (see Table 10.1). Extra-musical meaning can emerge from the act of referencing a musical sign to an (extra-musical) referent by virtue of three different types of sign quality (iconic, indexical, and symbolic). Beyond *conceptual* extra-musical meaning, I will propose two classes of *non-conceptual meaning* (and, notably, meaning beyond a single sign). One of these two classes is intra-musical meaning, which emerges from the act of referencing a structural musical element to another structural musical element. The other of these two classes is musicogenic meaning, which emerges from physical processes (such as actions), emotions, and personality-related responses (including preferences) evoked by music. That is, in contrast to how the term meaning is used in linguistics, the term meaning as it is used here is not confined to conceptual meaning, and can thus also refer to *non-conceptual* meaning.[1]

The dimensions of musical meaning are abstractions of musical meaning, and it is important to consider that interpreting and understanding music with regard to its meaning (as well as interpreting and understanding meaning information communicated by other individuals in general) is usually multi-dimensional. That is, during music listening or music performance, meaning can emerge from several sources simultaneously. For example, while listening to a symphonic poem, meaning may emerge from the interpretation of extra-musical sign qualities, from the processing of the intra-musical structure, and from music-evoked (musicogenic) emotions.

Moreover, the term *musical semantics* is used in this chapter (instead of simply using the terms 'musical meaning' or 'musical semiotics') to emphasize that musical meaning extends beyond musical sign qualities. For example, with regard to intra-musical meaning, musical meaning can emerge from structural

---

[1] To illustrate this: An individual with a receptive (Wernicke's) aphasia does not understand the conceptual meaning of words anymore. Yet, pain, or encountering a loved person, has meaning for this individual.

relations between successive elements. Another example, with regard to extra-musical meaning, is that during listening to *programme music*, processing of extra-musical meaning usually involves integration of meaningful information into a semantic context. Note, however, that the term *musical semantics* does not refer to binary (true–false) truth conditions. I agree with Uli Reich (2011) that no musical tradition makes use of quantifiers (e.g., 'all', 'some', 'none', 'always'), modals (e.g., 'must', 'may', necessary'), or connectives (e.g., 'and', 'if...then', 'if and only if', 'neither...nor'), unless music imitates language (such as drum and whistle languages; Stern, 1957). Hence, the term 'musical semantics' should not be equated with the term 'propositional semantics' as it is used in linguistics (this issue is discussed further at the end of this chapter).

10.2   Extra-musical meaning

Extra-musical meaning emerges from the interpretation of musical sign qualities with reference to the extra-musical world. Leonard Meyer (1956) referred to this class of musical meanings as *designative meaning*, Nattiez (1990) as *extrinsic referring*.[2] Here, the view is advocated that extra-musical meaning comprises three dimensions: Meaning due to (1) iconic, (2) indexical, and (3) symbolic sign quality of music. These sign-qualities are reminiscent of those introduced by Charles Sanders Peirce (although not with reference to music; Peirce, 1931/1958), and were, to my knowledge, first applied to music by Vladimír Karbusický (1986).

10.2.1   Iconic musical meaning

Iconic musical meaning emerges from musical information[3] resembling sounds of objects, qualities of objects, or qualities of abstract concepts. Examples for the use of iconic sign quality in Western music include the imitation of the sound of an animal's voice (e.g., dog barking in the 2nd movement of the 'Spring' of Vivaldi's *Le quattro stagioni*), of the qualities of an animal (e.g., large and heavy sounding tones resemble an elephant in the 5th movement of Saint-Saëns' *Le carnaval des animaux*), weather-related sounds (e.g., a thunderstorm in the 4th movement of Beethoven's *Pastorale*, or in the 'Summer' of Haydn's *Die Jahreszeiten*), sounds related to a landscape (e.g., 'Die Moldau' of Smetana's *Má vlast*), qualities of a fictional being (e.g., a gnome in Mussorgsky's *Pictures at an Exhibition*), etc. Even a single tone may sound 'warm,' 'round,' 'sharp,' 'colourful,' etc.[4] The use of iconic sign quality in music is one form of *tone painting*, the other form of

---

[2]   For an overview see Koopman & Davies (2001).
[3]   Musical information can be musical sounds, musical patterns, or musical forms.
[4]   Up to now, there is a sparsity of empirical research specifying the acoustic properties of such sound qualities.

tone painting uses indexical sign quality (see below).[5] Note that Susanne Langer (1942, 1953) used the term 'iconic' for what is referred to in the next section as 'indexical musical meaning'. With regard to language, the iconic sign-quality of the sound of a word that imitates the sound of the object or action the word refers to, is denoted as *onomatopoetic*.

10.2.2 Indexical musical meaning

Indexical musical meaning emerges from signals indicating the inner state of an individual. That is, in contrast to iconic musical meaning, indexical musical meaning refers to signs that do not signal the presence of something that is inherent in the sign itself (like 'smoke indicating the presence of fire'; Peirce, 1931/1958). In Western music, indexical musical meaning usually emerges from the imitation of expressions signalling the psychological state of an individual, such as an emotion, a mood, or an intention (imitation of such expressions is the other form of tone painting). Such expressions are inherently action-related (for example, vocalizations are produced by the vocal apparatus, gestures by skeletal muscles).

Stephen Davies (1994) uses the term 'intentional use of natural significance' (pp. 29–32) for indexical sign quality. This distinction is theoretically valid and useful for the description of how a musical system can serve to convey meaning information, but as mentioned above, this chapter will also consider the psychological reality that musical information is interpreted by a listener, regardless of whether or not a producer of a signal intended to convey a specific meaning emerging from indexical (or any other) sign quality. Cross (2008b) refers to this dimension of musical meaning as *motivational–structural* (in reference to Owings & Morton, 1998) due to the relationship between affective–motivational states of individuals on the one side, and the structural–acoustical characteristics of (species-specific) vocalizations on the other. For example, in a stressed individual, the vocal tract contracts, leading to a higher degree of acoustic roughness in vocalizations as compared to a vocalization uttered in a relaxed state. Susanne Langer used the term 'iconic' for what is referred to here as indexical musical meaning (Langer, 1942, 1953).[6]

Historically, concrete descriptions of emotional expression in music compositions were already formulated in the *Affektenlehre* (Theory of Affections) of the seventeenth century. The *Affektenlehre* prescribed certain musical methods and figures for imitating, or portraying (and thus, according to the *Affektenlehre*, summoning) individual emotions, and aimed at differentiating the relationships between the various individual elements of musical form and technique as

---

[5]   Tone painting in Western music can be traced back to the 13th century (e.g., *Sumer is icumen in*), other early examples include the caccia of the 14th century, and the madrigals of the 16th century.

[6]   The archaeologist Steven Mithen uses the term 'manipulative' (Mithen, 2006); for iconic meaning he uses the term 'mimetic'.

well as individual affections such as joy, sadness, love, hate, despair, desire, and admiration.[7]

Juslin & Laukka (2003) compared in a meta analysis the acoustical signs used for the expression of some 'basic' emotions (anger, fear, happiness, sadness, and tenderness) in music on the one hand, and in affective prosody on the other, showing that the acoustic properties that code emotional expression in prosody are highly similar to those coding these expressions in music. Correspondingly, a study by Fritz *et al.* (2009) showed that the expression of joy, sadness, and fearfulness in tonal music can be recognized universally; in that study, participants from the Mafa people (a native African population living in North Cameroon), who presumably had never heard Western tonal music before they participated in the experiment, were presented with musical stimuli (tonal music) expressing joy, sadness, and fearfulness. The Mafa participants recognized the emotional expression of the stimuli clearly above chance level, indicating that the expression of at least of some 'basic' emotions in tonal music can be recognized universally.[8] These findings parallel the universal recognition of facial (Ekman, 1999) and vocal (Scherer, 1995) expression of emotion. Interestingly, the music of the Mafa does not imitate emotions (although music is always joyful for them). Thus, the conception of music imitating emotions is not universal (but dependent on the use of music in a culture).

The Mafa also use music to index the physical fitness of a player (in addition to the use of music for group coordination in rituals, as well as for other social functions): Playing their flutes (usually along with dancing or running) is strenuous, involving rapid in- and exhalation, 'and a good physical fitness is regarded as a precondition to engaging in this activity which can last for several hours' (supplemental material of Fritz *et al.*, 2009). The duration of playing indexes the physical fitness of a player (and the Mafa appreciate powerful, long-lasting music performances; Fritz *et al.*, 2009). The indexical quality of musical signals with regard to the physical fitness of the producer of these signals has motivated one hypothesis for an adaptive function of music: The sexual selection model. This model dates back to Darwin (1874), who suggested, by analogy with birdsong, that 'musical notes and rhythm were first acquired by the male or female progenitors of mankind for the sake of charming the opposite sex' (p. 477). Darwin's supposition has often been repeated, despite lack of clear evidence supporting this

---

[7]   In contrast to the concept of *imitation* of emotion, Hanslick (1854) introduced in the 19th century the term *expression*, with the controversial claim that instrumental music cannot express definite emotions (for an extensive discussion see, e.g., Davies, 1994; see also last section of this chapter). After Hanslick, several theorists dealt with the issue of explaining 'art's expressiveness as arising from artists' expressing their concurrent emotions or feelings in the production of art' (although rarely with regard to music; see Davies, 1994, p. 170; Davies also provides a detailed account on what he refers to as 'expression theory').

[8]   Performance of Western listeners was significantly better than performance of Mafa, suggesting that the recognition of emotion in (tonal) music due to the understanding of indexical musical sign quality is also influenced by cultural experience.

supposition for humans (for reviews see Fitch, 2005, 2006, for other accounts on adaptive functions of music see the *Seven Cs* on p. 169).

Psychological states that can be indexed by sound information also include intentions.[9] An fMRI study by Steinbeis & Koelsch (2008c) showed that listeners automatically engage social cognition during listening to music, in an attempt to decode the intentions of the composer or performer (this was indicated by activations of the cortical *theory-of-mind* network). That study also reported activations of posterior temporal regions implicated in semantic processing, reflecting that the decoding of intentions also has meaning quality (see also Excursion on the decoding of intentions during music listening, and Excursion on posterior temporal cortex and processing of meaning in this Chapter).

Excursion: Decoding of intentions during music listening

When humans listen to music, they automatically attempt to figure out the intentions (perhaps even the desires, and beliefs) of the individuals who create(d) the music. That is, as soon as we listen to music, our brains engage in mental state attribution (*mentalizing*, or *adopting an intentional stance*, also often referred to as establishing a *Theory of Mind*, TOM).

An fMRI study by Steinbeis & Koelsch (2008c) showed that the perceived intentionality of musical information activates the typical theory-of-mind network in the listener, including brain structures involved in the processing of meaning information. In that study, non-tonal music (from Arnold Schönberg and Anton Webern) was presented to non-musicians. The same pieces of music were played – counterbalanced across subjects – either with the cue that they were written by a composer or with the cue that they were generated by a computer. Participants were told that this experiment was about emotion and music (that is, they were not informed about the real purpose of the study), and their task was to rate after each excerpt how pleasant or unpleasant they found each piece to be. Data of this behavioural task showed that valence ratings of participants did not differ between the two conditions (that is, whether participants were informed that the piece was from a composer or from a computer did not influence their perceived pleasantness of the piece). Interestingly, pieces were rated as moderately pleasant (that is, the non-tonal music was not rated as unpleasant). After the experiment, participants were presented with a post-imaging questionnaire, in which they answered to items such as 'imagining an agent' during the two conditions, 'visual imagery,' 'daydreaming,' etc. However, the only item in which a difference between conditions was found, was the

---

[9] For example, the prosody of a spoken sentence such as 'I would like you to mow the lawn' can convey a range of degrees of intentionality.

item about how strongly participants felt that intentions were expressed by the music.

The fMRI data showed that, when contrasting the brain activity of the composer condition against the computer condition, there was an increase in precisely the neuroanatomical network dedicated to mental state attribution, namely the anterior medial frontal cortex (aMFC), the left and right superior temporal sulcus, as well as left and right temporal poles. Notably, the brain activity in the aMFC correlated with the degree to which participants thought that an intention was expressed in the composed pieces of music. Thus, the data showed that listening to music automatically engages areas dedicated to mental state attribution (in the attempt to understand the composer's intentions).

The recognition of a composer's intentions also has a quality of meaning (and following the intentions of a composer might evoke musicogenic meaning; see Section 10.5). This was also reflected in the fMRI data of the mentioned study (Steinbeis & Koelsch, 2008c): In the left hemisphere, the activation of the posterior temporal cortex (posterior STS, extending into the posterior superior temporal gyrus) covered part of the so-called *Wernicke's area*, an area that is taken to host conceptual knowledge (see also Excursion on p. 135). Thus, these data suggest that meaning of music is derived in part from the understanding that every note reflects an intentional act, which signals personal relevance to the artist (or to the player or singer making music together with another individual). This represents a communication between the individual making, and the individual perceiving the music (naturally, this also holds if individuals make music together).

10.2.3   Symbolic musical meaning

Symbolic musical meaning emerges from arbitrary extra-musical associations; the symbolic sign quality of musical information can be conventional (e.g., any national anthem, soundtracks of movies and television series, music specific for a particular ritual, commercial jingle music, etc.), or idiosyncratic (e.g., the association between musical information and the memory of an event, a musical signal used as a personal telephone ringtone, etc.). Symbolic musical meaning also includes social associations such as associations between music and social or ethnic groups (for the influence of such associations on behaviour see Patel, 2008).

For conventional symbolic sign quality, Davies (1994) uses the term 'arbitrary stipulation of stand-alone meaning' (p. 34).[10] The meaning of the majority of words is due to this sign quality. For idiosyncratic symbolic sign quality, Koopman & Davies (2001) used the term 'meaning for the subject' (p. 268).

---

[10]   Davies also provides a number of interesting examples for symbolic musical meaning (Davies, 1994, pp. 40–47).

Cross (2008b) refers to symbolic musical meaning as *culturally enactive*, emphasising that symbolic qualities of musical practice are shaped by (and shape) culture: 'This dimension is rooted in the conventional and institutional use of music and is likely to be dependent in part on the frequency of co-occurrence of music and personal or social situations, in part on the trajectories of individual and social histories, and subject to the contingencies of cultural formation and change'.

10.3   Extra-musical meaning and the N400

This section will present evidence showing that processing of extra-musical meaning is reflected in the N400. As was described in Section 5.5.1, the N400 component is an electrophysiological index of the processing of meaning information, particularly conceptual/semantic processing or lexical access, and/or post-lexical semantic integration. The N400 elicited by words is highly sensitive to manipulations of semantic relations, being attenuated for words that are preceded by a semantically congruous context, compared to when preceded by a semantically incongruous context. That is, when a word is preceded by a semantic context, the amplitude of the N400 is inversely related to the degree of semantic fit between the word and its preceding semantic context (for an example see upper panel of Figure 10.1).

There is ample evidence that the N400 elicited by words is sensitive to conceptual analysis, and it is assumed that the N400 reflects that readers and listeners immediately relate the word to a semantic representation of the preceding contextual information. That is, the N400 elicited by words is particularly sensitive to the processing of meaning information, both in prime-target and in sentential contexts (Kutas & Federmeier, 2000; Friederici, 1999). More generally, an N400 effect can be elicited by the processing of almost any type of semantically meaningful information, such as faces and pictures (e.g., Lau *et al.*, 2008), environmental sounds (Van Petten & Rheinfelder, 1995; Cummings *et al.*, 2006; Orgs *et al.*, 2006, 2007; Aramaki *et al.*, 2010), and odours (Grigor *et al.*, 1999). Importantly, the N400 can also be elicited by music, and the N400 elicited by a word can be modulated by the meaning of musical information preceding that word.

An initial study investigating this issue (Koelsch *et al.*, 2004a) used a classical semantic priming paradigm in which target words (presented visually) were preceded by (auditorily presented) prime stimuli that were either sentences or musical excerpts (Figure 10.1). The prime stimuli were semantically either related or unrelated to the target word. In the language condition, for example, participants heard the sentence: 'The gaze wandered into the distance' and then saw the target word 'wideness,' which is semantically more closely related to the prime sentence than to a prime sentence like 'The manacles allow only little movement' (Figure 10.1). Target words were both concrete (e.g., needle, river, staircase, blossom, king, bird, pearls, sun) and abstract words (e.g., wideness, limitedness, devotion, mischief, reality, illusion, arrival, leave).

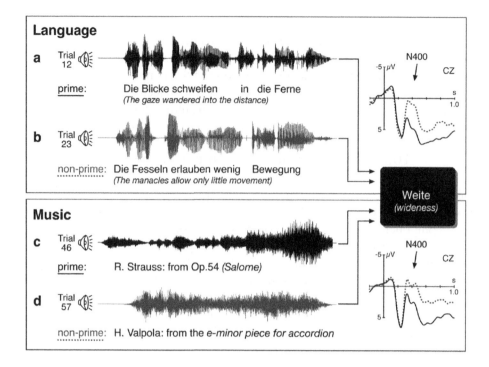

Figure 10.1 Left: Examples of the four experimental conditions preceding a visually presented target word. Top panel: Prime sentences that are semantically related (a), and unrelated (b) to the target word *wideness*. The upper diagram on the right shows grand-averaged ERPs elicited by target words after the presentation of semantically related (solid line) and unrelated prime sentences (dotted line), recorded from a central scalp electrode. Unprimed target words elicited a clear N400 component in the ERP (compared to the primed target words). Bottom panel: Musical excerpts that are semantically related (c), and unrelated (d) to the same target word. The diagram on the right shows grand-averaged ERPs elicited by target words after the presentation of semantically related (solid line) and unrelated musical excerpts (dotted line). As after the presentation of sentences, unprimed target words elicited a clear N400 component (compared to primed target words). Each trial was presented once, conditions were distributed in random order, but counterbalanced across the experiment. Note that the same target word was used for the four different conditions. Thus, condition-dependent ERP effects elicited by the target words can only be due to the different preceding contexts. Adapted from Koelsch *et al.* (2004a).

The upper panel of Figure 10.1 shows the ERPs of the target words following prime sentences to which the target words were semantically either related or unrelated. ERPs elicited by the target words showed the classical N400 priming effect: The N400 was significantly larger when elicited by semantically unrelated target words compared to when elicited by semantically related target words (as to be expected). Importantly, the conditions of primary interest in that study were the conditions in which the target words were preceded by musical excerpts.

These excerpts were believed to be meaningfully (or 'semantically') un/related to the target words, and it was of interest whether a musical excerpt could have the same effect on the semantic processing of a target word as a sentence. In the example presented in the lower panel of Figure 10.1, the target word 'wideness' was in one trial preceded by a musical excerpt in which the intervals of chords were set in wide position (therefore assumed to prime the concept of wideness), and in another trial by a musical excerpt in which the intervals were often dissonant, and set in close position (therefore used as a non-prime for the word wideness). The ordering of related and unrelated primes was balanced across the experiment, and any emotional content (such as pleasantness/unpleasantness) was controlled for.

As in the language condition, the ERPs elicited by target words that were meaningfully unrelated to the preceding musical excerpt showed an N400 (compared to the ERPs of target words that were meaningfully related to the preceding excerpt). That is, these results showed a modulation of the N400 elicited by the target words as a function of the semantic fit between musical excerpt and target word. That was the first empirical evidence that musical information can prime representations of meaningful concepts, and that music can systematically influence the semantic processing of a word (for probably the first behavioural experiment on this topic see Hevner, 1936). The priming of meaning by musical information was due to iconic sign qualities (e.g., ascending interval steps priming the word staircase), indexical sign quality (e.g., saxophone tones sounding like derisive laughter), or symbolic sign quality (e.g., a church anthem priming the word devotion). Unfortunately, it was not investigated in that study whether N400 responses differed between these dimensions of musical meaning, and, thus far, no subsequent study has investigated this issue. However, the results still allow us to conclude that processing of extra-musical meaning is associated with N400 effects.

Results of that experiment were obtained when participants judged the relatedness of prime-target pairs. In an additional experiment, participants performed a memory task on the stimulus items (thus being oblivious of the true purpose of the experiment). Even with this task, the N400 was clearly modulated by the semantic fit between musical excerpts and target-words. This showed that the musical information activated representations of meaningful concepts, and not simply covert verbalization of words (which could also have led to the observed N400 effects). That is, these results indicate that musical information can prime representations of meaningful concepts even when participants do not judge the semantic relatedness between prime and target word, and that the minds of individuals relate musical information to extra-musical concepts even without conscious intention (in a way that appears to be at least partially automatic).

The N400 effects elicited in the language condition (in which the target words followed sentences) did not differ from those observed in the music condition (in which the target words followed musical excerpts), with regard to amplitude, latency, and scalp distribution (and N400 effects did not differ between concrete and abstract words). In addition, a source analysis localized the main sources of the N400 effects, in both conditions, in the posterior part of the medial temporal gyrus bilaterally (Brodmann's areas 21/37), in proximity of the superior temporal sulcus. These regions have been implicated in the processing of semantic information during language processing (Lau *et al.*, 2008; see also Excursion on posterior temporal cortex and processing of meaning below).

Excursion: Posterior temporal cortex and processing of meaning

The so-called *Wernicke's area* is not a precisely defined anatomical region. It usually refers to left-hemispheric posterior superior temporal cortex (BA 22p), but often also to adjacent middle temporal cortex (BA 21p, and parts of BA 37). Activity in this region is a necessary condition for semantic processing, as evidenced by lesion studies with Wernicke's aphasics: These patients show semantic deficits in that they fail primarily on semantic processes in production as well as in acceptability judgements (Friederici, 1982). Functional neuroimaging studies indicate that the posterior superior temporal sulcus (STS) and the mid-posterior middle temporal gyrus host conceptual/semantic representations (representations of non-linguistic visual object features seem be stored in ventral parts of the inferior temporal cortex; Lau *et al.*, 2008). With regard to language production, the (left) MTG is also activated for tasks that require lexical selection (Indefrey & Levelt, 2004). The (left) posterior superior temporal cortex is also involved in the retrieval of lexical-syntactic information (Indefrey, 2004). With regard to the superior temporal gyrus (STG), several authors (e.g., Wernicke, 1874; Lau *et al.*, 2008) suggested that the posterior STG (as well as posterior MTG; Hickok & Poeppel, 2007, p. 398) is involved in early (auditory) stages of the 'sound-to-meaning transformation' (Lau *et al.*, 2008, note that neuroimaging studies typically show posterior STG activation for speech and other spectrotemporally complex stimuli such as music).

It should also be noted that semantic processing does not only involve posterior temporal cortex. As described later in this chapter, semantic processing includes (a) the activation of meaningful representations, (b) the selection of meaningful representations, and (c) the integration of the information of these representations with the previous semantic context. It appears that conceptual (including lexical) information is stored in temporal regions, and that strategic processes, including integrating and relating contextual meaningful representations, involve the inferior frontal cortex (all with left-hemispheric weighting in the case of language; Lau *et al.*, 2008; Friederici, 2009). Moreover, frontal areas have been implicated in semantic processing with regard to binding of phonological, syntactic and semantic aspects, and linearization of argument hierarchy during thematic role assignment based on syntactic and semantic features (for a review see Friederici, 2009).

Note that in humans, (non-verbal) vocalizations (Belin *et al.*, 2004), as well as (non-verbal) voice processing (e.g., von Kriegstein *et al.*, 2003) activate STS-regions. Petkov *et al.* (2008) reported that, in macaque monkeys, species-specific vocalizations activate posterior auditory parabelt regions (perhaps including the posterior STS), as well as anterior supra-temporal auditory regions. Animal vocalizations (which usually have a quite specific meaning) represent entries of a species-specific 'lexicon,' and it appears that auditory parabelt regions represent phylogenetically old areas for the storage of acoustic-semantic information. Due to the increase in the number of concepts that humans can store and communicate, the areas serving the storage of such information perhaps extended posteriorly, and along the superior and inferior convexity into the posterior superior and middle temporal regions (of which particularly the superior region is often referred to as 'Wernicke's area'). It is interesting to note that the left posterior STS, a region involved in abstracting categories from meaningful auditory input, is also involved, in violinists, in the processing of violin music (Dick *et al.*, 2011). This shows that meaning-related processes originating from this region are not specific for speech, or language, but also serve the processing of musical information.

Due to the length of musical excerpts (~10 s), musical information could not be used as target stimulus in the study by Koelsch *et al.* (2004a), thus only words were used as target stimuli. Hence, a question emerging from that study was whether musical information can also elicit N400 responses when presented as target stimulus. One study addressing this issue (Daltrozzo & Schön, 2009b) used short musical excerpts (duration was ~1 s) that could be used either as primes (in combination with word targets), or as targets (in combination with word primes). Figure 10.2 shows that when the musical excerpts were used as primes, meaningfully unrelated target words elicited an N400 (compared to related target words, as in the study by Koelsch *et al.*, 2004a). Importantly, when musical excerpts were used as target stimuli (and words as primes), an N400 was observed in response to excerpts that the participants rated as meaningfully unrelated to the preceding target word (compared to excerpts that were rated as related to the preceding word). This was the first evidence that musical information can also elicit N400 responses (when used as a target stimulus). Note that the musical excerpts were composed for the experiment, and thus not known by the participants. Therefore, musical meaning was not due to symbolic meaning, but due to indexical (e.g., 'happy') and iconic (e.g., 'light') meaning. In the data analysis used in that study (Daltrozzo & Schön, 2009b), the relatedness of prime-target pairs was based on the relatedness judgements of participants. In another article (Daltrozzo & Schön,

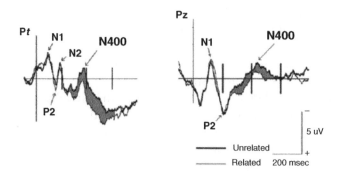

Figure 10.2 Data from the experiments by Daltrozzo & Schön (2009b). The left panel shows ERPs elicited by target words (primed by short musical excerpts), the right panel shows ERPs elicited by musical excerpts (primed by target words). The thick line represents ERPs elicited by unrelated stimuli, the thin line represents ERPs elicited by related stimuli. Note that the difference in N1 and P2 components is due to the fact that words were presented visually, and musical excerpts auditorily. Both meaningfully unrelated words and meaningfully unrelated musical excerpts elicited N400 potentials (see grey area). Reprinted with permission from Daltrozzo & Schön (2009b).

2009a) it was shown that even if the data are analysed based on the un/relatedness of prime-target pairs as pre-defined by the experimenters, a significant N400 was elicited (in the latter study a lexical decision task was used, whereas in the former study participants were asked to judge the conceptual relatedness of prime and target stimuli).

Further studies investigated the processing of musical meaning using only single chords or single tones. One study (Steinbeis & Koelsch, 2008a) used an affective priming paradigm with single chords (presented auditorily) and words (presented visually). Chords were either consonant or dissonant, and perceived by the participants (all musicians) as either more pleasant or more unpleasant. Words were either pleasant or unpleasant due to their positive or negative affective content (for example 'love' and 'hate'). Unrelated targets elicited an N400 effect, both when chords were primes and when words were targets, as well as when words were primes and chords were targets. The N400 elicited by words had a posterior scalp distribution, whereas the N400 elicited by chords had an anterior scalp distribution. The difference in scalp distribution perhaps reflects different semantic processes, such as more strategic processes (including integrating and relating the meaning between target and prime) during the music-evoked targets; such strategic processes have been shown to involve (inferior) frontal areas (Lau *et al.*, 2008; Friederici, 2009). However, the different scalp distributions of the observed N400 effects could also be due to the fact that words were presented visually, and chords auditorily.

In any case, the study by Steinbeis & Koelsch (2008a) revealed that a single musical stimulus (a chord that is more or less pleasant) can influence the semantic processing of a word (in this study presumably due to a chord's indexical sign qualities). Note that participants were musicians; using the same experimental paradigm with non-musicians also showed an N400 when words were targets (Steinbeis, 2008), but not when chords were targets.

We (Steinbeis & Koelsch, 2008a) also obtained fMRI data using the same experimental paradigm, and found that the semantic processing of words was related primarily to activity in temporal lobe structures, namely the posterior portion of the middle temporal gyrus (MTG) extending into the superior temporal sulcus (STS), again corresponding to Brodmann's areas 21/37 (there were also sub-threshold activations of inferior frontal cortex in both conditions that were presumably due to controlled selection and retrieval of conceptual representations; see also above and Lau *et al.*, 2008). As mentioned before, these temporal regions play a role for the storage and activation of lexical representations. Nearby (presumably within the same cytoarchitectonic areas), activation maxima were observed for the semantic processing of the chords. The topographical difference of activations between the processing of chords and words in this study was perhaps due to the fact that words (such as 'love') have different lexico-conceptual entries than, for example, a consonant major chord. It is likely that spreading activation is a mechanism underlying these priming effects: Single chords may activate affective representations, which spread onto affectively related lexical representations.[11]

Further experiments using the affective priming paradigm developed in the study by Steinbeis & Koelsch (2008a) used chords with pleasant and unpleasant timbres, as well as chords in major and minor mode as prime stimuli (i.e., target stimuli were always words; Steinbeis & Koelsch, 2011). Moreover, that study investigated both musicians and non-musicians. N400 effects for target words that did affectively not match with the prime chords were observed in the timbre, as well as in the major/minor mode experiment, with no statistical difference between musicians and non-musicians. These results showed that for both musicians and non-musicians a single chord is sufficient to communicate affective meaning (the meaning of the chords was presumably due to a mixture of indexical and symbolic sign qualities, but this was not specified in that study).

A recent ERP study by Goerlich *et al.* (2011) corroborated the results obtained by Daltrozzo & Schön (2009b), Steinbeis & Koelsch (2008a) and Steinbeis & Koelsch (2011). In that study (Goerlich *et al.*, 2011), short (660 ms) musical or prosodic stimuli with emotional valence (sad, neutral, happy) were used either as prime stimuli for visual word targets, or as targets following visual word primes. Results showed N400 effects for music as well as for prosodic target stimuli that did not match with the affective valence of word primes (as compared to affectively matching stimuli). Moreover, N400 effects were also found for word targets that did not match with the affective valence of music or prosodic prime stimuli (the

---

[11] This hypothesis of purported underlying mechanisms could be tested by varying the SOA and observing the persistence or absence of the effects.

N400 effect elicited by words following music primes was clearly visible in the ERP waveforms, but statistically not significant).[12]

In another study (Grieser-Painter & Koelsch, 2011), it was investigated whether similar semantic priming effects can also be observed when single tones are used as primes and/or targets. Tones had different timbres, and target words were adjectives (for example, 'tense,' 'open,' 'fast,' 'strong,' 'colourful'). Unrelated target words elicited an N400 effect, both when primed by tones, and when primed by words. Likewise, unrelated target tones elicited an N400 effect, both when primed by words, and even when primed by tones (none of the participants had received any professional musical training). That study (Grieser-Painter & Koelsch, 2011) also compared N400 effects between stimulus pairs that were pre-defined by the experimenters as related or unrelated, and stimulus pairs that were judged by the participants as related or unrelated. Although there was no statistically significant difference between these two conditions, the N400 was visible much more clearly in the ERPs of sound targets when the ERPs were analysed according to the ratings of each participant. Results of that study are supported by a recent study by Schön *et al.* (2010), in which similar N400 effects were reported for sounds following word primes, and words following sound primes (meaning of sounds was due to iconic and indexical sign quality). Neither the study by Grieser-Painter & Koelsch (2011), nor the study by Schön *et al.* (2010) differentiated between different sign qualities of sounds. Therefore, future studies could investigate N400 effects elicited by different extra-musical sign qualities.

In summary, the above-mentioned studies show that musical information (musical excerpts, single chords, and single tones), can systematically prime representations of meaningful concepts (as indicated by modulatory effects on the N400 elicited by words). Moreover, the studies show that musical excerpts, single chords, and single tones can elicit N400 effects that are modulated by the semantic fit with a preceding word. The N400 effects are due to extra-musical meaning, that is, meaning emerging from musical information referring to the extra-musical world of concepts.

10.4   Intra-musical meaning

The previous section dealt with extra-musical meaning and the N400. However, musical meaning can also emerge from one musical element (or group of elements) pointing structurally to another musical element (or group of elements). That is, musical semantics extends beyond extra-musical sign qualities in that musical meaning can also emerge from the interpretation of intra-musical structural

---

[12] A nice additional feature of that study was that alexithymia scores of participants were associated with the N400 amplitudes elicited by affectively incongruent music and prosody targets. The N400 of individuals with higher scores of alexithymia was smaller than the N400 of individuals with lower alexithymia scores. Alexithymia was assessed using the Toronto Alexithymia Scale (TAS-20).

relations, that is, from the structural reference of one musical element, or unit, to at least one other musical element, or unit (without reference to the extra-musical world).[13] Meyer (1956) used the term *embodied meaning*,[14] the term *intra-musical* was, to my knowledge, introduced later by Malcolm Budd (1996). Other theorists have used the terms *intrinsic referring* (Nattiez, 1990),[15] *formal meaning* (Alperson, 1994; Koopman & Davies, 2001), or *formal significance* (Davies, 1994).

10.4.1   Intra-musical meaning and the N5

In the following, data will be reported suggesting that the processing of intra-musical meaning is reflected electrically in the N5 (or N500). It will first be described that the N5 is sensitive to harmonic context build-up, and to harmonic incongruity. Then, evidence is presented that the N5 specifically interacts with the processing of the semantic incongruity of a word (which indicates that processes underlying the generation of the N5 are related to semantic processes).

Harmonic context buildup   As was described in Chapter 9, the N5 was first observed in experiments in which chord sequences consisting of five chords ended either on a music-syntactically regular or a music-syntactically irregular chord function (see Figure 9.3 on p. 90). Figure 10.3 shows ERPs elicited by regular (!) chords at positions one to five. Each of these chords elicits an N5 (see arrow in Figure 10.3), with the amplitude of the N5 declining towards the end of the chord sequence. This amplitude decline is taken to reflect the decreasing amount of harmonic integration required with progressing chord functions during the course of the cadence. The small N5 elicited by the (expected) final tonic chord presumably reflects that only a small amount of harmonic integration is required at this position of a chord sequence.[16] This phenomenology of the N5 is similar

---

[13]   For example, in major–minor tonal music, intra-musical meaning can emerge from chord functions pointing to the harmonic context, and listeners interpreting the function of a chord. Both the final tonic chord and the final double dominant of the sequences shown in Figure 9.5A (p. 95) are major chords (with identical superposition of intervals). The functional difference between these two chords, however, is that one points as a tonic to the previous harmonic context, and the other one as a double dominant.

[14]   Note that the term 'embodied' has nowadays a different meaning, often being associated with the role that the body plays for cognition and emotion.

[15]   I do not use the term 'intrinsic referring' here, because 'intrinsic referring' refers to the referencing of one musical element to another; by contrast, I use the term 'intra-musical meaning' to emphasize that (non-conceptual) meaning emerges from such referencing, and that structural references have a meaning quality.

[16]   Note also the clear differences in P2 potentials; these differences do not appear to be as systematic as those of the N5 (e.g., chords at the third and fourth positions elicit the smallest P2 amplitude, followed by the P2 potentials elicited by the second and the fifth chord). Nevertheless, perhaps these potentials are also related to syntactic structure building, e.g., with regard to the sensory consequences of predictions, and with regard to comparison processes; see also Chapter 9.

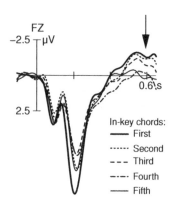

Figure 10.3 Grand-average ERPs elicited by regular chords, separately for each of the regular chords (first to fifth position) of a five-chord cadence (such as the one shown in the left panel of Figure 9.3). The amplitude of the N5 (indicated by the arrow) was dependent on the position of a chord in the cadence: The amplitude of the N5 decreases with increasing harmonic context build-up.

to that of the N400 elicited by open-class words: With progressing position of words in a sentence, the N400 amplitude also declines towards the end of a sentence (Van Petten & Kutas, 1990; cf. Figure 5.4 on p. 52). That is, during sentence processing, a semantically correct final open-class word usually elicits a rather small N400, whereas the open-class words preceding this word elicit larger N400 potentials. This is due to the semantic expectedness of words, which is rather unspecific at the beginning of a sentence, and which becomes more and more specific towards the end of the sentence (where readers already have a hunch of what the last word will be). Thus, a smaller amount of semantic integration is required at the end of a sentence, reflected in a smaller N400. If the last word is semantically unexpected, then a large amount of semantic integration is required, which is reflected in a larger amplitude of the N400.

Harmonic incongruity   Chapter 9 described ERPs elicited by regular and irregular chords at the final position of chord sequences such as those shown in Figure 9.3. Compared to regular chord functions, irregular chord functions typically elicit an early right anterior negativity (ERAN, usually maximal around 150–200 ms, and taken to reflect neural mechanisms related to syntactic processing). In addition, irregular chords elicit an N5 with a larger amplitude than the N5 elicited by regular chord functions (for studies reporting such N5 effects for melodies see Miranda & Ullman, 2007; Koelsch & Jentschke, 2010; see also Chapter 9). This increase of the N5 amplitude is taken to reflect the increased amount of harmonic integration, reminiscent of the N400 reflecting semantic integration of words. That is, at the same position within a chord sequence the N5 is modulated by the degree of fit with regard to the previous harmonic context, analogous to the N400 (elicited at the same position within a sentence), which is modulated by the degree of fit with regard to the previous semantic context (see also Figure 10.1). This suggests a relation between the N5 and the processing of musical meaning, or semantics (this relation will be elaborated in the next section).

Note that the N5 can be elicited even when individuals do not pay attention to the music (e.g., while listeners read a book, or a play a video game; see Chapter 9). That is, the neural mechanisms underlying the generation of the N5 operate partly automatically. Thus, similar to the processing of extra-musical meaning reflected in the N400 priming effects, intra-musical meaning is processed, at least in part, even when individuals do not pay attention to musical information (an exception is the synchronous processing of words and chords under the instruction to ignore the music and to pay attention to the syntax and semantics of the words; see Chapter 9).

N5 and N400   The following will provide data indicating that the N5 is related to the processing of meaning information. As explained in Section 9.6, two ERP studies (Koelsch *et al.*, 2005b; Steinbeis & Koelsch, 2008b) investigated interactions between music- and language processing by presenting chord sequences and sentences together synchronously to participants (see Figure 9.14 on p. 113).[17] Both studies showed that the left anterior negativity (LAN) elicited by morphosyntactic violations in language was influenced by the music-syntactically irregular chord functions, whereas the N400 was not affected by such irregularities. In addition, the study by Steinbeis & Koelsch (2008b) also found that the ERAN was smaller when elicited on syntactically wrong words (compared to the ERAN elicited on syntactically correct words; cf. Figure 9.16 A on p. 117).

Moreover, and more importantly with regard to the processing of musical meaning, results of that study (Steinbeis & Koelsch, 2008b) also showed an interaction between the N5 and the semantic cloze-probability of words (in the absence of an interaction between the N5 and the syntactic regularity of words; Figure 9.16 B,C).[18] The N5 was smaller when elicited on words with a semantic low-cloze probability (e.g., 'He sees the cool beer') compared to when elicited on words with a semantic high-cloze probability (e.g., 'He drinks the cool beer'). Importantly, the N5 did not interact with the LAN (i.e., the N5 did not interact

---

[17]   The chord sequences ended either on a regular tonic, or on a music-syntactically irregular Neapolitan sixth chord, and the last word of each sentence was (a) either syntactically correct, with a semantic high-cloze probability (a semantic high-cloze probability means that the last word of the sentence was semantically strongly expected, for example the word 'beer' in the sentence 'He drinks the cool beer'), or (b) syntactically correct, but with a semantic low-cloze probability (which means that the last word of the sentence was semantically less expected, for example the word 'beer' in the sentence 'He sees the cool beer'; note that both semantic high-cloze and low-cloze probability words were semantically correct words), or (c) it was syntactically incorrect, with semantic high-cloze probability (e.g., 'Er trinkt den kühlen Bier' / 'He drinks the$_{masc}$ cool$_{masc}$ beer$_{neuter}$'). As was described in Section 9.6 results showed three main effects (see also Figure 9.15 on p. 115): (1) music-syntactic irregularities elicited the ERAN; (2) language-syntactic irregularities elicited the left anterior negativity (LAN); (3) semantically less expected words elicited an N400 (compared to the semantically highly expected words).

[18]   No N5 was elicited in the previous study by Koelsch *et al.* (2005b), because participants focussed on the words without any task related to the chords (for further explanation see Chapter 9).

with the syntactic processing of words), indicating that the N5 is not simply modulated by any type of deviance, or incongruity, but that the N5 is specifically modulated by neural mechanisms underlying semantic information processing. That is, the N5 potential can be modulated by semantic processes, namely by the activation of lexical representations of words with different semantic fit to a previous context. This modulation indicates that the N5 is related to the processing of meaning. Note that the harmonic relation between the chord functions of a harmonic sequence is an intra-musical reference (i.e., a reference of one musical element to another musical element), and not a reference to anything belonging to the extra-musical world.

It appears that intra-musical meaning is modulated by musical expression (e.g., emphasizing an irregular musical event by playing it with an accent), and that such modulation is also reflected in N5 potentials: In a study in which excerpts of classical piano sonatas were played either with musical expression (played by a pianist) or without expression (played by a computer without dynamics and agogics), chords played in the condition with expression elicited a larger N5 than the same chords played in the condition without expression (Koelsch *et al.*, 2008b).

Neural generators of the N5   The locations of the neural generators of the N5 have remained elusive. This is in part due to the difficulty that the N5 is usually preceded by the ERAN (for an exception see Poulin-Charronnat *et al.*, 2006), making it difficult to differentiate the neural correlates of N5 and ERAN in experiments using functional magnetic resonance imaging. The N5 usually has a frontal scalp distribution, thus the scalp distribution of the N5 is more anterior than that of the N400, suggesting at least partly different neural generators. Perhaps the N5 originates from combined sources in the temporal lobe (possibly overlapping with those of the N400 in BAs 21/37) and the frontal lobe (perhaps in the posterior part of the inferior frontal gyrus). This needs to be specified, for example using EEG source localization in a study that compares an auditory N400 with an auditory N5 within subjects. Such research could also aim to clarify whether the N5 consists of several sub-components, and whether one of these sub-components is a (late) N400.

N5, CNV, and RON   It is unlikely that the N5 is related to processes reflected in the O-wave of the contingent negative variation (CNV), or the reorienting negativity (RON, Schröger & Wolff, 1998). The O-wave of the CNV is a long-latency, long-duration ERP component that usually (a) occurs in experimental settings in which motor responses are required, (b) has a frontally negative and parietally positive scalp distribution (with mastoid reference), and (c) is elicited under attend conditions (Rugg & Coles, 1995). By contrast, the N5 can be elicited under pre-attentive listening conditions (i.e., when no responses are required; see Chapter 9), and does not usually show a parietally positive scalp distribution with mastoid reference (e.g., Koelsch *et al.*, 2000). The RON (which also occurs at around 500 msec) is elicited when participants turn back attention to the primary task after being oriented away, and is thought to reflect the reorienting back to task-relevant information. In contrast to the N5, the RON is confined to conditions in which a deviant is task-irrelevant (the N5 is also present under attend conditions in which harmonically irregular chords are task-relevant, Koelsch *et al.*, 2000; Koelsch & Friederici, 2003). Moreover, the N5, but not the RON, shows a clear polarity inversion at mastoidal sites with nose reference (e.g., Koelsch *et al.*, 2007).

Further intra-musical phenomena of tonal music   The previous sections described that, in tonal music, intra-musical meaning can emerge from harmonic context build-up, or – on a more abstract level – from the *build-up* of structure. Such a structure has a certain *stability* (for example, a harmonic structure consisting of chords of the harmonic core is more stable compared to a structure consisting mainly of chords not belonging to the harmonic core). With regard to syntactic processes (second row of Table 10.2), structural stability is related to the *hierarchy of stability* of tones and chords established based on *knowledge-free structuring*. Moreover, a musical structure has a certain *extent* (for example, a harmonic structure can be confined to a single key, or span several keys). Extent and stability of structures are further phenomena that give rise to intra-musical meaning. Moreover, as described in the previous sections, intra-musical meaning can emerge from harmonic incongruity, or – again, on a more abstract level – from a *breach* of an (expected) structure. With regard to the model of music perception (summarized in the first row of Table 10.2), a structural breach is related to the disruption of *structure building*. In tonal music, a structural breach is usually followed by a *post-breach structure* (e.g., chord functions following a deceptive cadence). The post-breach structure leads to the *resolution* of the structural breach (for example, the tonic resolving a structural breach at the end of a harmonic sequence). Post-breach structure and resolution are also phenomena that give rise to intra-musical meaning: The chord functions of a post-breach structure have a different intra-musical meaning than the same chord functions in a context without breach, and the tonic resolving a structural breach has a different intra-musical meaning (and is harmonically integrated in a different way) than a tonic at the end of a harmonic sequence without a structural breach. The differentiation between post-breach structure and resolution is important because these structural phenomena give rise to different emotional responses, namely tension due to the anticipation of resolution during the post-breach structure, and relaxation during the resolution. Finally, in tonal music intra-musical meaning can also emerge from *large-scale structural relations* (for example, the chord functions of a second theme of a sonata are integrated differently during the exposition, where it is in a different key than the first theme, compared to the recapitulation, where it is in the same key as the first theme). The principles of structural phenomena that give rise to meaning, in relation to processes of music perception (Chapter 8) and the syntactic processes described in Chapter 9, are summarized in Table 10.2.

Table 10.2   Principles of structural phenomena that give rise to intra-musical meaning (bottom row), in relation to processes of music perception and syntactic processing. Note that, with regard to *syntactic processes*, the *stability of structure* is related to a hierarchy of harmonic stability which can be established based on *knowledge-free structuring*. Moreover, a structural *breach* is related to the disruption of *structure building*, and meaning emerging from *large-scale relations* is related to *large-scale structuring*.

music perception	feature extraction	Gestalt formation	interval analysis	structure building	structural re-analysis	vitalization	premotor, immune system
syntactic processing	element extraction	knowledge-free structuring	musical expectancy formation	structure building	structural re-analysis	syntactic integration	large-scale structuring
intra-musical	build-up	stability	extent	breach	post-breach	resolution	large-scale

These principles are not confined to music. On the contrary, they reflect rather general principles, parts of which have been described, for example, with regard to lyric poetry, rhetoric, aesthetics, visual arts, and linguistics (e.g., Jakobson, 1960).[19] Use of the N5 as dependent variable enables music psychologists to further test and investigate these principles.

With regard to tonal music, it is important to understand that the meaning emerging from these principles is not the iconic meaning (or a metaphorical meaning) of 'build-up,' 'extent,' 'stability' etc., but the meaning emerging from *harmonic integration* due to the establishment of a structural model, its modifications, etc. This does, however, not exclude that representations of such extra-musical concepts are activated during the processing of musical structure. For example, in music analysis, music theorists often use such *extra*-musical concepts metaphorically to describe *intra*-musical structural principles (see also Cook, 1992), suggesting that (at least in those individuals) such intra-musical phenomena might also give rise to extra-musical meaning.

Furthermore, it is important to note that harmonic integration is implicitly related to musical expectancy, and that the violation, or fulfilment of expectancies has emotional effects (such as tension, suspense, and relaxation; for extensive accounts on this issue see Huron, 2006; Lerdahl, 2001b). For example, Leonard Meyer (1956) stated that 'as soon as the unexpected, or for that matter the surprising, is experienced, the listener attempts to fit it into the general system of beliefs relevant to the style of the work. . . . Three things may happen: (1) The mind may suspend judgement, so to speak, trusting that what follows will clarify the meaning of the unexpected consequent. (2) If no clarification takes place, the

---

[19]   Roman Jakobson defined the *poetic function* as the projection of 'the principle of equivalence from the axis of selection into the axis of combination. Equivalence is promoted to the constitutive device of the sequence' (p. 358). Notably, according to Jakobson, the poetic function gives rise to meaning emerging from structure.

mind may reject the whole stimulus and irritation will set in. (3) The expected consequent may be seen as a purposeful blunder. Whether the listener responds in the first or third manner will depend partly on the character of the piece, its mood or designative content. The third response might well be made to continuing music whose character was comic or satirical' (pp. 29–30). 'From this point of view what a musical stimulus or series of stimuli indicate and point to are not extra-musical concepts and objects but other musical events which are about to happen. This is, one musical event (be it a tone, a phrase, or a whole section) has meaning because it points to and makes us expect another musical event' (p. 35).

The structural relations of musical events can also lead to emotional responses (such as surprise, increase in tension, relaxation, etc.; see also Lerdahl, 2001b; Bigand *et al.*, 1996; Lerdahl & Krumhansl, 2007), which can, in turn, have meaning for the individual. I will differentiate between these two dimensions (intra-musical meaning, and emotional responses to music), and thus deal with the dimension of emotional meaning in the following section on musicogenic meaning.[20]

10.5   Musicogenic meaning

The previous sections dealt with meaning emerging from the interpretation of musical information; this section deals with meaning emerging from the interpretation of physical, emotional, and personality-related effects elicited by music. That is, listeners do not only interpret musical information expressed by another individual, but also the effects evoked by the music in themselves. There is scarcity of empirical data on these dimensions of meaning, therefore the following sections on musicogenic meaning mainly provide theoretical considerations that need empirical testing.

10.5.1   Physical

Individuals tend to move to music (singing, playing an instrument, dancing, clapping, conducting, head-nodding, tapping, swaying, etc.), that is, individuals tend

---

[20]   With regard to the relation of expectancy, emotion, and meaning, Meyer's basic hypothesis states that affect is aroused when an expectation, a tendency to respond activated by the musical stimulus situation, is temporarily inhibited or permanently blocked. Meyer suggests that in musical experience the same stimulus (the music) activates tendencies, inhibits them, and provides meaningful and relevant resolutions for them. Meyer notes that formalist and expressionist views may see the meaning of music as being essentially intra-musical ('non-referential'). His studies are concerned 'with an examination and analysis of those aspects of meaning which result from the understanding of and response to relationships inherent in the musical progress rather than with any relationships between the musical organization and the extra-musical world of concepts, actions, characters, and situations. The position adopted admits both formalist and absolute expressionist viewpoints' (Meyer, 1956, p. 3).

to show physical activity in response to, and in synchrony with, music. Merely the fact that an individual shows such activity has meaning for the individual; in addition, the way in which the individual moves expresses meaning information. These movements are 'composed' by the individual (and should therefore be differentiated from the motor-effects as one aspect of the emotional effects of music, such as smiling during emotional contagion when listening to joyful music; see next section). That is, it is important to understand that there is no objectivity of music (in the sense that music would direct movements of individuals like a puppeteer). Music can elicit an impetus to move, and this impetus can be intended by the composer or player, but it is the individual that decides to move (i.e., to follow something that was intended by another individual), and it is the individual that 'composes' the movements while moving to music (including dance movements, clapping, vocalizations, and singing or playing along).

In a social situation, that is, when more than one individual moves to (or plays) music, meaning also emerges from joint, coordinated activity. For example, an action-related effect that becomes apparent in music based on an isochronous pulse (that is, a pulse to which we can easily clap, sing, and dance) is that individuals synchronize their movements to the external musical pulse. In effect, this leads in a group of individuals to coordinated physical activity. Notably, humans are one of the few species that are capable of synchronizing their movements to an external beat (non-human primates apparently do not have this capability, although some other species appear to have this capability as well; for a detailed account see Patel *et al.*, 2009). In addition, humans are unique in that they can understand other individuals as intentional agents (see also Excursion on decoding of intentions during music listening on p. 131), share their intentionality, and act jointly to achieve a shared goal. In this regard, communicating and understanding intentions, as well as inter-individual coordination of movements is a pre-requisite for cooperation (see also Excursion on *The Seven Cs* on p. 169). Cross (2008b) stated that, in a social context, musical meaning can emerge from such joint performative actions. Cross referred to this dimension as *socio-intentional* (Cross, 2008b, p. 6). According to Cross, this dimension of musical meaning would therefore 'be oriented towards attributions and interpretations of intentions and intentionality in engagement with music,' and socio-intentional meaning 'would be rooted in performative actions and sound structures that afford cues about shared intentionality that direct attention in interaction rather than the sharing of attention *per se*: These may be, e.g., declarative and disclosural (making manifest), concerned with the direction of another's attention to an object or event distinct from the individuals involved in the interaction; they may even be dissimulative, concerned with the *mis*direction of another's attention' (Cross, 2008b, p. 6). Note that mental state attribution (such as decoding intentions of another individual) is not specific for musicogenic meaning, but is engaged in response to any (man-made) musical signal. Therefore, mental state attribution is a meta-phenomenon of musical meaning, and not conceptualized here as a separate dimension of musical meaning. However, the thoughts by Cross on socio-intentional musical meaning are relevant here because they describe how social interaction modulates physical musicogenic meaning. Also note that, as soon as (isochronous) music becomes a *symbol* for calling upon to move (in synchrony), then the meaning quality is symbolic (and not emerging from one's own physical activity in response to, and in synchrony with, the music).

Especially in non-Western cultures, musical meaning is grounded in social interaction (see also Cross, 2011; Seifert, 2011). Giving consideration to the concepts of symbolic and musicogenic meaning aims at incorporating cultural phenomena in which 'music is embedded in the fabric of everyday lives' (Cross, 2011). For example, we (Fritz *et al.*, 2009) studied music perception in the Mafa people in northern Cameroon, who have different pieces of instrumental music, each being associated with a certain ritual. Therefore, each song has a clear symbolic meaning. With regard to musicogenic meaning, I have emphasized that, especially during music making in a group, communicating and understanding intentions, as well as inter-individual coordination of movements and actions is a pre-requisite for cooperation. Cross, 2011, refers to these functions as 'relational' in terms of 'involving the formation, maintenance or restructuring of connections and affiliations between participants.' These social functions are part of what makes us human, and engaging in these social functions has meaning for the individual (for a more detailed description of these social functions see *The Seven Cs* on p. 169). With regard to a comparison between meaning in music and language, Cross (2011) noted that the participatory, and therefore social, nature of music (allowing us to experience coordination, cooperation, group cohesion, spirituality, and the feeling to belong) represents a realm which 'might best be thought of not as an autonomous realm but as a mode of human communication that is homologous with aspects of linguistic interaction'.

10.5.2   Emotional

Musicogenic meaning can also emerge from emotions evoked by music. This view considers that feeling one's own emotions is different from the recognition of emotion expressed by the music (Gabrielson & Juslin, 2003), the latter usually being due to indexical sign quality of music. The different principles by which music may evoke emotions are discussed in Chapter 12; here, the meaning emerging from (music-evoked) emotions is discussed.

A priori musical meaning   The evocation of emotions with music has important implications for the specificity of meaning conveyed by music as opposed to language. The use of language for the communication of emotions faces several problems. In the paragraphs about rule following and the argument against the idea of a 'private language,' Ludwig Wittgenstein (1984) demonstrates that 'inner' states (like feelings) cannot be directly observed and verbally denoted by the subject who has these states. His argument shows that the language about feelings functions in a different mode than the grammar of words and things. Wittgenstein argues that it is not possible (1) to identify correctly an inner state, and (2) to guarantee the correct language use which is not controlled by other speakers. This means (3) that it is impossible for the speaker to know whether his or her use corresponds to the rules of the linguistic community, and (4) whether his or her use is the same in different situations. According to Wittgenstein, correct use of the feeling vocabulary is only possible in specific language games. Instead of assuming a direct interaction of subjective feelings and language, Gunter Gebauer (in press) proposes that feeling sensations (Wittgenstein's *Empfindungen*) are *reconfigured* by linguistic expressions (although reconfiguration is not obligatory for subjective feeling). This means that there is no (direct) link or translation between feelings and words, posing fundamental problems for any assumption of a specificity of verbal communication about emotions.

However, affective prosody, and perhaps even more so music, can evoke feeling sensations (*Empfindungen*) which, *before* they are reconfigured into words, bear greater inter-individual correspondence than the words that individuals use to describe these sensations. In other words, although music seems semantically less specific than language (e.g., Slevc & Patel, 2011; Fitch & Gingras, 2011), music can be more specific when it conveys information about feeling sensations that are problematic to express with words because music can operate *prior* to the reconfiguration of feeling sensations into words. Note that, in spoken language, affective prosody also operates in part on this level, because it elicits sensational processes in a perceiver that bear resemblance to those that occur in the producer. I refer to this meaning quality as *a priori musical meaning*.

The reconfiguration of a feeling sensation into language involves the activation of representations of meaningful concepts (such as 'joy', 'fear', etc.; Zentner *et al.*, 2008, report a list of 40 emotion words typically used by Western listeners to describe their music-evoked feelings). Such activation presumably happens without conscious deliberation, and even without conscious (overt or covert) verbalization, similarly to the activations of concepts by extra-musical sign qualities, of which individuals are often not consciously aware.

10.5.3 Personal

Feeling sensations evoked by a particular piece of music, or music of a particular composer, can have a personal relevance, and thus meaning, for an individual in that they touch, or move, the individual more than feeling sensations evoked by other pieces of music, or music of another composer. This is in part due to inter-individual differences in personality (both on the side of the recipient and on the side of the producer). Due to the fact that an individual has a personality (be it a receiver or producer of music), and that personalities differ between individuals, there are also inter-individual differences among receivers in the particular preference for, or connection with, a particular producer of music. For example, one individual is touched more strongly by Beethoven than by Mozart, another one vice versa. That is, music-evoked emotions can also be related to one's inner self, sometimes leading to the experience that one recognizes oneself in the music in a particular, personal way. Personality-specific characteristics, notably, are not confined to emotional aspects. They are also reflected in different extra-musical subjects (such as subjects chosen for programme music – it is not a coincidence that Saint-Saëns' chose *Le carnaval des animaux* as subject, and Rachmaninoff *The Isle of the Dead*), and in personality-characteristic construction of musical structures (think of the statics of Bruckner's symphonic constructions as opposed to the dynamic constructions of Dvořák).

10.6   Musical semantics

Musical semantics is the theory of musical meaning, including the description of how musical meaning is conveyed, and how musical information is interpreted by an individual (musical semiotics, on the other hand, is the theory of musical signs). Semantic processing (that is, processing of meaning) includes (a) storage of meaningful information, (b) activation of representations of meaningful information, (c) selection of representations of meaningful information, and (d) integration of the semantic information of these representations with the previous semantic context. As for language, it remains to be specified where in the brain these processes are located, and how they are reflected electrically in ERPs. However, the reported N400 studies indicate that these processes can be activated by musical information with regard to extra-musical meaning information, and the reported N5 studies suggest that semantic processes also emerge from (intra-musical) harmonic integration.

10.6.1   Neural correlates

The studies investigating semantic processes with music indicate that posterior temporal (neo)cortical regions store *conceptual* features, and not (only) lexical representations (unless musical information elicits implicitly lexical access, which, however, does not seem likely because N400 priming effects are also observed when stimuli are not relevant for semantic tasks). It appears that the conceptual representations stored in posterior temporal cortex interface with a semantic network that is distributed across different brain regions, particularly frontal regions (Lau *et al.*, 2008; Friederici, 2009). This is consistent with the notion that the neural generators of the N5 are located in both temporal and frontal regions, and consistent with the affective priming studies showing (sub-threshold) activations of frontal areas during the semantic processing of musical information (Steinbeis & Koelsch, 2008a).[21]

---

[21]   Besson *et al.* (2011) raised the possibility that the N400 also reflects intra-musical meaning (as indicated by N400 effects elicited by musical targets primed by musical stimuli): This is a valid point,

10.6.2   Propositional semantics

As mentioned in the Introduction, it appears that no musical tradition makes precise use of propositional codes (involving quantifiers, modals, or connectives), unless the music imitates language (such as drum and whistle languages). This feature partly distinguishes language from music: The degrees of freedom differ between language and music with regard to the construction of propositions (as well as the ambiguity of meaning). Note that, in reality, there is a transitional zone between 'propositional' and 'non-propositional': For example, instrumental music can prime representations of concepts such as 'some' and 'all,' and modifiers, modals, or connectives are often used imprecisely in everyday language (think of the 'logical and', or the 'logical or'). However, anyone interested in how listening to music with propositional semantics might feel like, just has to listen to a song with lyrics containing propositional semantics.

The interesting phenomenon in this regard is, in my view, that humans appreciate music as a communicative medium, *even though* (or perhaps *because*) there are no operators such as quantifiers, modals, or connectives. Cross (2011) offers one interesting hypothesis in this regard: In music, due to relational goals that involve 'the formation, maintenance or restructuring of connections and affiliations between participants,' the truth of meanings is 'not required to be made mutually explicit'. In music (particularly in Western music), we are often dealing with a description (not with a verification), and with certainty (not with knowledge). Moreover, in the section on emotional musicogenic meaning, it was argued that music has the advantage of defining a feeling sensation without this definition being biased by the use of words (think of Mahler's assertion that 'if a composer could say what he had to say in words, he would not bother trying to say it in music'). With regard to propositional semantics and binary (true–false) truth conditions, this also means that there is no 'true' or 'false' with regard to the inner application of a rule for the usage of a musical concept.

10.6.3   Communication vs. expression

Meaning conveyed by music is often referred to as 'expressive,' as opposed to the 'communicative' nature of language. For example, Slevc & Patel (2011) mentioned that 'linguistic, but not musical, semantics exists for communicative reasons,' and that 'instrumental music might better be conceived of as a form of expression rather than of communication'. Apart from the fact that this depends on how music is used in a particular culture (see also Cross, 2011; Seifert, 2011), it should also be noted that the two terms 'communication' (in the sense of conveying specific, unambiguous information) and 'expression' (in the sense of conveying rather unspecific, ambiguous information) are normally used as if there

---

which requires future investigation. One could argue that both musical prime and target sounds evoke representations of (extra-musical) meaningful concepts, but this remains to be specified.

were two separate realms of conveying meaningful information (communication and expression) with a clear border between them. However, as outlined above, this does not seem to be the case. Rather, there is a continuum of the degree of specificity of meaning information, with 'expression' being located towards one end, and 'communication' towards the other (due to the lack of clear borders between music and language, I will propose the concept of a *music-language continuum* in Chapter 13).

10.6.4 Meaning emerging from large-scale relations

Musical meaning can also emerge from large scale relations (such as relations between phrases, parts, and movements of a work). This holds for all three classes of musical meaning (extra-musical, intra-musical, and musicogenic meaning). With regard to extra-musical meaning, one example is *programme music* featuring a number of story components that are related to each other (e.g., symphonic poems such as Strauss' *Till Eulenspiegels lustige Streiche*, Dukas's *L'Apprenti sorcier*, and many others). With regard to intra-musical meaning, one example is that the second theme in the sonata form has a different intra-musical meaning in the exposition than in the recapitulation due to the second theme being in the same key as the first theme during the exposition, but not during the recapitulation. That is, the structural relation between first and second theme, and thus the intra-musical meaning emerging from the second theme, differs between exposition and recapitulation.[22] Davies (1994) stated in this regard that 'to understand the musical work is to understand how it is put together' (p. 48); another example is that the last movement of a symphony, suite, cantata, etc., has a different meaning in relation to the previous movements compared to when heard in isolation, perhaps because 'musical ideas fit together – as complementary, or as variations, or as repetitions – so that there is a development or progress of ideas' (p. 368).

With regard to the integration of meaningful information into a larger semantic context, a study by Krumhansl (1996) explored parallels between music and linguistic *discourse*. According to Krumhansl (1996), 'music and discourse both consist of units that have well-defined beginnings and ends. Topics [or musical ideas] are introduced and developed within these units, with various devices used to move the argument forward. Acoustic cues, such as pauses, pitch contour, dynamic stress, and rhythmic patterning, serve to define these units and highlight certain elements within them' (p. 405). So far, however, such methods have rather investigated how listeners segment musical information. Krumhansl (1996) stated that beginnings of new segments are associated with the presence of 'new musical ideas' (p. 427), and that 'musical ideas [are] marked by a variety of surface characteristics, such as changes in rhythmic and pitch patterns, register,

---

[22] For a critical account of the perception of large-scale structures see Cook (1987).

and texture' (p. 427). The specification of the semantic processing of such musical ideas remains to be developed.

10.6.5   Further theoretical accounts

With regard to other theoretical accounts on musical meaning, different theorists often deal with different meaning dimensions when referring to musical meaning, and therefore it is not surprising that several theories are not compatible with each other, or that even within one theory different musical meaning dimensions are confused. For example, Hanslick's argument that absolute music is not about expression of emotion has created confusion, because he implicitly wrote about intra-musical meaning, whereas expression of emotion due to (extra-musical) indexical sign quality belongs to a different dimension of musical meaning (for details about Hanslick's argument see, e.g., Davies, 1994; Cumming, 1994). Another example is that Scruton (1983) aims at making statements about *intra-musical* meaning, but his thoughts about the perception of successive pitches with increasing and decreasing fundamental fre-quency as upward- and downward-movement refer to *extra-musical* sign quality.[23] This has important theoretical implications, for two reasons. Firstly, meaning emerging from extra-musical sign qualities does not require a 'metaphorical transference of ideas' (as claimed by Scruton, 1983, 1999), but mainly mere semantic priming (for a critical account on Scruton's use of the term 'metaphor' see Cumming, 1994). For example, hearing the words 'doctor,' 'nurse,' 'scalpel' activates representations of related concepts such as 'hospital' due to semantic priming (i.e., not due to a metaphorical transference of ideas); similarly, successive tones with increasing pitch prime the representation of the concept 'upward'. Secondly, the use of the term *metaphorical meaning* (or *metaphorical transference of ideas*) is appropriate with regard to intra-musical structural properties, such as 'breach' and 'conflict,' because intra-musical structural properties might prime representations of such concepts, although the music does not *sound*, e.g., like a breach (that is, there are no cracking noises and the like). However, 'breach' and 'conflict' are extra-musical concepts, and therefore their meaning is different from intra-musical meaning (and so are their musicogenic effects, such as perceived tension).

---

[23]   The association of high/low pitches with the concepts 'high'/'low' is presumably a mixture of (a) iconic sign quality, mainly due to the position of the larynx being higher (if the individual is standing or sitting) during the production of tones with higher pitches compared to lower pitches, and (b) possibly symbolic sign quality, due to the cultural shaping of the association of high/low pitches with the concepts 'high'/'low': Lawrence Zbikowski (1998) noted that Greek music theorists of antiquity used the terms 'sharpness' and 'heaviness,' that in Bali and Java the terms 'small' and 'large' are used, and that the Suyá of the Amazon basin use the terms 'young' and 'old' to refer to high and low pitches (Zbikowski, 1998, p. 5). This does not rule out, however, that those peoples would also agree that high/low pitches are 'high'/'low,' and that in Western listeners high and low pitch information also primes representations of concepts such as 'sharp/heavy,' 'small/large,' and 'young/old'.

The neurobiological theory of musical meaning presented in this chapter suggests that music can communicate meaning, notably not only meaning related to
emotion, or affect, but iconic, indexical, and symbolic meaning (with regard to
extra-musical meaning), as well as intra-musical meaning. The data presented in
this chapter indicate that extra-musical meaning is at least partly processed with
the same mechanisms as meaning in language (as reflected in the N400). Therefore, the notion that language and music are strictly separate, non-overlapping
domains with regard to the processing of meaning does not seem to make sense.
The fact that music can communicate intra-musical meaning (as reflected in the
N5) also implies that musical meaning is not only a matter of semiotics (i.e., not
only a matter of the sign qualities of music), but a matter of semantics that includes
integration of meaningful information with a semantic context. Notably, neural
correlates of such meaning processing have so far not been shown for language;
therefore, music – and not only language – is important for understanding how
the human brain processes meaning information.

11

Music and Action

11.1 Perception–action mediation

The perception of events can give rise to action-related processes. With regard to music, simply listening to music can automatically engage action-related processes. In his *common coding* approach to perception and action, Wolfgang Prinz (1990) described how actions are represented in terms of their perceptual consequences, and that the late stages of perception overlap with the early stages of action in the sense that they share a common representational format.[1] Such a common format can, for example, be a common neuronal code. 'Common coding' is supposed to be involved both when an individual perceives a movement, as well as when an individual perceives the effects of an action produced by another individual (and due to common coding, such perception evokes movement representations). Similarly, Liberman & Mattingly (1985) proposed in their *motor theory of speech perception* that, during speech perception, speech is decoded in part by the same processes that are involved in speech production. With regard to the observation of movements, Giacomo Rizzolatti and colleagues published in the 1990s reports about neurons located in the area F5 of the premotor cortex of macaque monkeys, which were not only active when the monkeys performed a movement, but also when the monkeys simply observed that movement

---

[1] The common coding approach follows the ideomotor approaches of Hermann Lotze (1852) and Willam James (1890). For a summary of the Lotze-James account and the ideomotor framework see, e.g., Prinz (2005), pp. 142–143.

Brain and Music, First Edition. Stefan Koelsch.
© 2013 John Wiley & Sons, Ltd. Published 2013 by John Wiley & Sons, Ltd.

(the so-called *mirror neurons*, reviewed in Rizzolatti & Sinigaglia, 2010). For example, when a monkey observed an experimenter grasping a piece of food with his hand, neural responses were evoked in neurons located in area F5. These neurons ceased to produce action potentials when the experimenter moved the food toward the monkey, and they produced action potentials again when the monkey grasped the food. The 'mirror function' of these premotor neurons is a physiological correlate of 'common coding', and fundamental for perception-action mediation. This section will provide an overview of studies on perception–action mediation during listening to music.[2]

The first neuroscientific study on auditory perception–action mediation was an MEG study by Jens Haueisen and Thomas Knösche (Haueisen & Knösche, 2001). In that study, both non-musicians and pianists were presented with piano melodies. Compared to non-musicians, musicians showed neuronal activity in (pre-)motor areas that was elicited simply by listening to music (the task was to detect wrong notes, and those trials with wrong notes were excluded from the data analysis).[3] Interestingly, the centre of neuronal activity for notes that would usually be played with the little finger was located more superiorly than activity for notes that would usually be played with the thumb (according to the somatotopic representation of the fingers), supporting the notion that the observed neural activity was premotor activity. Similar activations were observed with fMRI when violinists listened to violin music (Dick *et al.*, 2011).

One year later, Evelyne Kohler *et al.* (2002) investigated neurons in the area F5 of macaque monkeys that discharged not only when the monkeys performed a hand action (such as tearing a piece of paper), but also when the monkeys saw, and simultaneously heard the sound of, this tearing action (similar to the mirror neurons mentioned above, that are active during both observation and execution of actions). Importantly, simply hearing the sound of the same action (performed out of the monkey's sight) was equally effective in evoking a response in these neurons. Control sounds that were not related to action (such as white noise, or monkey calls) did not evoke excitatory responses in those neurons. Thus, this study showed that (in monkeys) some premotor neurons are active during both hearing and execution of actions.

As mentioned above, the study by Haueisen & Knösche (2001) showed perception-action mediation in musicians (pianists). Music-related perception-action mediation in non-musicians was shown by Callan *et al.* (2006). In that study, activation of premotor cortex was observed not only when participants (non-musicians) were singing covertly, but also when they simply listened to song. Interestingly, premotor activity in the same area was also observed during both covert speech production and listening to speech (Figure 11.1a). This

---

[2]   For an fMRI study investigating pianists and non-musicians *observing* finger-hand movements of a person playing piano see Haslinger *et al.* (2005).
[3]   Neural sources were located on the crown of the precentral gyrus, thus presumably in the premotor cortex, rather than in the motor cortex.

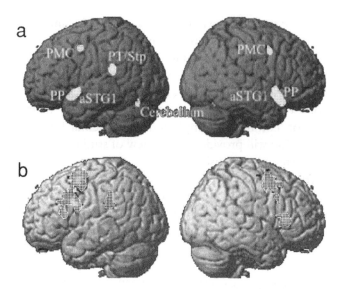

Figure 11.1 Premotor activation during listening to musical information in non-musicians. The top panel (a) shows areas that were activated during listening to singing, covert singing, listening to speech, and covert speech (conjunction analysis) in the study by Callan *et al.* (2006). Both left and right premotor activity was observed in all four conditions, showing that this area is active not only during perception of speech or music, but also during the production of speech or music. (b) shows areas that were more active during listening to trained melodies compared to listening to melodies consisting of untrained (and different) tones in the study by Lahav *et al.* (2007). aSTG1 = anterior superior temporal gyrus; PMC = premotor cortex; PP = planum polare; PT = planum temporale; Stp = superior temporal plane. Modified with permission from Callan *et al.* (2006) and Lahav *et al.* (2007).

showed that neural correlates of mirror mechanisms overlap strongly for music and speech perception.

In a study on the effects of musical training on perception-action mediation in non-musicians by Lahav *et al.* (2007), non-musicians were trained over the course of five days to play a piano melody with their right hand. After this training period, simply listening to the trained melody activated premotor activity (Figure 11.1b). Listening to an untrained melody did not activate premotor cortex, suggesting that in the early stages of learning, perception-action mediation relies on fairly specific learned patterns. Dick *et al.* (2011) reported that in trained musicians, on the other hand, such activity does not differ between familiar and unfamiliar music (and similar training effects were observed in trained actors listening to dramatic speech). Bangert *et al.* (2006) measured BOLD signals during both listening to melodies and producing simple melodies with the right hand on a keyboard (without auditory feedback). In pianists, activation was observed during both perception and production of melodies in the premotor cortex, the pars opercularis (corresponding to BA 44), the planum temporale, and the supramarginal gyrus

(BA 40).[4] Activations within the premotor cortex (PMC) and BA 44 during both perception and production of melodies were clearly left lateralized.

Interestingly, perception-action mediation appears to be modulated by emotional processes. In an fMRI experiment on music and emotion (in which pleasant and unpleasant music was presented to the participants; Koelsch *et al*., 2006a)[5] the contrast of listening to pleasant versus listening to unpleasant music showed an increase in BOLD signal in premotor areas, as well as in the Rolandic, or 'central', operculum during listening to pleasant music. During listening to unpleasant music, a decrease of BOLD signal in these areas was found. That is, premotor activity during listening to music was modulated by the emotional valence of the music, suggesting that perception-action mediation is modulated by emotional processes. It is likely that the Rolandic operculum contains, at least partly, the representation of the larynx, and therefore it seems that participants were quasi-automatically (without being aware of this, and without intentional effort) singing subvocally along with the pleasant, but not with the unpleasant music. The activation of the Rolandic operculum during singing is different from the one reported by Callan *et al*. (2006), perhaps because the former study (Koelsch *et al*., 2006a) used instrumental music, whereas the latter (Callan *et al*., 2006) used songs. The notion that mirror mechanisms can be modulated by emotional factors is consistent with findings showing that auditory mirror mechanisms elicited by emotional vocalizations can be modulated by the emotional valence of these vocalizations (Warren *et al*., 2006).[6]

With regard to temporal aspects of music, both cortical (supplementary motor area, SMA, and PMC) and subcortical structures (basal ganglia and cerebellum) are active during both perception and production of tactus, metre, and rhythm (e.g., Grahn & Brett, 2007; Grahn & Rowe, 2009; Grahn, 2009). Moreover, functional connectivity between the basal ganglia, SMA, and PMC increases during the perception of tone sequences based on an isochronous pulse (Grahn & Rowe, 2009). Finally, patients with Parkinson's disease show increased difficulties in discriminating changes in such sequences (compared to healthy controls; Grahn, 2009), corroborating the notion that the basal ganglia (in addition to SMA, PMC, and cerebellum) play an important role for both the perception and the generation of rhythm and metre.

11.2   ERP correlates of music production

The last section described action-related neural processes activated by music perception. This section will report studies investigating neural correlates of the

---

[4]   In addition, *performing* the melodies on a piano (without auditory feedback) elicited activity in auditory areas.

[5]   See also Chapter 12.

[6]   That study used vocalizations such as 'yuck' or 'yippee' expressing triumph, amusement, fear, or disgust.

execution of actions during music production. The term *action* as used here implies that an action (a) consists of at least one movement, or a chunk of movements (such as playing triad arpeggios across several octaves), (b) has a goal, (c) can be voluntarily executed (or inhibited), (d) can be corrected during execution (if necessary), and (e) is modulated by the anticipation of a specific action effect.[7] Actions can be chained into action sequences, each action having a sub-goal (and related action effects), and the action sequence having a superordinate goal (and a related action effect).

When playing a musical instrument (or when singing), alone or together in a group, a player continuously establishes action goals, forms the corresponding motor programs to execute the right movements at the right time with the right strength, monitors ongoing movements by relating information such as proprioceptive feedback of the actual movements to the planned movements, and initiates corrective movements (when necessary). Corrective movements are also required when synchronizing movements with movements of other players. While playing, such corrections are memorized, and integrated with the execution of simultaneous movements. The perception of the action effects completes the action, and can modify selection, programming, execution, and control of new actions (see also Table 11.1 for a summary of the processes involved in the selection, programming, execution, and control of an action). All these processes overlap in time, making the investigation of these different processes challenging. One approach to investigating music production with ERPs is to examine neural correlates of error-related processes.

Questions that arise in this context are: At what point in time errors are actually detected by the sensorimotor system, whether they are detected already prior to execution, and, if so, at what point in time potential errors can still be corrected. Using music, an ERP-study by Maidhof *et al.* (2009) investigated whether errors are detected already *before* a movement is fully executed (for a similar study see Herrojo-Ruiz *et al.*, 2009a). That study (Maidhof *et al.*, 2009) investigated expert pianists playing scales and scale patterns (bimanually) in a relatively fast tempo (Figure 11.2). These stimuli were chosen to provoke pianists into committing errors, with the aim to compare the brain-electric potentials related to incorrect, with those related to correct, keystrokes (in time intervals preceding, and following the onsets of keystrokes).

Results showed that, behaviourally, pianists pressed incorrect and correct keys with different velocities: Participants pressed incorrect keys with a lower velocity than (a) correct keypresses, and (b) the simultaneous correct keypresses (and the velocity of these simultaneous correct keypresses was not influenced by the lower velocity of the erroneous keypresses of the other hand). Moreover, correct and incorrect keypresses were produced with different inter-onset intervals (IOIs): The IOI between an incorrect keypress and the preceding keypress was prolonged

---

[7]   An action effect can, e.g., be a single tone, or the sounds of a triad arpeggio across several octaves.

Table 11.1 Overview of the processes involved in the production of actions (bottom row), in relation to processes of music perception and syntactic processing: Establishment of an action goal, formation of a motor program, motor command and establishment of efference copies to predict outcomes of movements, differentiation of predicted outcomes and actual movements (by virtue of sensory feedback), correction of motor commands, integration of movements performed by different effectors (such as left and right hand), and (perception of the) action effect. Note that feedback loops lead to modulations of motor commands during correction and integration. Predictive processes during *program formation* parallel predictive processes during *Gestalt formation* (e.g., with regard to automatic change detection). The *efference copy*, on the other hand, presumably contains additional knowledge-based predictive information about sensory consequences of external objects. Thus, this sensorimotor *prediction* parallels the predictions due to *musical expectancy formation*. The *differentiation* of sensory feedback (that is, of information about the actual consequences of a movement) from the information of the efference copy, parallels processes of *structure building* which include the differentiation between predicted (or expected) sound and actual musical information. Processes of *correction* parallel processes of *structural reanalysis* (and revision).

music perception	feature extraction	Gestalt formation	interval analysis	structure building	structural re-analysis	vitalization	premotor, immune system
syntactic processing	element extraction	knowledge-free structuring	musical ex-pectancy formation	structure building	structural re-analysis	syntactic integration	large-scale structuring
action	action goal	program formation	motor-commands, efference copies	differenti-ation (relating)	correction	integration	perception of action effect

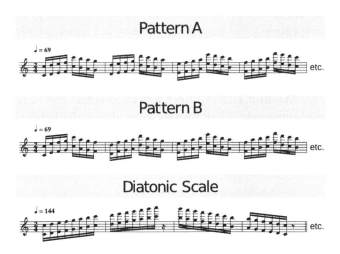

Figure 11.2 Illustration of the patterns used in the study by Maidhof *et al.* (2009). Patterns are shown in C major (in the experiment, the stimuli had to be produced in different major keys). The instructed tempo for the scales was 144 bpm, and for the patterns 69 bpm.

(compared to the IOI between successive correct keypresses), indicating that an upcoming error slowed down the keypresses (*pre-error slowing*). IOIs of simultaneous keystrokes performed by the other hand (without errors) were influenced by the error; that is, in contrast to the velocity, both hands showed a similar pre-error slowing during an error (even if the error occurred only in one hand), presumably due to the integration of several movements during executing an action.[8] In addition to pre-error slowing, Herrojo-Ruiz *et al.* (2009a) also reported *post-error slowing* following the commission of an error. Those data (Herrojo-Ruiz *et al.*, 2010) also indicate that pre-error and post-error slowing effects are limited to the trials directly preceding and following the errors (i.e., these temporal disruptions are not progressive phenomena, and slowing does not occur several events before or after an error).

The ERPs of the study by Maidhof *et al.* (2009) showed that, compared to correct keypresses, incorrect keypresses elicited an increased negativity already *before* a wrong key was actually pressed down (Figure 11.3). This ERP effect was maximal at central leads and peaked around 100 ms before a key was pressed down (left of Figure 11.3B). This *pre-error negativity* was followed by a later positive deflection with an amplitude maximum at around 280 ms after the onset of an incorrect note. This potential had a fronto-central scalp topography and resembles the early *Error Positivity* (Pe) or the P3a (see also right of Figure 11.3B). Virtually the same ERP pattern was reported in the study by Herrojo-Ruiz *et al.* (2009a); that study also employed a condition in which pianists played without auditory feedback. Under that condition, the error positivity was considerably reduced. In addition, the pre-error negativity elicited in the motor condition (without auditory feedback) was identical to the pre-error negativity elicited in the audiomotor condition. This finding is consistent with findings showing that, after extensive learning of a sequence, auditory feedback is irrelevant for music performance with regard to error-monitoring (Finney, 1997; Finney & Palmer, 2003; Pfordresher, 2003, 2005, 2006). In those studies, even the complete absence of auditory feedback had mostly no effects on the performance of extensively trained piano pieces.[9]

In the study by Maidhof *et al.* (2009), left-hand errors and right-hand errors were also analysed separately. That analysis showed that the ERPs elicited by the errors were not lateralized. Therefore, left- or right-hand errors were probably not

---

[8]   The integration of bimanual movements is also referred to as *bimanual coupling*: Bimanual movements begin and end synchronously, even when they have different parameters (e.g., amplitudes), and even when movement times differ when the respective movements are performed in isolation by one hand (Marteniuk *et al.*, 1984; Spijkers *et al.*, 1997; Swinnen & Wenderoth, 2004; Diedrichsen *et al.*, 2010). In addition, musicians specifically train to play synchronously with both hands. Such integrative processes (including bimanual coupling) are not only due to low-level processes of motor execution, for 'the symmetry constraint observed in bimanual coordination ... depends on perceptual variables and task demands ... More generally, many demonstrations of constraints in bimanual coordination appear to reflect limitations in the simultaneous estimation of high-level, task-relevant states ..., rather than hard-wired coordination constraints between the two hands. The

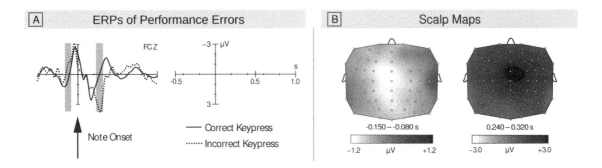

Figure 11.3   Grand-average ERPs (recorded from ten pianists) elicited by correctly and incorrectly performed keypresses, time-locked to the onset of keypresses. (A) The arrow indicates the note onset and thus the onset of the auditory feedback. The grey areas highlight the pre-error negativity (occurring prior to the keypress), and the later positive effect (occurring after the incorrect keypress). (B) Scalp topographies for the difference potentials for correct keypresses subtracted from incorrect keypresses in the two time windows marked grey in (A). Modified with permission from Maidhof *et al.* (2009).

simply due to some right- or left-hemispheric neural disturbance that caused the erroneous movement (such hemispheric disturbances would then occur bilaterally when averaged across left- and right-hand errors). That is, it does not appear that the early ERP difference occurring before a key was pressed down was the cause for the error. Instead, it appears that this early ERP difference reflects cognitive processes of error detection, error correction, and/or movement integration. With regard to movement integration, note that IOIs were prolonged before incorrect keypresses in *both* hands (in contrast to velocities, which differed between synchronous erroneous and correct key presses). Such integrative adjustment of bimanual movements perhaps contributed to this ERP effect.

The fact that the early ERP elicited by incorrect movements occurred *prior* to keypresses indicates that errors were detected *before* they were fully executed (and before auditory feedback was available). Such an error detection process is based on internal forward models: Probably during the formation of a motor program, a forward model is prepared which includes an efference copy, or 'corollary discharge'. With regard to the model of music perception (summarized in the top row of Table 11.1), the forward model based on the formation of a motor program parallels the predictive forward model established based on operations of the

---

human coordination system has evolved to achieve single goals flexibly using many effectors rather than to achieve multiple goals simultaneously' (Diedrichsen *et al.*, 2010, p. 38).

[9]   Specific alterations of auditory feedback, on the other hand, profoundly disrupt performance. For example, disruptive effects of pitch manipulations (false auditory feedback) occur during learning, or when the perceived feedback resembles an intended sequence (reviewed in Pfordresher, 2006). However, if auditory feedback is random, that is, when the feedback sequence is highly dissimilar to the intended sequence, the auditory feedback does not disrupt performance (presumably because players perceive the feedback as being unrelated to the planned actions).

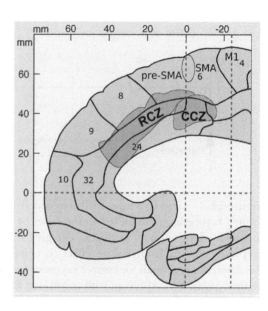

Figure 11.4 Anatomical map of the medial frontal cortex. SMA: Supplementary motor area, RCZ: Rostral cingulate zone, CCZ: Caudal cingulate zone. The vertical line through the anterior commissure (left vertical dashed line) indicates the approximate border between SMA and pre-SMA. The oval indicates the presumed location of the supplementary eye field according to Amiez & Petrides (2009). The numbers indicate Brodmann areas. Modified with permission from Ridderinkhof *et al.* (2004).

auditory sensory memory (summarized under the concept of *Gestalt formation*). The formation of a motor program takes into account the action goal, as well as the initial movement conditions, such as the respective locations and movements of body, arm, hand, and target. The formation of a motor program appears to involve several areas, including the pre-supplementary motor area (pre-SMA) and SMA proper (see Figure 11.4), the (pre-)supplementary eye field, premotor cortex, primary motor cortex (M1), basal ganglia, and parietal areas (e.g., Hoover & Strick, 1999; Middleton & Strick, 2000; Nachev *et al.*, 2008; Desmurget & Sirigu, 2009; Schubotz, 2007).[10] Studies investigating the activity of neurons in M1 of non-human primates showed that the latency between the first activity in M1 and movement onset is variable and can range up to several hundred milliseconds, with the typically assumed latency being around 100 to 150 ms (Evarts, 1974; Porter & Lewis, 1975; Thach, 1978; Holdefer & Miller, 2002; Hatsopoulos *et al.*, 2007).

Then, at the same time as the motor command is sent from M1 to the periphery, an *efference copy* is created (supposedly in brain structures that are also involved in the generation of the movement), and sent to sensory structures. The efference copy is not used to generate the ongoing motor activity, but can be used to predict the outcome (i.e., the sensory consequences) of the motor command (information

---

[10]   The basal ganglia are either part of the motor planner itself or are in a loop with planning structures. Note that cortico – basal ganglia – thalamo – cortical and cerebellar loops contribute to the programming, initiation, execution, and control of movements. For the role of the basal ganglia (as well as other motor structures) in the perception of tactus and metre see Grahn & Brett (2007). For the role of the basal ganglia in sensorimotor synchronization see Schwartze *et al.* (2010).

from efference copies interacts at several levels of the central nervous system, and usually modulates sensory processing; see also Poulet & Hedwig, 2007; Crapse & Sommer, 2008).[11] Particularly when actions are carried out quickly, it is likely that the efference copy contains more predictive information than the preparation of the efference copy during program formation, e.g., with regard to those sensory consequences of movements that are due to external objects (such as objects that are touched, grazed, or moved during the movement). That is, in the course of the establishment of the efference copy, predictive information about sensory consequences is probably added to the forward model. Because these consequences are based on knowledge about the nature of external objects (their weight, temperature, surface texture, etc.) this predictive information parallels the predictive information due to *musical expectancy formation* during music perception (Table 11.1).

During the execution of an action, information about the actual consequences of a movement is *differentiated* from the information of the efference copy and related to the predicted consequences of planned movements (Wolpert *et al.*, 1995; Miall & Wolpert, 1996; Wolpert *et al.*, 1998; Desmurget & Grafton, 2000; Wolpert & Ghahramani, 2000). Information about the actual consequences originates from (a) somatosensory feedback (such as proprioceptive and tactile information), (b) visual feedback, and (c) efferent information (motor outflow).[12] Whenever there is a mismatch between actual and predicted consequences, an error signal is generated. It is likely that such signals are also generated when the sensory consequences of a movement deviate from the predicted consequences, even though the movement itself was correct (for example, when a key of the piano got stuck, or when an experimenter provides false auditory feedback; see also below). The occurrence of the pre-error negativity in the studies by Maidhof *et al.* (2009) and Herrojo-Ruiz *et al.* (2009) reflects that this detection mechanism operates *prior* to the full execution of the movement, and thus *before* the perception of the auditory feedback, that is, before the perception of the tone produced by the movement (action effect). With regard to the processes of music perception and syntactic processing (first two rows of Table 11.1), the *differentiation* of sensory feedback (that is, of information about the actual consequences of a movement) from the information of the efference copy, parallels processes of *structure building* which include the differentiation between predicted (or expected) sound and actual musical information.

---

[11]   The information from the efference copy also tells the sensory areas about the upcoming sensory perceptions and allows them to prepare for the sensory consequences of the movement, for details see Crapse & Sommer (2008).

[12]   Note that, despite some overlap, the networks underlying somatosensory feedback on the one hand, and visual feedback on the other, differ significantly from each other (see, e.g., Swinnen & Wenderoth, 2004).

The error signal can, in turn, lead to the *corrective modulation* of motor commands (for details see Desmurget & Grafton, 2000). Note, however, that the execution of movements is sometimes faster than the propagation of sensory information to the cortex, and that, thus, sensory feedback cannot always be used to correct movements (Desmurget & Grafton, 2000).[13] Also note that, without any erroneous movement, movements can also be re-programmed during execution (Leuthold & Jentzsch, 2002).[14] With regard to the processes of music perception and syntactic processing (Table 11.1), processes of *correction* parallel processes of *structural reanalysis* and revision.

In the studies by Maidhof *et al.* (2009) and Herrojo-Ruiz *et al.* (2009), the corrective modulation of the motor command might have resulted in the lower velocities of incorrect keypresses. Note that the IOIs were prolonged before incorrect keypresses in both hands (although only one hand performed an erroneous movement). This indicates that the movement of the hand playing the correct key was integrated with the erroneous movement of the other hand. Neural mechanisms of such integration processes during the performance of actions remain to be specified.[15]

The error positivity (Pe) following incorrect keypresses in the study by Maidhof *et al.* (2009) is probably related to the conscious recognition of a committed error (see also the 'error-awareness hypothesis', e.g., Nieuwenhuis *et al.*, 2001). Conscious recognition might also involve the adaptation of response strategy after an error has been perceived, involving remedial performance adjustments following errors ('behaviour-adaption hypothesis', Hajcak *et al.*, 2003). Such adaptive processes might include making up for delays due to pre- and post-error slowing (Herrojo-Ruiz *et al.*, 2010). Recognition of errors might also result in affective processes following the committed error or its consequences, including autonomic responses (such as changes in heart rate and sweat production). Finally, recognition of errors has effects on learning, to avoid similar errors in the future when aiming to obtain a similar action goal.

Notably, in addition to the experimental design described so far, the above-mentioned study by Maidhof *et al.* (2009) provided the participants during playing every so often with manipulated (false) feedback when correct notes were played: Randomly between every 40th to 60th produced note, the auditory feedback of the digital piano was manipulated in a way that the pitch of one tone was lowered by one semitone (for a similar study see Katahira *et al.*, 2008). This was done to investigate the time course of the neural mechanisms underlying

---

[13]   Or, depending on the movement and the speed of a movement, sensory feedback loops might allow corrections only at the very end of the trajectory.

[14]   In that study (Leuthold & Jentzsch, 2002), re-programming was reflected electrically in a negative centro-parietal potential that was maximal at around 370 ms after the onset of a cue that required participants to re-program a movement that had already been commenced.

[15]   It is likely that a network of numerous (sensori-)motor structures mediates bimanual movement integration. For a study suggesting involvement of the SMA in this network see Steyvers *et al.* (2003).

the processing of (manipulated) feedback during music performance, and thus to study the processing of auditory effects of actions intended by a player or singer. Musicians expect to perceive the auditory feedback of their action, and the intention of musicians to produce specific auditory effects by executing certain actions is a fundamental aspect of music performance. Skilled piano players are trained to produce specific auditory effects with highly accurate movements (Ericsson & Lehmann, 1996; Palmer, 1997; Sloboda, 2000). Accordingly, results of behavioural (Drost *et al.*, 2005a,b), electrophysiological (Bangert & Altenmüller, 2003), and neuroimaging studies (see the first section of this chapter) consistently show pronounced coupling of auditory and motor systems in individuals with musical training.[16]

ERPs of correct (!) keypresses with and without feedback manipulation are shown in the top of Figure 11.5 (reported in Maidhof *et al.*, 2010). Feedback manipulations (i.e., the sound of a wrong note, although the correct key was pressed) elicited a negativity that was maximal at around 200 ms and had a fronto-central scalp distribution. As will be discussed further below, this negativity was presumably mainly a *feedback error-related negativity* (feedback-ERN), with additional contributions of MMN/ERAN potentials. The negativity was followed by a P3a, and a P3b.[17]

The feedback ERN is a type of *error-related negativity* (ERN or Ne, Botvinick *et al.*, 2001; Yeung *et al.*, 2004; Van Veen & Carter, 2006; Falkenstein *et al.*, 1990). Classically, the ERN (or *response-ERN*) is an ERP peaking shortly after participants commit an error in a variety of speeded response tasks (although the ERN often begins to emerge already before a button press, shortly after the onset of electromyographic potentials). The ERN typically peaks around 50–100 ms after incorrect responses, regardless of the modality in which the stimulus is presented, and regardless of the modality in which the response is made. The *feedback ERN* is elicited after negative performance feedback (compared to positive feedback), and after feedback stimuli indicating loss (or punishment) in time estimation tasks, guessing tasks, and gambling tasks (Miltner *et al.*, 1997; Hajcak *et al.*, 2005, 2007). The feedback ERN is generally taken to reflect expectancy-related mechanisms, probably irrespective of whether the outcome of an event is worse or better than expected (Oliveira *et al.*, 2007; Ferdinand *et al.*, 2008).

Note that musical feedback manipulations most presumably elicit, in addition to action-related processes, cognitive processes related to the perception of acoustic deviants. Wrong notes (whether produced by the player him/herself, or whether due to false feedback) probably elicit ERPs such as ERAN or MMN, which overlap with ERPs such as the ERN (or the N2b). A study by Maidhof *et al.* (2010) offered one approach to deal with this difficulty: In that study, ERPs elicited during music

---

[16]   Notably, in the EEG study by Bangert & Altenmüller (2003), musically naïve participants showed auditory-sensorimotor co-activity already within 20 minutes of piano learning.

[17]   Behavioural results showed that feedback-manipulated tones did not cause longer IOIs with regard to succeeding tones.

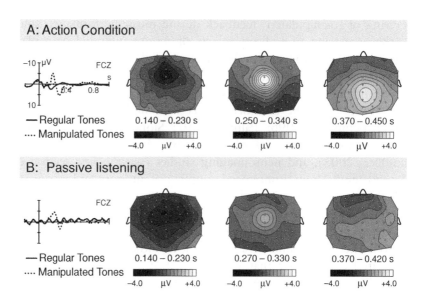

Figure 11.5 (A) Grand-average ERPs (recorded from twelve pianists) elicited by correct keypresses with correct (solid line) and manipulated (dotted line) auditory feedback, time-locked to the onset of tones (i.e., when a key was pressed down). Feedback-manipulated tones elicit a feedback ERN (presumably overlapping with MMN/ERAN potentials), followed by a P3a, and a P3b (best to be seen in the isopotential maps). (B) Grand-average ERPs (recorded from the same twelve pianists) elicited while passively listening to the auditory stimuli of the action condition. Here, the deviants ('manipulated tones') also elicit a negativity peaking around 200 ms (possibly in part due to ERN potentials), and a P3a; no P3b was elicited in this condition (reflecting that the deviants were not task-relevant). Note that in both the action and the passive listening condition, ERN potentials presumably overlapped with MMN/ERAN potentials, making it challenging to disentangle action-related from perception-related brain potentials. Modified with permission from Maidhof *et al.* (2010).

performance (*action condition*) were compared with ERPs elicited when musicians merely listened to such stimuli (*perception condition*). Such comparisons are also interesting because they can perhaps inform us about action-related processes evoked during music *perception*. Figure 11.5 (bottom panel) shows ERPs elicited in a condition in which pianists passively listened to the tone patterns with and without wrong (feedback-manipulated) tones produced in the action condition. As in the action condition, incorrect tones (compared to correct tones) elicited a negativity that was maximal around 200 ms, but with smaller amplitude than in the action condition. This negativity was followed by a small P3a (being maximal at frontal leads; no P3b was elicited in this condition).

That is, manipulated tones elicited during both the production and the perception of tones negative potentials with maximal amplitudes around 200 ms, with larger amplitude in the action compared to the perception condition. Similarly, the P3a elicited by (wrong) tones was more pronounced during the action condition (when participants were playing) compared to the perception condition (i.e., when participants only listened to the stimuli). The absence of a P3b during the perception condition reflects that the pitch manipulations were task-irrelevant for the participants. Because the N2b, or the ERN, is usually observed in combination with a P3b, it is likely that the observed negative potential is not simply an N2b or ERN, but consists to a considerable degree of MMN/ERAN potentials.

Therefore, although the early negativity observed during the *action* condition is presumably in part due to a feedback ERN, it might well be that this ERN effect overlaps with MMN/ERAN potentials related to the processing of acoustic or harmonic-syntactic irregularity. On the other hand, whereas the negativity elicited during the *perception* condition presumably reflects at least in part an MMN/ERAN, it might well be that this potential overlaps in part with ERN/N2b potentials due to the simulation of action during the perception of music. This illustrates the difficulty of disentangling the different contributions of these components during music performance and music perception. However, there are several aspects that can be addressed to distinguish ERN- from MMN- or ERAN-potentials:

1. A comparison of feedback-manipulated tones with wrong tones produced by players themselves. In contrast to the ERPs elicited by feedback manipulations, ERPs of self-performed errors (Figure 11.3) did not show a significant negative effect around 200 ms after the onset of erroneous keypresses (although a small negativity is visible in the ERPs of wrong tones in this time range; cf. Figure 11.3). Because the auditory deviance is comparable between self-generated errors and feedback manipulations, the ERPs of self-performed errors provide an estimate of possible MMN contributions. Thus, because no clear MMN is visible in the ERPs of self-performed errors, it is unlikely that the negativity elicited by feedback manipulations is simply an MMN or ERAN.

2. A localization of the sources of ERPs. Using current source density, the study by Maidhof *et al.* (2010) localized the sources of the negativities elicited in both action and perception condition in the rostral cingulate zone (RCZ) of the posterior medial frontal cortex (see also Figure 11.4). These results are consistent with an explanation in terms of a feedback ERN: Studies on action monitoring and cognitive control indicate that the RCZ plays a key role in the processing of expectancy violations, performance monitoring, and the adjustment of actions for the improvement of task performance (van Veen & Carter, 2002; Ridderinkhof *et al.*, 2004; Nieuwenhuis *et al.*, 2004; van Veen *et al.*, 2004; Folstein & Van Petten, 2008). Therefore, the ERN, the feedback ERN, and the N200/N2b presumably all receive (main) contributions from neural generators located in the RCZ (this also supports the notion that feedback ERN and N2b are subcomponents of the N200, with feedback

ERN and N2b possibly being synonymous labels for effects with very similar functional significance).

3. A comparison between musical experts and musical beginners. A study by Katahira *et al.* (2008) reported an ERN to feedback-manipulated tones in musical experts (similar to the study by Maidhof *et al.*, 2009), but no ERN was observed in participants who had only moderate musical training.[18] This is consistent with results by Maidhof *et al.* (2010), which showed that the ERN amplitude correlated negatively with the duration of musical training. An absence of ERN potentials (as in the study by Katahira *et al.*, 2008) renders it unlikely that the effect observed in expert musicians is an MMN or ERAN, because frank violations elicit such potentials also in non-experts (although the amplitude of these potentials is also modulated by musical training; see Sections 5.3.2 and 9.8).

4. A comparison between diatonic (in-key) and non-diatonic (out-of-key) feedback manipulations. If ERN potentials partly overlap with ERAN potentials, then non-diatonic feedback manipulations should evoke larger negative effects than diatonic manipulations (because the ERAN amplitude is related to the degree of violation; see Chapter 9). The study by Katahira *et al.* (2008) reported that the ERN amplitude did not differ between diatonic and non-diatonic feedback manipulations, suggesting that ERAN-related potentials did not, or only minimally, contribute to the ERN potentials.

It is also worth noting that the MMN is not influenced by the anticipation of deviant tones, nor by prior knowledge of deviant stimuli (e.g., Rinne *et al.*, 2001; Waszak & Herwig, 2007; Scherg *et al.*, 1989). Moreover, the MMN amplitude does not differ between a condition in which participants trigger the presentation of tones, or listen to the same sequence of tones (Nittono, 2006).[19] Therefore, different ERN amplitudes in the absence of P3b potentials (as observed in the study by Maidhof *et al.*, 2009) indicate that possible MMN contributions could have been only minor.

The discussed methods provide approaches to illuminate to which extent evoked negativities following the perception of feedback manipulations reflect ERN, N2b, or MMN/ERAN potentials. As mentioned above, the manipulated tones elicited in both the action and the perception condition an early negativity that strongly resembled the ERN (in terms of latency, distribution, and neural generators). In both the study by Katahira *et al.* (2008) and by Maidhof *et al.* (2010), this feedback-ERN effect was more pronounced during the performance of music compared to the mere perception of music. Thus, it seems likely that similar

---

[18]   Participants played unfamiliar melodies on a keyboard, and in five percent of the keypresses, the tone was shifted a semitone upwards.

[19]   In that study (Nittono, 2006), participants triggered the presentation of a tone (which was either a standard tone, or one of two pitch-deviants) by pressing buttons. That is, participants had control over the timing of the stimuli, but not over the pitch of the stimuli.

expectancy-related mechanisms operate during both performance and perception of music.[20] Importantly, the feedback ERN in the mentioned music studies is influenced by the expectancies generated by the intention and action of the pianists to produce a certain auditory effect. In contrast to these action-related expectancies, pianists could also build expectancies during the perception of the sequences based on the preceding musical context and its underlying regularities. Consequently, the manipulated tones during piano performance were more unexpected than the manipulated tones during the perception of the sequences, resulting in the enlarged feedback ERN in the action compared to the perception condition.

If the feedback ERN reflects the processing of violations of action-related predictions, how are these predictions established during the production and perception of musical sequences? It appears that, during the production of action sequences, pianists anticipate the tone mapped to the particular keypress they are about to perform. After having learned these associations (due to extensive training), the formation of an action plan leads to the establishment of predictions of the sensory feedback using the internal forward model described above. This implies that predictions are formed before a motor command is sent. According to the common coding theory (Prinz, 1990), it also seems likely that, when making music, actions are selected, and controlled, using an inverse model of the intended effect, leading to an expectation for a certain effect (the *ideomotor principle*; see e.g., Hommel *et al.*, 2001). As mentioned above, the common coding theory assumes that coding of perception overlaps with the coding of action in the sense that they share a common representational format. Therefore, the anticipated effects of an action should influence its planning, control, and execution (the *action-effect principle*). The notion that the prediction of action effects is related to the training of the participants is supported by the correlation between ERN amplitude and amount of training. In the study by Katahira *et al.* (2008), no ERN was elicited in non-expert players, and in the study by Maidhof *et al.* (2010) pianists with longer training showed larger ERN amplitudes. While listening to the sequences without playing, predictive mechanisms probably extrapolate from the regularities of the preceding auditory input, and thus generate a prediction towards a specific sound to follow. This expectancy (or prediction) seems to be a fundamental aspect of perception, which is most likely not under the strategic control of individuals (Schubotz, 2007; see also Chapter 9).

Even more importantly, the combined data show that the processing of expectancy violations is modulated by the action of an individual. During music performance, players (or singers) expect, based on their intention and their act of performing, to perceive a specific auditory effect. In addition, the preceding musical context induces expectancies for specific tones. Hence, when an unexpected

---

[20]   Note that it is conceivable that a feedback ERN can also be elicited during perception (without action), because feedback ERN-like waveforms are also observed when no actions, or responses, are required on the part of the participants (Donkers *et al.*, 2005), and when rules (i.e., expectations) are violated in tasks without overt responses (Tzur & Berger, 2007, 2009).

tone is encountered following an action, the detection of the violation of such expectancies elicits a brain response similar to the feedback ERN/N200. A similar effect, although with smaller amplitude, is elicited when pianists merely perceive an unexpected tone (without performing). It is tempting to speculate that this effect is in part due to action-related mechanisms, such as an effect of simulated action during the perception of music.

12

Emotion

Using music to investigate the neural correlates of emotion has several benefits:
(1) Music is capable of evoking *strong emotions* (usually more powerful than, for
example, static images of faces). Strongly pleasurable responses to music can involve, e.g., goose bumps or shivers down the spine. (2) Music can be used to
investigate *mixed emotions* (such as 'pleasant sadness'). (3) Music can evoke a
wide *variety of emotions*. For example, with regard to positive emotions, music
can evoke joy, amusement, amazement, feelings of vitalization, consolation, spirituality, calmness, triumph, etc.[1] (4) Studying the neural correlates of emotions
with music has direct relevance for applications of music in *therapy*. (5) Both
listening to music and making music can evoke emotions, enabling investigators
to study interactions between emotion and *action*. (6) Music can be used to study
the *time course* of emotional processes, with regard to both short-term emotional
phenomena (in the range of seconds) and longer-term emotional phenomena (in
the range of minutes, or even hours). (7) It appears that, with regard to human
evolution, music is originally a social activity.[2] Therefore, music is well suited to
study interactions between emotion and *social factors*.

---

[1] Zentner *et al.* (2008) report a list of 40 emotion words typically used by Western listeners to
describe their music-evoked feelings.
[2] This assumption is based on the observations of musical practice in non-industrialized cultures,
which is almost exclusively observed in social contexts (e.g., Cross, 2008a; Cross & Morley, 2008).

However, using music in the study of emotion also bears several difficulties, as follows. (1) Musical preferences often differ substantially between individuals (a death-metal enthusiast can utterly despise thrash-metal), and the necessary control over the stimulus material might result in quite different emotional responses in different subjects. (2) Participants have to be equally familiar with different musical pieces, or styles, used in different experimental conditions (to avoid differences in neural activity between different emotion conditions not being simply due to differences in familiarity). (3) It is often challenging to control for the musical and acoustical parameters that differ between pieces used for different experimental conditions. For example, sad music is usually slower than joyful music, and when comparing effects of joyful and sad music, differences between conditions observed in functional neuroimaging data might simply be due to differences in arousal and corresponding cardiovascular responses related to the tempo of the pieces (that is, at least in some brain structures contrasting responses to slower joyful and faster joyful music might lead to the same results). (4) Some emotions are better investigated with other stimuli than music. For example, music is presumably not optimal to investigate emotional phenomena that involve a high load of cognitive appraisal (such as jealousy, regret, and the like).[3] Moreover, although individuals might feel disgusted by certain music, disgust can probably better be studied with odours and images.

12.1 What are 'musical emotions'?

Some researchers advocate that music can evoke a wide spectrum of 'real' emotions, whereas others argue that emotions evoked by music are artificial (and not real). For example, Zentner *et al.* (2008) state that 'musical antecedents do not usually have any obvious material effect on the individual's well-being and are infrequently followed by direct external responses of a goal-oriented nature' (p. 496) and thus reject the notion that music can evoke joy, sadness, anger, or fear. Similarly, Scherer (2004) writes that 'music is unlikely to produce basic emotions' (p. 244), Noy (1993) states that 'the emotions evoked by music are not identical with the emotions aroused by everyday, interpersonal activity' (p. 126), and Konečni (2003) claims that 'instrumental music cannot directly evoke genuine emotions in listeners' (p. 333). The assumption that music would not have material effects on the individual's well-being is also taken as argument for the notion that music cannot evoke basic emotions related to survival functions (e.g., Kivy, 1991; Scherer, 2004; Zentner *et al.*, 2008). Based on these assumptions, Scherer (2004) proposed to distinguish between goal-oriented *utilitarian* emotions (elicited in order to 'adapt to specific situations that are of central significance

---

[3]   Nevertheless, music can in part evoke the feeling sensations of such emotions, see also the *a priori musical meaning*, p. 145.

to the individual's interests and well-being', Scherer & Zentner, 2008, p. 595), and *aesthetic* emotions (elicited without 'obvious material effect on the individual's well-being and only rarely leading to specific goal-oriented responses').

However, the assumption that music-evoked emotions are not goal-relevant conflicts with the assumption that making music in a group supports the emergence of several evolutionarily adaptive social functions; engagement in such functions is associated with goal-relevant motivations to fulfil social needs (such as the 'need to belong'; Baumeister & Leary, 1995), and the fulfilment of such needs evokes emotions (such as feelings of reward and pleasure, and possibly attachment-related emotions such as love and happiness). Moreover, the assumption that music cannot evoke 'real' emotions conflicts, for example, with the experience that music can 'hale souls out of men's bodies' (William Shakespeare)[4], and with therapeutic effects of music-evoked emotions. In the following, several arguments will be offered that speak for the notion that music making and music listening have material effects, that it can serve the fulfilment of basic needs and thus serve the achievement of goals related to the survival of the individual.

Music and 'real' emotions   Juslin & Västfjäll (2008) argued that several psychological mechanisms underlying the evocation of emotion with music are shared with mechanisms underlying the evocation of 'a wide range of both basic and complex emotions in listeners' (see also Section 12.2 for details). Moreover, listening to music can evoke changes in the three major reaction components of an emotion, namely in *physiological arousal* (as reflected, for example, in changes in autonomic and endocrine activity; e.g., Steinbeis *et al.*, 2006; Sammler *et al.*, 2007; Grewe *et al.*, 2007b; Koelsch *et al.*, 2008; Orini *et al.*, 2010; Koelsch *et al.*, 2011), *subjective feeling* (such as feelings of pleasantness, happiness, sadness, etc.), and *motor expression* (e.g., smiling or frowning; Grewe *et al.*, 2007a). Moreover, music listening often elicits *action tendencies* (dancing, foot tapping, clapping, etc.). Finally, music can modulate activity in all so-called limbic and paralimbic brain structures (that is, in those structures that generate emotions; see also Section 12.6), indicating that music-evoked emotions are not merely illusions of the mind, but that music can indeed evoke 'real' emotions.

Music and 'basic' emotions   Music can evoke 'basic' emotions (Ekman, 1999) such as joy, fear, anger, sadness, and disgust. Many individuals experience joy and happiness while making music, or while listening to music (this is a frequent motivation for listening to music, e.g., Juslin *et al.*, 2011). It has also been shown that music can elicit surprise (Koelsch *et al.*, 2008),[5] and some participants of an earlier study (Koelsch *et al.*, 2006a) reported to us that continuously highly dissonant stimuli (used in that study) evoked a feeling of disgust and vertigo.

---

[4] Much Ado About Nothing, Act II, Scene 3.

[5] See also Meyer (1956), Huron (2006), and Section 12.5.

Most people get quite angry when they have to listen to music that they utterly dislike, and music is sometimes used to stimulate anger and aggression (e.g., the 'hate music' of neo-Nazis, Messner *et al.*, 2007, and partly also military music). Music-evoked sadness is dealt with in the next section.

The fact that music does usually not evoke all basic emotions equally often does not mean that music is not capable of evoking these emotions (e.g., music usually evokes happiness more often than disgust). Moreover, even if music does not evoke a basic emotion in an individual with the same intensity as a particular 'real life' situation, the underlying brain circuits are nevertheless presumably the same (whether I feel strong or moderate fear – the emotion is still fear). The assumption that music has no goal relevance, and no relation to the survival of an individual, will be dealt with further below.

'Real' and music-evoked sadness   Another argument against the assumption that music can evoke 'basic' emotions is that music cannot evoke 'real' sadness, because it is assumed to have no 'real implications for the individual's well-being' (Zentner *et al.*, 2008), and because in 'real' life, sadness tends to be experienced as a negative state, which most people try to avoid. By contrast, some individuals 'do not usually turn off the radio when a sad song hits the air' (Zentner *et al.*, 2008; other individuals, however, do turn off the radio when a sad song is played). Thoughts about why humans seek negative emotional experiences in the arts date back to Aristotle's *Poetics*, and have been summarized for the musical domain by Levinson (1990). Here, suffice it to say that the experience of sadness during art reception (or production) can have several rewarding effects (such as emotional catharsis, identifying expression, empathic emotional responses, understanding one's own feelings, emotional simulation, distraction from the extra-musical world, reward of expressive potency, emotional communication, realizing that no true loss occurred, etc.). Such rewarding effects are experienced as feelings of pleasure, or fun, and they correlate with activity of reward circuits in the brain (described in Section 12.6.2). A crucial point is that some of such rewarding effects (particularly emotional catharsis, empathic emotional responses, understanding one's own feelings, and emotional simulation) are only possible *because* 'real' sadness occurred prior to the experience of reward – otherwise there would be no reason for the elicitation of such rewarding effects. Therefore, music-evoked sadness must be congruent, at least for brief episodes, with the sadness evoked by a 'real' event.

Music and goal relevance: Survival functions   As mentioned above, it is assumed by some that music cannot evoke real emotions because it has no material effects (such as food, or money). However, music can have effects on the well-being of an individual, often including regenerative autonomic, endocrine and immunological effects. Although such effects are not material in the sense that the individual can eat, drink, or even touch them, they nevertheless involve matter (hormone- and immune-molecules and cells which also modulate expenditure of glucose, fat, and minerals).[6] Moreover, as will be described in more detail in Section 12.3, music activates, and facilitates, social functions such as communication, cooperation, and group cohesion. Engaging in such social functions fulfils basic needs of an individual, and is vital for the well-being of an individual. Moreover, for humans, engaging in social functions was (and is) critical for the survival of the human species. Therefore, the pleasure of engaging in these social functions is indeed related to survival functions.

12.2   Emotional responses to music – underlying mechanisms

Juslin & Västfjäll (2008) suggested several mechanisms underlying the evocation of emotions with music. These mechanisms include *brain stem reflexes* (due to basic acoustic properties of music such as timbre, attack time, intensity, and consonance/dissonance), *evaluative conditioning* (the process of evoking emotion with music that has been paired repeatedly with other positive, or negative, stimuli), *emotional contagion* (where a listener perceives an emotionally relevant feature or expression of the music, and then copies this feature or expression internally; see also Juslin & Laukka, 2003), *visual imagery* (where music evokes images with emotional qualities), *episodic memory* (where music evokes a memory of a particular event, also referred to as the 'Darling, they are playing our tune' phenomenon; Davies, 1978), and *musical expectancy* (where a specific musical feature violates, delays, or confirms expectancies of listeners, leading to feelings of tension and suspense; see Section 12.5).

Additional factors suggested by other researchers include the repeated *mere exposure* (which can contribute to, and modify the liking of music; Moors & Kuppens, 2008), extra-musical *semantic associations* with emotional valence (the association with an emotionally positive concept has different emotional effects than the association with a negative concept; Fritz & Koelsch, 2008), *rhythmic entrainment* ('a biological mechanism synchronizing body oscillators to external rhythms, including music'; Scherer & Zentner, 2008),[7] and *engaging in social functions*. The latter factor will be described in more detail in the next section, because it is of particular relevance for the argument of goal relevance, for the evolutionarily adaptive value of music, and for the therapeutic potential of music.

---

[6]   For a review see Koelsch & Stegemann (in press). For example, a study by Koelsch *et al.* (2011) showed a decrease of cortisol levels due to reduced stress before and during surgery. Release of cortisol increases glucogenesis in the liver (leading to higher levels of blood sugar), enhances lipolysis (leading to increased fat metabolism) and protein catabolism.

[7]   However, while it is highly plausible that emotions involve a synchronization of biological systems involved in that emotion (for details see Section 12.6.6), it is unclear why simply the entrainment of biological oscillation(s) to an external isochronous pulse should evoke an emotion. Entrainment of brain oscillations to the temporal properties of external stimuli, for example, often occur, without any apparent emotional component.

Table 12.1 Synopsis of social functions engaged by music (bottom row), in relation to processes of music perception, syntactic processing, and the dimensions of musical meaning. Analogous to *auditory feature extraction* (which is a prerequisite for the other processes of music perception), social *contact* is the prerequisite for the participation in the other social functions (therefore listed in the outermost left dimension). With regard to the domain of musical meaning, *social cognition* parallels the interpretation of *indexical* sign quality of music (both are related to the recognition of an inner state of an individual). Moreover, *coordination* of movements is related to the physical activation from which *physical musicogenic meaning* can emerge. *Cooperation* parallels the co-operated activity of syntactic features (e.g., of melody, harmony, metre, and rhythm) during *syntactic integration*. Cooperation leads to increased *social cohesion* which, in turn, is related to increased health of the individuals of a group (which relates to the effects of music perception on the *immune system*). Moreover, the inter-individual nature of social relations implied by social cohesion parallels *large-scale* structural relations, and the inter-individual liking component of social cohesion parallels *personal musicogenic meaning*.

music perception	feature extraction	Gestalt formation	interval analysis	structure building	structural re-analysis	vitalization	premotor, immune system
syntactic processing	element extraction	knowledge-free structuring	musical expectancy formation	structure building	structural re-analysis	syntactic integration	large-scale structuring
musical meaning	iconic	indexical	symbolic	intra-musical	physical	emotional	personal
social functions	contact	social cognition	co-pathy	communi-cation	coordina-tion	cooper-ation	social cohesion

12.3 From social contact to spirituality – The Seven Cs

Music making is an activity involving several social functions. The ability, and the need, to engage in these social functions is part of what makes us human, and emotional effects of engaging in these functions include experiences of reward, fun, joy and happiness. Exclusion from engaging in these functions represents an emotional stressor, leads to depression, and has deleterious effects on health and life expectancy (Cacioppo & Hawkley, 2003).[8] Therefore, engaging in such social functions is important for the survival of the individual and the species. These functions can be categorized into seven areas (summarized, and related to other domains, in Table 12.1).

(1) When individuals make music, they come into **contact** with each other. Being in contact with other individuals is a basic need of humans (as well as of numerous other species; Harlow, 1958), and social isolation is a major risk factor for morbidity as well as mortality (House *et al.*, 1988; Cacioppo & Hawkley, 2003). As will be outlined in Section 12.6.3, a plausible hypothesis is that social

[8] Negative mood effects are described in the phenomenon of 'Appression' by Siebel & Winkler (1996), Section 9.6.5 in that book.

isolation results in damage of the hippocampal formation and that, on the other hand, contact with other individuals promotes hippocampal integrity. With regard to the model of music perception (summarized in the top row of Table 12.1), social *contact* is the prerequisite for the participation in the other social functions, analogous to *auditory feature extraction*, which is a prerequisite for the other processes of music perception.

(2) Music automatically engages **social cognition**. During music listening, individuals automatically engage processes of mental state attribution ('mentalizing', or 'adopting an intentional stance'), in an attempt to figure out the intentions, desires, and beliefs of the individuals who actually created the music (also often referred to as establishing a 'theory of mind', TOM). As described in Section 10.2.2, a study by Steinbeis & Koelsch (2008c) showed that listening to music automatically engages brain structures dedicated to social cognition (i.e., a network dedicated to mental state attribution in the attempt to understand the composer's intentions). These processes are also required when making music together in a group, for example, when varying tempo and/or loudness, during improvisation, etc. Interestingly, individuals with Autistic Spectrum Disorder (ASD) seem to be surprisingly competent in social cognition in the musical domain (in contrast to their problems with social cognition in other social contexts; see also Allen *et al.*, 2009). This supports the notion that music therapy can aid the transfer of socio-cognitive skills in the musical domain to non-musical social contexts in individuals with ASD. With regard to the domain of musical meaning (summarized in the third row of Table 12.1), *social cognition* parallels the interpretation of *indexical* sign quality of music, because both social cognition and the interpretation of indexical extra-musical sign quality are related to the recognition of an inner state of an individual.

(3) Music making can engage **co-pathy** in the sense that inter-individual emotional states become more homogeneous (e.g., reducing anger in one individual, and depression or anxiety in another), thus decreasing conflicts and promoting cohesion of a group (Huron, 2001). With regard to positive emotions, for example, co-pathy can increase the well-being of individuals during music making or during listening to music.[9] The term 'co-pathy' is used here (instead of 'empathy') because co-pathy refers to the social function of empathy. Moreover, empathy has many different connotations, due to various definitions of empathy provided by different researchers. By using the term *co-pathy* I do not only refer to the phenomenon of thinking what one *would* feel if one were in someone else's position. Instead, I refer to the phenomenon that one's own emotional state is actually affected in the sense that it occurs when one perceives (e.g., observes or hears), or imagines, someone else's affect, and that this perception or imagination evokes a feeling in the perceiver which bears strong congruency with what the other individual is feeling (for a review of the concept of empathy see Singer &

---

[9]   For a study showing an increase of positive mood due to music making in a group see Koelsch *et al.* (2010a).

Lamm, 2009). Co-pathy should be differentiated from (a) *mimicry* (a low-level perception–action mechanism of imitating another individual's emotional expression, e.g., contraction of the musculus zygomaticus major when looking at a smiling face),[10] and (b) *emotional contagion* (a short-term spread of a behaviour which is presumably a precursor of co-pathy; e.g., children laughing because other children laugh).[11] Both mimicry and emotional contagion contribute to co-pathy. They may occur outside of awareness, and do not require a self/other concept. By contrast, co-pathy requires self-awareness and self/other distinction, i.e., the capability to make oneself aware that the affect may have been evoked by music made by others, although the actual source of one's emotion lies within oneself. Moreover, co-pathy should be differentiated from (c) *sympathy*, *empathic concern*, and *compassion*, which do not necessarily involve shared feelings (e.g., feeling pitiful for a jealous person, without feeling jealous oneself; for details see Singer & Lamm, 2009).

(4) Music always involves **communication** (notably, for infants and young children, musical communication during parent–child singing of lullabies and play-songs appears to be important for social and emotional regulation, as well as for social, emotional, and cognitive development; Trehub, 2003; Fitch, 2006). As described in Chapters 9 and 10, neuroscientific and behavioural studies revealed considerable overlap of the neural substrates and cognitive mechanisms underlying the perception of music as well as of language, with regard to syntax, and with regard to the processing of meaning. With regard to speech and music production, a study by Callan *et al.* (2006) also showed a strong overlap of the neural substrates of speaking and singing. Because music is a means of communication, particularly active music therapy (in which patients make music) can be used to train skills of (nonverbal) communication (Hillecke *et al.*, 2005).

(5) Music making also involves **coordination** of actions. This requires individuals to synchronize to a beat, and to keep a beat. The coordination of movements in a group of individuals appears to be associated with pleasure (for example, when dancing together), even in the absence of a shared goal (apart from deriving pleasure from concerted movements; see also Huron, 2001). Interestingly, a study by Kirschner & Tomasello (2009) reported that children as young as 2.5 years synchronized more accurately to an external drum beat in a social situation (i.e., when the drum beat was presented by a human play partner) compared to nonsocial situations (i.e., when the drum beat was presented by a drumming machine, or when the drum sounds were presented via a loudspeaker). This effect might have originated from the pleasure that emerges when humans coordinate their movements with each other (see also Overy & Molnar-Szakacs, 2009; Wiltermuth & Heath, 2009). The capacity to synchronize movements to an external beat appears to be uniquely human among primates, although other mammals

---

[10]   For a study on EEG correlates of emotional mimicry during viewing facial expressions see, e.g., Achaibou *et al.* (2008).

[11]   Some (e.g., Hatfield *et al.*, 2009) assume that mimicry, in turn, is a component of contagion.

(such as seals) and some song birds (such as cockatoos) might also possess this capacity. A current hypothesis (Patel, 2006, 2008) is that this capacity is related to the capacity of vocal learning, which might depend (in mammals) on a direct neuronal connection between the motor cortex and the nucleus ambiguus. The nucleus ambiguus is located in the brainstem and contains motor neurons innervating the larynx (the motor cortex also directly projects to brainstem nuclei innervating the tongue, jaw, palate, and lips; e.g., Jürgens, 2002). Coordination of movements between individuals also increases likelihood of future cooperation between these individuals (even if future situations of cooperation require personal sacrifice; Wiltermuth & Heath, 2009). With regard to the dimensions of musical meaning (third row of Table 12.1), coordination of movements is related to the physical activation from which *physical musicogenic meaning* can emerge.

(6) A convincing musical performance by multiple players is only possible if it also involves **cooperation** between players. Cooperation implies a shared goal, and engaging in cooperative behaviour is an important potential source of pleasure. For example, Rilling *et al.* (2002) reported an association between cooperative behaviour and activation of a reward network including the nucleus accumbens (NAc). Cooperation between individuals increases inter-individual trust, and increases the likelihood of future cooperation between these individuals. It is worth noting that only humans have the capability to communicate about coordinated activities in order to cooperatively achieve a joint goal (Tomasello *et al.*, 2005). With regard to the dimensions of syntactic processes (second row of Table 12.1), cooperation parallels the co-operated activity of syntactic features (melody, harmony, metre, etc.) during *syntactic integration*.

(7) As an effect, music leads to increased **social cohesion** of a group (Cross & Morley, 2008). A wealth of studies show that humans have a 'need to belong' and a strong motivation to form and maintain enduring interpersonal attachments (Baumeister & Leary, 1995). Meeting this need increases health and life expectancy (Cacioppo & Hawkley, 2003). Social cohesion also strengthens the confidence in reciprocal care (see also the caregiver hypothesis; Trehub, 2003; Fitch, 2005), and the confidence that opportunities to engage with others in the mentioned social functions will also emerge in the future. Social cohesion is related to increased health of the individuals of a group (Cacioppo & Hawkley, 2003). This relates social cohesion to the effects of music perception on the *immune system* (first row of Table 12.1). Moreover, with regard to the dimension of musical meaning (third row of Table 12.1), the inter-individual liking component of social cohesion parallels *personal musicogenic meaning*. Finally, with regard to syntactic processes (second row of Table 12.1), the systemic nature of inter-individual relations implied by social cohesion parallels the systemic nature of *large-scale* structural relations.

Although it should be clearly noted that music can also be used to manipulate other individuals, and to support non-social behaviour (e.g., Brown & Volgsten, 2006), music is still special – although not unique – in that it can engage all of these social functions at the same time (similar, e.g., to cooperative forms of play). This is presumably one explanation for music's potential to evoke strong emotions. Therefore, music *does* serve the goal of fulfilling social needs (the human need to be in contact with others, to belong, to communicate, etc.). In this regard, music-evoked emotions are related to survival functions and to functions that are of vital importance for the individual (for a discussion on the role of other factors, such as sexual selection, for the evolution of music see Huron, 2001; Fitch, 2005).

It is also worth mentioning that the experience of engaging in these social functions, along with the experience of emotions evoked by participating in social functions, is a *spiritual experience* (such as the experience of communion; note that neither 'spiritual' nor 'communion' are used here as religious terms). This is probably one reason why religious practices usually involve music.

Engaging in social functions during music making evokes activity of neural 'reward circuits',[12] and we (Koelsch *et al.*, 2010) have previously suggested that activity of these reward circuits is subjectively experienced as 'fun' (see also Section 12.6.2). In addition to experiences of fun, music making can also evoke attachment-related emotions (due to the engagement in the mentioned social functions), such as love, joy and happiness. As will be described in Section 12.6.3, the latter emotions presumably involve activity of the hippocampal formation. In this regard, music can not only be fun, it can also make people happy. As will be described in more detail further below, the capacity of music to evoke such emotions is an important basis for beneficial biological effects of music, and thus for the use of music in therapy.

12.4 Emotional responses to music – underlying principles

The notion of 'mechanisms' that 'induce' emotions with music (e.g., Scherer, 2004; Juslin & Västfjäll, 2008) gives the impression that specific musical antecedents determine a specific emotional effect. This, however, does not seem to be the case (otherwise, depressive patients could easily be healed with happy music). Therefore, instead of using the term 'mechanism', the term 'principle' will be used in the following. Likewise, the term 'music-evoked emotion' will be used (instead of 'music-induced emotion'), to emphasize that some emotional effects cannot be caused (or intended) in a deterministic way. In the following, several principles underlying the evocation of emotion with music (as well as relations to music perception, syntactic processes, musical meaning, and social functions) will be suggested (see also Table 12.2). These principles include, and partly expand, the mechanisms underlying the evocation of emotions with music suggested by Juslin & Västfjäll (2008).

---

[12] These circuits include projections from the lateral hypothalamus via the medial forebrain bundle to the mesolimbic dopamine pathway. The mesolimbic dopamine pathway includes projections of dopaminergic neurons located in the ventral tegmental area (VTA) to the nucleus accumbens (NAc). Details of this pathway will be reported in Section 12.6.2.

Table 12.2 Synopsis of principles underlying the evocation of emotions with music (bottom row), in relation to the processes of music perception, syntactic processing, musical meaning, intra-musical features, and social functions. Processes of *evaluation* that elicit emotions can happen at the earliest stages of *feature extraction* (therefore listed in the outermost left dimension). Emotional *contagion* is related to *social cognition* and the processing of *indexical* sign quality. Note that contagion is a precursor of full-blown empathic responses whose social function is *co-pathy* (therefore, contagion is listed left of co-pathy). Musical information can activate an episodic *memory* representation which, in turn, can evoke an emotion. In this case, the musical information has (usually idiosyncratic) *symbolic* sign quality. Moreover, in the case of musical information with (more general) *symbolic* sign quality (semantic memory), musical information might evoke a concept with emotional valence (which might lead to an emotional response). With regard to syntactic processes, long-term memory is also related to the (implicit) knowledge required for *musical expectancy formation*. The emotion principle of *expectancy* is related here to the *communication* of *intra-musical* meaning, that is, to intra-musical features that give rise to meaning, as well as to emotional responses. *Imagination* is related to the resourceful processes of *structural reanalysis and revision*. Engaging in *social functions* has positive effects on human health, related to the potentially regenerative effects of music perception (and music making) on the *immune system*.

music perception	feature extraction	Gestalt formation	interval analysis	structure building	structural re-analysis	vitalization	premotor, immune system
syntactic processing	element extraction	knowledge-free structuring	musical expectancy formation	structure building	structural re-analysis	syntactic integration	large-scale structuring
musical meaning	iconic	indexical	symbolic	intra-musical	physical	emotional	personal
intra-musical	build-up	stability	extent	breach	post-breach	resolution	large-scale
social functions	contact	social cognition	co-pathy	communi-cation	coordin-ation	cooper-ation	social cohesion
emotion principles	evaluation	contagion	memory	expectancy	imagina-tion	under-standing	social functions, aesthetics

(1) Conceptually, 'brain stem reflexes' (Juslin & Västfjäll, 2008) to music are the result of an *evaluation* (on the level of the brainstem; see also Chapter 1). Other evaluative processes may occur on a number of other levels: For example, Scherer (2001) noted that evaluative processes can occur on a sensory-motor, a schematic, and a conceptual level (Scherer, 2001, p. 103).[13] Note that evaluative processes can be (a) automatic and non-cognitive (e.g., evaluative processes occurring on the level of the brainstem or the thalamus; see also Chapter 1), (b) automatic and cognitive, but without awareness (processes on the level of the

---

[13] The sensory-motor level represents reflex-systems responding to stimuli that are innately preferred or avoided. The schematic level includes learned preferences/aversions, and the conceptual level includes recalled, anticipated, or derived positive–negative estimates.

orbitofrontal cortex; Siebel *et al.*, 1990), or (c) cognitive with involvement of conscious awareness (processes on the level of the neocortex). On each of these levels, several evaluative processes can be carried out. In his *Sequential Check Theory of Emotion Differentiation*, Scherer (2001) proposed several *sequential checks* underlying the evaluation (appraisal) of stimuli.[14]

Evaluative processes are major antecedents for emotions (Scherer, 2001), and are therefore considered here as one principle underlying the evocation of emotion in response to music. Scherer & Zentner (2001) have outlined a number of appraisal processes with regard to music (referred to as 'production rules' by the authors). These appraisal processes are determined by the musical structure, the quality of the performance, the expertise and current mood or motivational state of the listener, as well as by contextual features such as location and the form of the event. With regard to the model of music perception (summarized in the top row of Table 12.2), processes of *evaluation* can happen at the earliest stages of *feature extraction*. Therefore, *evaluation* is listed in the outermost left dimension of Table 12.2.

(2) Another principle is emotional ***contagion***, i.e., the evocation of an emotion due to an individual perceiving an emotional expression (facial, vocal, gestural, and/or postural), and then copying this expression internally in terms of motor expression and physiological arousal. For instance, music might express joy (due to faster tempo, large pitch variation, etc.), and this expression is copied by the listener in terms of smiling, (covert or overt) vocalization, and/or bouncing. The (peripheral) feedback of these motor acts (and related physiological changes) then evokes an emotion (e.g., Hatfield *et al.*, 1993). As mentioned above, contagion may contribute to full-fledged co-pathic phenomena (also involving self-awareness and self/other distinction). With regard to *social functions* (fifth row of Table 12.2), emotional contagion is related to *social cognition*. With regard to *musical meaning* (third row of Table 12.2), emotional contagion is related to the processing of *indexical* sign quality. Moreover, because contagion is a precursor of full-blown empathic responses whose social function is *co-pathy*, contagion is listed left of co-pathy in Table 12.2.

(3) Emotions, and stimuli associated with emotions, can be memorized. With regard to music, a musical stimulus might be associated with a memory of an autobiographical event, and the perception of the music might evoke an emotional memory representation leading to an emotional response (Juslin's 'episodic memory' mechanism; for an fMRI study on music-evoked autobiographical memories and emotional effects see Janata, 2009). In this case, the musical information has (usually idiosyncratic) *symbolic* sign quality (see third row of Table 12.2). Musical information with (more general) symbolic sign quality (semantic *memory*), might evoke a concept with emotional valence, which in turn might also lead to an

---

[14]   These checks include relevance detection (including a novelty check), implication assessment, coping potential determination, and normative significance evaluation. Note that some of these checks can only be performed by cortical structures, such as normative significance evaluation.

emotional response (similar to emotional responses to the affective valence of a word; Võ *et al.*, 2009).

Moreover, a musical stimulus might repeatedly be paired with a certain emotion elicited by another stimulus, so that music can become a conditioned stimulus and trigger an emotional response (Juslin's 'evaluative conditioning'). Although episodic memory, semantic memory, and evaluative conditioning involve different learning processes, and rely on different neural correlates,[15] they are all (long-term) memory functions, and thus categorized here under a *memory* principle. Note that, with regard to syntactic processes (second row of Table 12.2), long-term memory parallels the (implicit) knowledge required for *musical expectancy formation*.

(4) The principle of musical *expectancy* states that the build-up, fulfilment, and violation of expectancies has emotional effects (such as surprise, tension, suspense, or relaxation). The next section will deal with this principle in more detail. Here, it is important to note that, in addition to emotional effects due to musical expectancy, emotional effects due to a 'mere exposure' (Moors & Kuppens, 2008) are also related to predictive processes; although the mechanisms underlying the *mere exposure effect* are unclear, it appears that the ability to predict upcoming events (thus facilitating processing fluency) correlates with perceived pleasure (Armstrong & Detweiler-Bedell, 2008). Complete predictability, on the other hand, can easily lead to boredom.[16] The emotion principle of *expectancy* is related here to the *communication* of *intra-musical* meaning, that is, to intra-musical features that give rise to meaning as well as to emotional responses (see also Table 12.2).

(5) Another principle is *imagination*, which refers to emotional effects of being resourceful, inventive, curious, or creative, and to emotional effects of trying something out. The imagination principle is related to Juslin's principle of 'visual imagery' (note, however, that visual imagery can also lead to processes of evaluation, for example because imagined objects or scenes usually have an emotional valence). With regard to the processes of music perception and syntactic processes (first and second row of Table 12.2), the imagination principle is related to *structural reanalysis and revision*, which often involves the (resourceful) establishment of possible new continuations of a musical sequence.

(6) Emotional effects also arise from *understanding*. With regard to music, an individual might understand an extra-musical meaning, the (intra-musical) meaning of a musical structure, the 'logic of musical ideas and their progress' (Davies, 1994, p. 48), the musical discourse, etc. Perlovsky (2007) argued that

---

[15] See, e.g., LeDoux (2000) for neural correlates of evaluative conditioning, and Platel *et al.* (2003) for possible neural correlates of a semantic and an episodic musical memory. See Groussard *et al.* (2010) for a comparison between a semantic musical memory, and a semantic language memory.

[16] The explanation of the mere exposure effects in terms of classical conditioning (assuming 'that the absence of aversive events constitutes the unconditioned stimulus'; Zajonc, 2001) is weak, because this does not explain the decrease of preference following over-exposure.

humans (and perhaps other species as well) have an inborn need to understand (or 'make sense of') how elements of contexts, or structures, are synthesized into coherent entities. This need is referred to as the *knowledge instinct* by Perlovsky (2007).[17] The fulfilment of this need to understand is experienced as rewarding (the 'aha moment', or 'eureka moment'), and presumably involves activity of the dopaminergic reward pathway, although this remains to be specified empirically (Section 12.6.2 will describe details on neural correlates of music-evoked feelings of reward). With regard to the *intra-musical* features that give rise to meaning and emotion, *understanding* is related to the *resolution* of a structural breach (a sequence with a structural breach can only be fully understood once it is resolved).

(7) Finally, as described in Section 12.3, *engaging in social functions* during music making (or during listening to music) may also elicit emotional responses. In Section 12.3, it was mentioned that engaging in social functions has positive effects on human health, which relates this emotion principle to the potentially regenerative effects of music perception (and music making) on the *immune system* (first row of Table 12.2). Note that such regenerative effects only emerge in the absence of violence. Thus, regenerative effects of engaging in social functions are inherently linked to experiences of *beauty*, and thus to *aesthetic experience* (Siebel *et al.*, 1990, posited that a defining feature of *beauty* is the absence of violence).[18,19] Beyond the aesthetic experience of social functions during music making (and in part also during music listening), it is assumed that the beauty of musical sounds, contents, and structures can also evoke emotions (for example, Kivy, 1999, argued that 'music moves us emotionally by its sheer musical beauty').[20]

Note that it is often difficult to investigate one particular emotion-evoking mechanism, or principle, because usually several principles are at work at the same time (making it difficult to tease apart emotional effects evoked by different principles). This might be one reason why only few studies have so far investigated the neural correlates of the principles, or mechanisms, underlying the evocation of emotion with music (see also next sections). However, one exception in this regard is the principle of musical expectancy, which will be dealt with in the next section.

12.5   Musical expectancies and emotional responses

As already mentioned in Chapter 10, Leonard Meyer (1956) proposed that emotions can be evoked on the basis of fulfilled or suspended musical expectancies. He proposed that the confirmation or violation of such musical expectancies

---

[17]   Perlovsky (2007) also argues that meaning emerges in part from such understanding.

[18]   Siebel *et al.* (1990) also posited that the *aesthetic* view asks '*how* something is beautiful' (as opposed to asking *if* something is beautiful).

[19]   A study by Istók *et al.* (2009) reported that Finnish students associated the 'aesthetic value of music' most strongly with the adjectives *beautiful* and *touching*.

[20]   Studies on the empirical aspects of aesthetics are extremely sparse. For neuroscience studies approaching this issue by investigating neural correlates of judgements of beauty see Jacobsen *et al.* (2006) and Müller *et al.* (2010).

produces emotions in the listener (for a more recent account see Huron, 2006). In accordance with this proposal, Sloboda (1991) found that specific musical structures were associated with specific psycho-physiological reactions (shivers, for example, were often evoked by new or unexpected harmonies).

A study by Steinbeis *et al.* (2006) tested the hypothesis that emotional responses can be evoked by unexpected chord functions. In that study, physiological measures including EEG, skin conductance responses (SCR), and heart rate were recorded while subjects listened to three versions of Bach chorales. One version was the original version composed by Bach with a harmonic sequence containing an irregular chord function (e.g., a deceptive cadence with a submediant instead of a tonic). In another version, that chord was rendered regular (e.g., by replacing the submediant with a tonic), and in a third version this chord was rendered very irregular (e.g., by replacing the submediant with a Neapolitan sixth chord). The SCRs elicited by regular (expected) chords clearly differed from those elicited by irregular (unexpected) chords.[21] Because the SCR reflects activity of the sympathetic branch of the autonomic nervous system, and because this system is intimately linked to emotional experiences, these data clearly corroborate the assumption that unexpected harmonies elicit emotional responses. The findings of the study by Steinbeis *et al.* (2006) were replicated in a study by Koelsch *et al.* (2008b), which also obtained behavioural data showing that irregular chords were perceived by listeners as more surprising, more arousing, and less pleasant, than regular chords.

Corroborating these findings, functional neuroimaging experiments using chord sequences with unexpected harmonies (originally designed to investigate music-syntactic processing; Koelsch *et al.*, 2005a; Tillmann *et al.*, 2006) showed activity changes in response to the unexpected chords in the amygdala (Koelsch *et al.*, 2008a), and the orbitofrontal cortex (Tillmann *et al.*, 2006; Koelsch *et al.*, 2005a). The orbitofrontal cortex (OFC), comprising the Brodmann areas 11, 47, and partly 10, is a paralimbic structure that plays an important role for a range of emotional processes such as the evaluation of 'breaches of expectation' (Nobre *et al.*, 1999), and the evaluation of the emotional significance of sensory stimuli (for reviews see Mega *et al.*, 1997; Rolls & Grabenhorst, 2008). Thus, in addition to the *expectancy principle*, irregular chord functions presumably also evoke emotional responses due to the *evaluation principle*.

Note that the findings of the studies reported above show that unexpected musical events do not only elicit responses related to the processing of the structure of the music, but also emotional responses (this presumably also holds for unexpected words in sentences, and any other stimulus which is perceived as more or less expected). Thus, research using stimuli that are systematically more or less expected should ideally assess the valence and arousal experience of the listener (even if an experiment is not originally designed to investigate emotion), so that these variables can potentially be used to explain variance in the data.

---

[21]   SCRs also differed between unexpected and very unexpected chords.

12.5.1   The tension-arch

Chapter 10 described structural principles that give rise to intra-musical meaning (listed in Table 12.3). These principles are also relevant for the elicitation of emotional phenomena. For example, as described in the previous section, a breach of expectancy might lead to surprise and an increase in tension. This increased tension extends for the post-breach period until the breach is resolved. During this post-breach structure, listeners familiar with tonal music anticipate a resolution. The resolution is perceived as relaxing, pleasurable, and rewarding. A reasonable working hypothesis is that the dorsal striatum is involved in emotional activity due to anticipation: In the study by Koelsch *et al.* (2008a) this region was activated during blocks of chord sequences with irregular chords (evoking the anticipation for resolution), and a study by Salimpoor *et al.* (2011) showed dopaminergic activity in this region while listeners anticipated a music-evoked frisson (an intensely pleasurable experience often involving goosebumps or shivers down neck, arms, or spine). The anticipated, and rewarding, frisson itself evoked dopaminergic activity in the ventral striatum, presumably the nucleus accumbens. Thus, another working hypothesis is that the pleasurable and rewarding experience of the resolution of a breach of expectancy involves activity of the mesolimbic dopaminergic reward pathway including the nucleus accumbens.

The increase and decrease in tension during the build-up of expectancy, the breach of expectancy, the anticipation for resolution, and the resolution is illustrated schematically by the *tension-arch* in Table 12.3. The increase and decrease in tension has been modelled, and investigated empirically, using the tonal pitch space theory (e.g., Lerdahl, 2001b; Bigand *et al.*, 1996; Lerdahl & Krumhansl, 2007), but note that the underlying fine-structure of emotional activity, including its neural correlates, cannot be grasped adequately by one-dimensional tension values. For example, the tension value of a tonic chord at the beginning of a harmonic sequence with a structural breach is identical to the tension value of a tonic chord at the end of a sequence (both tonic chords have low tension values). However, the underlying affective phenomenon is different (build-up vs. resolution), and so are the neuro-affective correlates (presumably involving, for example, reward-related dopaminergic activity during the resolution of a harmonic expectancy violation).

Particularly in Western music, numerous tension-arches are usually interweaved into large-scale tension archs. For example, imagine a simple sonata form with

Table 12.3   The tension arch and its relation to the principles of structural phenomena that also give rise to musical meaning.

intra-musical	build-up	stability	extent	breach	post-breach	resolution	large-scale

one tension arch spanning the first theme, one the transition, one the second theme, and one the codetta. These four archs are spanned by an open arch (thus spanning the entire exposition, the arch is somewhat open because, although the exposition comes to an end, it does not close with the tonic). Development and recapitulation also comprise several arches, the entire first movement is spanned by a tension arch, and so might be the entire sonata or symphony. The large-scale construction of tension-arches is a major component of the aesthetic, and thus emotional, experience of a musical piece. There is a lack, however, of empirical studies investigating how the interweaving of tension-arches contributes to the aesthetic experience of music, and the neural correlates of the emotional effects of musical tension-arches are not well known.

12.6   Limbic and paralimbic correlates of music-evoked emotions

So far, most neuroscientific studies on music and emotion had the primary aim to investigate neural correlates of emotion, rather than to investigate a particular mechanism, or principle, underlying the evocation of that emotion. For example, some studies investigated pleasantness/unpleasantness (Blood *et al.*, 1999; Gosselin *et al.*, 2006; Koelsch *et al.*, 2006; Sammler *et al.*, 2007; Ball *et al.*, 2007; Khalfa *et al.*, 2008), emotional responses to music paired with images (Baumgartner *et al.*, 2006a,b) or film clips (Eldar *et al.*, 2007), neural correlates of music-evoked sadness (Mitterschiffthaler *et al.*, 2007), or of intensely pleasurable experiences during music listening (Blood & Zatorre, 2001; Salimpoor *et al.*, 2011). The following sections will provide an overview of these studies, relate them to the emotion principles, and describe how these studies advance the understanding of the neural correlates of emotion in general.

The first functional neuroimaging study on music and emotion was a study by Anne Blood and colleagues (Blood *et al.*, 1999). Using PET, they investigated the emotional dimension of pleasantness/unpleasantness with sequences of harmonized melodies. The stimuli varied in their degree of (continuous) dissonance, and were perceived as less or more unpleasant (stimuli with the highest degree of continuous dissonance were rated as the most unpleasant). Stimuli were presented under computerized control without musical expression. Therefore, it is likely that the primary principle underlying the evocation of emotional responses was *evaluation* (on the *sensory-motor* level due to the sensory dissonance, and on the *schematic*, possibly also on the *conceptual* level, due to the cultural experience of listeners; see also section 1.4 and Fritz *et al.*, 2009). Variations in pleasantness/unpleasantness modulated activity in the (posterior) subcallosal cingulate cortex, as well as in a number of paralimbic structures: Increasing unpleasantness correlated with activations of the (right) parahippocampal gyrus, while decreasing unpleasantness of the stimuli correlated with activations of frontopolar and orbitofrontal cortex (for anatomical illustrations see Figure 12.1).

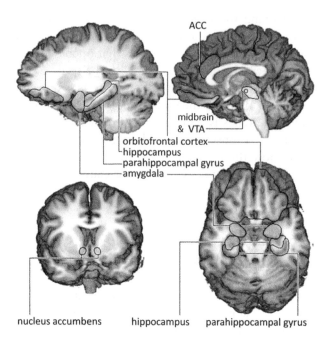

Figure 12.1 Illustration of some limbic/paralimbic structures, of which activity modulation due to music-evoked emotions was reported in functional neuroimaging studies (top left: View of the right hemisphere; top right: Medial view; bottom left: Rostral view; bottom right: Ventral view). VTA: Ventral tegmental area (approximate location is indicated by the circle). ACC: Anterior cingulate cortex.

In another PET study, Blood & Zatorre (2001) investigated neural correlates of intensely pleasurable responses to music involving, e.g., goosebumps and shivers down neck, arms, or spine. As mentioned above, this emotional experience is also referred to as *musical frisson* (e.g., Levinson, 2004), other researchers used terms such as *chills* (e.g., Panksepp, 1995; Grewe *et al.*, 2007b), *thrills* (e.g., Goldstein, 1980; Sloboda, 1991), or *skin orgasm* (Panksepp, 1995). However, Huron (2006) noted that listeners can find music 'thrilling' without necessarily experiencing goosebumps, and that the term 'chills' is best reserved for the phenomenological feeling of coldness, which often, but not necessarily, accompanies a frisson. Frissons can presumably be evoked by each of the principles listed in Section 12.4, although it seems likely that frissons are usually the effect of the simultaneous workings of several of these principles.[22,23]

---

[22]   For overviews see Sloboda (1991), Panksepp & Bernatzky (2002), Huron (2006), and Grewe *et al.* (2007b).

[23]   The principles underlying the evocation of frissons in the study by Blood & Zatorre (2001) were not addressed by the authors.

In the study by Blood & Zatorre (2001), frissons were evoked when participants were presented with a piece of their own favourite music (using normal CD recordings; as a control condition, participants listened to the favourite piece of another participant). Increasing intensity of frissons correlated with increases in rCBF in the insula, orbitofrontal cortex, the ventral medial prefrontal cortex, and the ventral striatum. Also correlated with increasing intensity of frissons were decreases in rCBF in the amygdala and the (anterior) hippocampal formation.[24] Thus, in that study, activity changes were observed in core structures of the limbic/paralimbic system (e.g., amygdala and hippocampal formation).

Although not well defined, limbic/paralimbic structures are considered as core structures of emotional processing, because their lesion or dysfunction is associated with emotional impairment (for an overview see, e.g., Dalgleish, 2004).[25] Moreover, these structures play a critical role for emotions that are assumed by some to have survival value for the individual and for the species (Dalgleish, 2004).[26] The study by Blood & Zatorre (2001) provided evidence that music can evoke activity changes in these brain structures, suggesting that at least some music-evoked emotions involve the very core of evolutionarily adaptive neuro-affective mechanisms, thus supporting the view that music can evoke 'real emotions'. Moreover, the finding that amygdala activity was modulated by music was important because affective disorders, such as depression and pathological anxiety, are related to amygdala dysfunction (Drevets et al., 2002; Stein et al., 2007), and the influence of music perception on amygdala activity strengthened the empirical basis for music-therapeutic approaches for the treatment of affective disorders.

The findings of limbic activations during listening to music were corroborated by fMRI experiments. An fMRI study by Koelsch et al. (2006a) used pleasant and unpleasant musical stimuli (similar to the study by Blood et al., 1999).[27] In contrast to the study by Blood et al. (1999), the pleasant musical excerpts were not computerized sounds, but joyful instrumental tunes played by professional musicians. Unpleasant stimuli were continuously dissonant counterparts of the original musical excerpts. The principles underlying the evocation of emotion in that study were thus presumably mainly *evaluation* (similarly to the study by Blood et al., 1999), and *contagion* (due to the expression of joy in the pleasant music). Unpleasant music elicited increases in BOLD signals in the amygdala, the

---

[24] For patient studies on music-evoked pleasure see Griffiths et al. (2004), Stewart et al. (2006), and Matthews et al. (2009).

[25] How limbic/paralimbic structures (such as amygdala, hippocampus, parahippocampal gyrus, temporal poles, thalamus, hypothalamus, midbrain, nucleus accumbens, anterior cingulate cortex, insular cortex, orbitofrontal cortex, etc.) interact, and which functional networks they form, is still not well understood.

[26] For example, the amygdala has been implicated in the initiation, generation, detection, maintenance, and termination of emotions that are assumed to be important for the survival of the individual and the species (Price, 2005).

[27] For further studies using consonant and dissonant stimuli see Ball et al. (2007), Sammler et al. (2007), and Gosselin et al. (2006).

(anterior) hippocampal formation, the parahippocampal gyrus, and the temporal poles (a decrease of BOLD signal was observed in these structures in response to the pleasant music). During the presentation of the pleasant music, an increase of BOLD signal was observed in the ventral striatum and the insula (and in cortical structures such as the auditory cortex and Broca's area). None of the participants reported experiences of musical frissons, indicating that activity changes in the amygdala, the hippocampal formation, and the ventral striatum can be evoked by music even when individuals do not experience frissons.

Activity changes in the amygdala in response to music were also reported in an fMRI study by Ball et al. (2007). That study used original (mainly consonant) piano pieces as pleasant stimuli, and electronically manipulated, continuously dissonant versions of these stimuli as unpleasant stimuli (similar to the study by Koelsch et al., 2006). Interestingly, signal changes in the amygdala in response to both consonant and dissonant musical stimuli were positive in a central region of the amygdala (referred to as *laterobasal* group by the authors), and negative in a dorsal region of the amygdala (referred to as *centromedial* group by the authors).[28] This indicates that different subregions of the amygdala show different response properties to auditory (musical) stimulation. No signal difference was observed in the amygdala between the consonant and the dissonant music conditions, although participants clearly rated the consonant pieces as more pleasant.[29]

The findings of different response properties in the amygdala reported by Ball et al. (2007) were corroborated by an fMRI study by Fritz & Koelsch (2005). That study used short pleasant musical excerpts and their manipulated (continuously dissonant) unpleasant counterparts (similar to the study by Koelsch et al., 2006a).[30] Two separate anatomical regions of the amygdala were shown to be selectively involved in the perception of stimuli with positive and negative valence. A correlation of BOLD signals with *decreasing* valence was observed within a central aspect of the amygdala (presumably lateral and/or basal nuclei), whereas a correlation of BOLD signal with *increasing* valence was observed in a dorsal aspect of the amygdala (including the substantia innominata). A functional connectivity analysis with seed voxels in these dorsal and central aspects of the amygdala revealed two different networks displaying BOLD signal synchronicities with the respective

---

[28]   Notably, the amygdala is not an anatomical unity, but is composed of several distinct groups of cells. These are usually referred to as the lateral, basal, and accessory basal nuclei (which are often collectively termed the basolateral amygdala), as well as of several surrounding structures, including the central, medial, and cortical nuclei. These surrounding structures, together with the basolateral amygdala, are often referred to as 'the amygdala', although the amygdala is clearly not a functional unity (Ball et al., 2007; Davis & Whalen, 2001; Fritz & Koelsch, 2005). Although the amygdala has become one of the most heavily studied brain structures, the functional significance of these nuclei, as well as their interaction with other structures, is not well understood (LeDoux, 2007).

[29]   Perhaps because the consonant pieces were not all happy dance tunes, as in the study by Koelsch et al. (2006a).

[30]   In additional conditions, all stimuli were also played backwards in the study by Fritz & Koelsch (2005).

amygdala regions. The central aspect of the amygdala (involved in the processing of stimuli with negative emotional valence) was functionally connected to the temporal pole, the hippocampus, and the parahippocampal gyrus. The coordinates of these activations were virtually identical to those observed in the study by Koelsch *et al*. (2006). Notably, the dorsal aspect of the amygdala (involved in the processing of stimuli with positive emotional valence) was functionally connected with the ventral striatum and the orbitofrontal cortex. The functional connectivity between these regions parallels the anatomical connections between them (these connections are illustrated in Figure 12.2). These results suggest that different aspects of the amygdala regulate activity in at least two networks of emotional processing: One network involved in the processing of stimuli with positive, and one in the processing of stimuli with negative, emotional valence.

An fMRI study by Baumgartner *et al*. (2006b) investigated emotional responses to negative (fearful or sad) pictures[31] and to pictures presented together with fearful or sad music. Thus, the principles underlying the evocation of emotion in that study were presumably *evaluation*, *contagion* and *imagination* (due to the imagination of what might happen, or what might have happened, to the individuals shown in the pictures). Brain activations were stronger during the combined presentation of pictures and music than during the presentation of pictures alone. For example, activation of the amygdala was only observed in the combined condition, but not in the condition where only pictures were presented. The combined presentation also elicited stronger activation in the hippocampal formation, the parahippocampal gyrus, and the temporal poles. As mentioned above, the network comprising amygdala, hippocampal formation, parahippocampal gyrus, and temporal poles was also observed in the studies by Koelsch *et al*. (2006) and Fritz & Koelsch (2005), suggesting that this network plays a consistent role in the emotional processing of music.

Similarly, an fMRI study by Eldar *et al*. (2007) showed that activity changes in response to music in both the amygdala and the (anterior) hippocampal formation increase significantly when the music is presented simultaneously with film clips (film clips were neutral scenes from commercials, positive music was also taken from commercials, and negative music mainly from soundtracks of horror movies). Activity changes in the amygdala were considerably larger for the combined (film and music) presentation than for the presentation of film clips alone, or music alone. Analogue response properties were observed in the ventro-lateral frontal cortex for both positive and negative, and in the hippocampal formation for negative, music combined with the film clips. Notably, emotional music without the film clips did not elicit a differential response in these regions (although all film clips were neutral). Importantly, activity changes in the amygdala were observed in response to both positive and negative stimulus combinations. This supports the view that the amygdala is not only involved in negative, but also in positive

---

[31] Images of the International Affective Picture System showing humans or human faces.

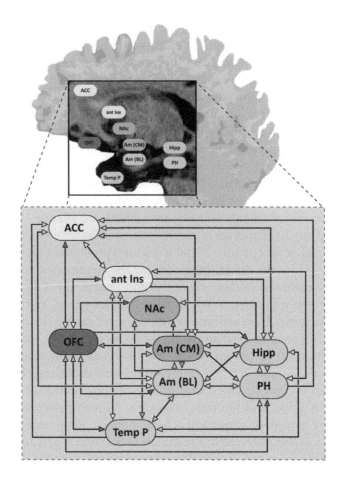

Figure 12.2   Schematic illustration of anatomical connections of limbic and paralimbic structures involved in the emotional processing of music (connectivity according to
Nieuwenhuys *et al.*, 2008; Öngür & Price, 2000; Barbas *et al.*, 1999; Augustine, 1996).
ACC: Anterior cingulate cortex; ant Ins: Anterior insula; Am (BL): Basolateral amygdala;
Am (CM) corticomedial amygdala (including the central nucleus), Hipp: Hippocampal
formation; NAc: Nucleus accumbens; OFC: Orbitofrontal cortex; PH: Para-hippocampal
gyrus; Temp P: Temporal pole. Reprinted with permission from Koelsch (2010).

emotions (see, e.g., Murray, 2007), clearly challenging the rather simplistic view
that the amygdala is primarily a fear centre of the brain.

Curiously, behavioural ratings obtained in that study (Eldar *et al.*, 2007) did
not significantly differ between the combined conditions on the one hand (positive music with neutral film, as well as negative music with neutral film), and
the condition in which music was presented alone on the other (note that film
clips played without music were rated as neutral). Therefore, the reasons for the
signal increase in amygdala and hippocampal formation during the combined
presentation of music and film clips remained unclear. Perhaps the combination

of emotional music with neutral film clips stimulated imagery about positive or negative events that might happen next, increasing the overall emotional activity without modulating the emotional valence of the stimuli. The findings that visual information modulates signal changes in the amygdala are corroborated by data showing that simply closing the eyes during listening to fearful music also leads to increased amygdalar activity (Lerner *et al.*, 2009), perhaps due to increased visual imagery during the eyes-closed condition.

Involvement of the amygdala in the emotional processing of music was not only reported in functional neuroimaging studies, but also in a lesion study by Gosselin *et al.* (2005), in which patients with medial temporal lobe resections that included the amygdala showed impaired recognition of fearful music. In addition, Griffiths *et al.* (2004) reported that a patient with a lesion of the left amygdala and the left insula showed a selective loss of intensely pleasurable experiences, and of autonomic responses, during music listening: The patient lost the capability to experience frissons in response to musical pieces that had elicited such responses in him before he had suffered the brain lesion.

12.6.1   Major–minor and happy–sad music

Several functional neuroimaging studies used music in major and minor modes to investigate 'happiness and sadness' (Khalfa *et al.*, 2005; Mitterschiffthaler *et al.*, 2007), 'musical beauty' (Suzuki *et al.*, 2008), or 'liking' (Green *et al.*, 2008). However, these studies do not yield a consistent picture yet, except perhaps activation of the anterior fronto-median cortex (BA 10m/9m) for minor contrasted to major music in two studies (Khalfa *et al.*, 2005; Green *et al.*, 2008). Problems in comparing these studies include: (1) different participant populations, e.g., only males in one study (Suzuki *et al.*, 2008) compared to eight males and five females in another (Khalfa *et al.*, 2005), (2) interpretation of unsystematic effects (such as rCBF decrease in a striatal region during 'beautiful major', increase during 'beautiful minor', but increase during 'ugly major', and decrease during 'ugly minor' music; Suzuki *et al.*, 2008), (3) use of 'true performances' (Mitterschiffthaler *et al.*, 2007; Khalfa *et al.*, 2005) on the one hand, and use of melodies (Green *et al.*, 2008) or chords (Mizuno & Sugishita, 2007; Suzuki *et al.*, 2008) played without musical expression on the other, and (4) different tasks: Participants were asked 'how well they liked it' (Green *et al.*, 2008), to 'rate the beauty of the chord sequence' (Suzuki *et al.*, 2008), to rate 'their mood state … from sad … to happy' (Mitterschiffthaler *et al.*, 2007), or to 'judge the emotion represented in the music … from sad to happy' (Khalfa *et al.*, 2005).

Moreover, whereas some studies aimed to match major and minor stimuli in tempo and timbre (Suzuki *et al.*, 2008; Green *et al.*, 2008; Mizuno & Sugishita, 2007), 'happy' and 'sad' stimuli differed considerably in their acoustic and musical properties in other studies (e.g., 'happy' excerpts having a faster tempo than 'sad' excerpts; Mitterschiffthaler *et al.*, 2007; Khalfa *et al.*, 2005). Thus, further studies are needed to provide more information about the neural correlates of happiness and sadness, how major and minor tonal features might contribute to emotional effects related to happiness and sadness, and how such effects are related to musical preference and cultural experience.

12.6.2   Music-evoked dopaminergic neural activity

Several studies showed that listening to pleasant music activates brain structures implicated in reward and experiences of pleasure. As mentioned above, Blood & Zatorre (2001) reported that the ventral striatum (presumably the nucleus accumbens, NAc; see Figure 12.1 for illustration) is involved in intensely pleasurable responses involving frissons. Similarly, another PET study by Brown *et al.* (2004) reported activation of the ventral striatum (in addition to the subcallosal cingulate cortex, the anterior insula, and the posterior part of the hippocampus) during listening to two unfamiliar, pleasant pieces contrasted with a resting condition. Activation of the ventral striatum in response to pleasant music was also observed in three studies using fMRI. One of these studies investigated the valence dimension (Koelsch *et al.*, 2006a), another one examined differences in pleasantness due to the predictability of music (Menon & Levitin, 2005), and the third investigated music-evoked autobiographical memories (Janata, 2009).[32] One of these studies (Menon & Levitin, 2005) reported that activation of the ventral striatum was connected to activity in the ventral tegmental area (VTA) and the hypothalamus. This suggests that the haemodynamic changes observed in the ventral striatum reflected dopaminergic activity: The NAc is innervated in part by dopaminergic brainstem neurons (located mainly in the VTA, as well as in the substantia nigra), and is part of the so-called 'reward circuit' (Berridge *et al.*, 2009; Björklund & Dunnett, 2007). This circuit includes projections from the lateral hypothalamus via the medial forebrain bundle to the mesolimbic dopamine pathway involving the VTA with projections to the NAc (this circuit is part of what Panksepp, 1998, refers to as the *SEEKING* system). Further support for the assumption that the activity changes in the ventral striatum reported in the above-mentioned studies involved dopaminergic neural activity stems from a recent PET study by Salimpoor *et al.* (2011) showing that strong music-evoked pleasure (including musical frissons) is associated with increased dopamine binding in the NAc.

Importantly, activity in the NAc (as well as activity in the ventral pallidum; Berridge *et al.*, 2009)[33] correlates with motivation- and reward-related experiences of pleasure, for instance during the process of obtaining a goal, when an unexpected reachable incentive is encountered, or when individuals are presented with a reward cue (reviewed in Berridge *et al.*, 2009; Nicola, 2007). In humans, NAc activity has been reported, e.g., for sexual activity, intake of drugs, eating of

---

[32]   Thus, with regard to the principles underlying the evocation of emotion (Section 12.4), emotional responses in the study by Menon & Levitin (2005) were presumably mainly due to *expectancy*, and in the study by Janata (2009) mainly due to *memory*.

[33]   The ventral pallidum is formed by rostral and ventral extensions of the globus pallidus.

chocolate, and drinking water when dehydrated (Berridge *et al.*, 2009; Nicola, 2007). It has, therefore, previously been suggested that NAc activity correlates with the subjective experience of *fun* (Koelsch *et al.*, 2010b), but more detailed information about the functional significance of the NAc is needed to determine the role that the NAc possibly plays for other emotions as well.

The NAc also appears to play a role in invigorating, and perhaps even selecting and directing, behaviour in response to stimuli with incentive value, as well as in motivating and rewarding such behaviour (Nicola, 2007). The NAc is considered as a 'limbic motor interface' (Nieuwenhuys *et al.*, 2008), because (1) the NAc receives input from limbic structures such as amygdala and hippocampus, (2) injecting dopamine in the NAc causes an increase in locomotion, and (3) the NAc projects to other compartments of the basal ganglia, which play an important role for the learning, selection, and execution of actions. This motor-related function of the NAc puts it in a key position for the generation of a drive to move to, join in, and dance to pleasant music, although the neural basis for this drive needs to be specified.

It is important to note that in three of the above-mentioned studies (Koelsch *et al.*, 2006a; Brown *et al.*, 2004; Menon & Levitin, 2005) participants did not report frissons during music listening. This indicates that dopaminergic pathways including the NAc can be activated by music as soon as it is perceived as pleasant (i.e., even in the absence of extreme emotional experiences involving frissons). Thus, results from the reviewed studies indicate that music can easily evoke experiences of pleasure, or fun, associated with activity of a reward pathway involving the hypothalamus, the VTA and the NAc. This emotional power of music needs to be explored further, to provide more systematic knowledge in support of the therapy of disorders related to dysfunctions involving the mesolimbic reward-pathway (such as depressive disorders or Parkinson's disease). Notably, in addition to subjective experiences of fun (involving the NAc), music can also evoke experiences of joy and happiness (Koelsch *et al.*, 2010b). The next section puts forward the hypothesis that the latter experiences involve different neural systems than those involved in experiences of fun.

12.6.3   Music and the hippocampus

Compared to studies investigating emotion with stimuli such as emotional faces, affective pictures, pain stimuli, or reward stimuli, the review of functional neuroimaging studies on music and emotion reveals a particularly noticeable feature: The proportion of studies reporting activity changes within the (anterior) hippocampal formation in response to music is remarkably high (such activity changes have been reported by Blood & Zatorre, 2001; Koelsch *et al.*, 2006a; Eldar *et al.*, 2007; Baumgartner *et al.*, 2006b; Mitterschiffthaler *et al.*, 2007; Brown *et al.*, 2004; Fritz & Koelsch, 2005; Koelsch *et al.*, 2007b; see also Figure 12.3). It is well established that the hippocampus plays an important role for learning and memory, spatial orientation, novelty, as well as expectedness (for reviews see, e.g.,

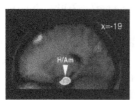 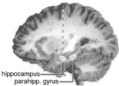 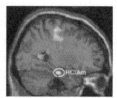

Blood et al., 2001   Koelsch et al., 2006   Eldar et al., 2007   Mitterschiffthaler et al., 2007

Figure 12.3 Activity changes in the anterior hippocampal formation in response to music. In the PET study by Blood & Zatorre (2001), decrease of rCBF correlated with increasing intensity of musical frissons (the x-coordinate refers to Talairach stereotaxic space). In the fMRI study by Koelsch *et al.* (2006a), activity changes in the anterior hippocampal formation were reflected in BOLD signal increase during unpleasant, and decrease during pleasant music. In both the study by Blood & Zatorre (2001) and by Koelsch *et al.* (2006), activity changes were stronger in the left than in the right hemisphere. In the study by Eldar *et al.* (2007), BOLD signal increase in the anterior hippocampal formation (see illustrated regions of interest) was stronger during the combined presentation of negative (scary) music and neutral film clips (compared to when only film clips or only music was presented). Activity changes did not differ between hemispheres. In the study by Mitterschiffthaler *et al.* (2007), activity changes in the right hippocampal formation were reflected in BOLD signal increase in response to sad, and decrease during neutral music. Modified with permission from Blood & Zatorre (2001), Koelsch *et al.* (2006a), Eldar *et al.* (2007), and Mitterschiffthaler *et al.* (2007).

Moscovitch *et al.*, 2006; Nadel, 2008). However, at least in some of the functional neuroimaging studies that used music to investigate emotion, it is unlikely that the hippocampal activations were simply due to such processes. For example, in the study by Mitterschiffthaler *et al.* (2007) sad (as compared to neutral) music elicited changes in the anterior hippocampal formation, although participants were probably comparably familiar with neutral and sad pieces. Similarly, participants were presumably equally unfamiliar with the happy and fearful musical pieces used in the study by Eldar *et al.* (2007). Finally, in the study by Blood & Zatorre (2001), rCBF changes in the anterior hippocampal formation were observed even when analyzing responses only to stimuli that participants brought themselves into the experiment (supporting Figure 5 of Blood & Zatorre, 2001), thus every subject was highly familiar with the music included in that analysis.

Therefore, studies on music and emotion indicate that the hippocampus plays an important role for emotional processes, a view that had been advocated already by Papez (1937) and MacLean (1990) several decades ago (but ignored by many neuroscientists). The hippocampus has dense reciprocal connections with structures involved in the regulation of behaviours essential for survival (such as ingestive, reproductive, and defensive behaviours), and with structures involved in the regulation of autonomic, hormonal, and immune system activity (Nieuwenhuys *et al.*, 2008). Such structures include the amygdala, hypothalamus, thalamic nuclei, the septal-diagonal band complex, the cingulate gyrus, the insula, and autonomic brain stem nuclei. Efferent connections project to the NAc, other parts of the striatum, as well as to numerous other limbic, paralimbic, and non-limbic structures (Nieuwenhuys *et al.*, 2008). The functional significance of these connections places the hippocampus (along with the amygdala and the orbitofrontal cortex) in a pivotal position for emotional processing, and it has previously been noted that the key to understanding the function of the hippocampus lies in the fact that it has major projections not only to cortical association areas, but also to subcortical limbic structures (Nieuwenhuys *et al.*, 2008).

The notion that the hippocampus is involved in emotional processes (in addition to its more cognitive functions such as memory and spatial representation) is supported by a wealth of empirical evidence. (1) Lesion of the hippocampus leads to impairment of maternal behaviour in rats (as indexed by less frequent and less efficient nursing, poorer nest building, increased maternal cannibalism, poorer retrieving, and fewer pups surviving to weaning; Kimble *et al.*, 1967). (2) Individuals with depression show structural as well as functional abnormality within the hippocampus (reviewed in Videbech & Ravnkilde, 2004; Warner-Schmidt & Duman, 2006). (3) The hippocampus is unique in its vulnerability to chronic emotional stressors: In animals, chronic stress related to helplessness and despair leads to death of hippocampal neurons and related hippocampal atrophy (Warner-Schmidt & Duman, 2006), consistent with studies on humans showing reduced hippocampal volume in individuals suffering from childhood sexual abuse (Stein *et al.*, 1997), and post-traumatic stress disorder (PTSD; Bremner, 1999). The loss of hippocampal volume during and after emotional traumatization, or during depression, is assumed to be due to the death of hippocampal neurons, and a down-regulation of neurogenesis in the dentate gyrus (Warner-Schmidt & Duman, 2006).[34] (4) Activity changes in the anterior hippocampal formation (as well as in the amygdala) in response to pleasant and unpleasant music are reduced in individuals with reduced capability of producing tender positive feelings (i.e., feelings that can be described as soft, loving, warm and happy) compared to individuals of a normal control group (Koelsch *et al.*, 2007b).

Although only little specific information about the involvement of the hippocampus in the processing of emotions is yet available, the results of the studies mentioned above motivate the hypothesis that the hippocampus is a critical structure for the generation of joy and happiness, and therefore for emotions that play a particular role for social attachments. We (Koelsch *et al.*, 2007) have denoted such emotions as *tender emotions*, referring to Charles Darwin's *The Expression of Emotions in Man and Animals* (Darwin, 1872), in which he wrote that 'tender feelings ... seem to be compounded of affection, joy, and especially of sympathy' (p. 247).[35] These feelings are 'of a pleasurable nature', and it is interesting to note

---

[34]   The dentate gyrus is one of three structures constituting the hippocampal formation, the other two structures are the hippocampus proper, and the subiculum (Nieuwenhuys *et al.*, 2008).
[35]   Note that Darwin means sympathy in the sense with which nowadays the word empathy is often used, e.g., 'feeling either pity for the grief of someone else, or feeling the happiness or good fortune' (Darwin, 1872).

that, in his chapter about love, joy, and devotion, Darwin also writes about 'the wonderful power of music' (p. 250, a topic which is elaborated in *The Descent of Man*, Darwin, 1874). The experience of social attachments is related to positive tender emotions (such as joy and happiness), whereas social loss is related to negative emotions such as sadness. Attachment-related behaviour includes licking, grooming, nest-building, and pup retrieval, and particularly in humans hugging, kissing, caressing, stroking, softly touching, and softly vocalizing (Panksepp, 1998, associates attachment-related emotions in part with a *CARE* system). In humans, another attachment-related emotion is love.[36]

Negative feelings such as anxiety and depression appear to be related to inhibition of hippocampal activity (as suggested by studies reporting reduced hippocampal activity in individuals with pathological depression; Warner-Schmidt & Duman, 2006). Notably, in healthy individuals, inhibition of neural pathways projecting to the hippocampus during the perception of unpleasant stimuli might well represent a sensitive neural mechanism that serves to prevent potential damage of hippocampal neurons (note that, as mentioned above, severe emotional stress leads to the death of hippocampal neurons; Warner-Schmidt & Duman, 2006). Therefore, it is important that researchers are more cautious in attributing activity changes observed in the amygdala and the hippocampus during the presentation of unpleasant (or threatening) stimuli simply to the generation of fear (or other unpleasant emotions), and that researchers consider the possibility that these activity changes reflect inhibitory processes activated automatically in response to stimuli that are potentially emotionally stressful.

It is also important to differentiate the feelings related to the activation of the 'reward circuit' (including the lateral hypothalamus, as well as the mesolimbic dopamine pathway involving the ventral tegmental area with projections to the NAc; see above) from the tender positive emotions that involve activity of the hippocampus, although both are naturally not mutually exclusive (usually, having joy is fun). Panksepp (1998) described oxytocin as one possible link between these two systems: Oxytocin plays a role in establishing social bonds, and thus in attachment-related emotions (Panksepp's CARE system), and activity of the mesolimbic reward-pathway (part of Panksepp's SEEKING system) can be elicited in part by endogenous opioid and oxytocin release in the VTA. We (Koelsch *et al.*, 2010) have previously noted that feelings arising from activity of the former circuit (involving the NAc) might perhaps best be referred to as *fun*, whereas attachment-related (tender positive) emotions involving hippocampal activity might best be referred to as *joy*, *love* and *happiness* (see also Siebel, 2009).

An important difference between reward-related and attachment-related emotions is that the former ones satiate; once an organism has satisfied bodily needs and achieved homeostasis, the organism is satiated, and stimuli that were

---

[36] According to my experience from experiments using music as stimuli to evoke emotions, emotional experiences related to hippocampal activity are often described by participants as *touching*, or *moving*.

previously incentive can become even aversive (for example, because too much of a chemical compound can be harmful for an organism). This stays in contrast to the hippocampus-centred emotions, which do not satiate. Note that a brain system for attachment-related affect that does not satiate is evolutionary adaptive, because, for example, feeling attached to a child, loving a child and feeling the joy of being together with the child are emotions that serve the continuous protection, and nurturing of the offspring. Similarly, the need to belong to a social group and the feeling of social inclusion (both of which do not appear to satiate), serve the formation and maintenance of social attachments, thus strengthening social cohesion.[37] Also note that the capability of music to evoke attachment-related emotions renders the comparison between music and other stimuli that evoke emotions that satiate, such as cheesecake, quite unfortunate: In contrast to cheesecake, one can consume music for hours and still be happy (and still want more).

Whether or not the present conception of the quality of hippocampus-centred emotions is already sufficient, or needs to be expanded–it is important to recognize the importance of the hippocampus for emotional processing in affective neuroscience. Future neuroimaging studies on emotion should carefully control for familiarity, novelty, and memory processes elicited by different stimulus categories to rule out the possibility that hippocampal activations are due to such factors. Notably, due to the capability of music to evoke activity changes in the hippocampus, it is conceivable that music therapy with depressed patients and with PTSD patients has positive effects on the up-regulation of neurogenesis in the hippocampus; this remains to be studied empirically.

12.6.4   Parahippocampal gyrus

The parahippocampal gyrus plays an important role for encoding and storage of memories of emotional events (e.g., Kilpatrick & Cahill, 2003; Rugg & Yonelinas, 2003). This presumably also holds for emotional events related to music. That is, the parahippocampal gyrus may enable an individual to remember the emotions of previously experienced music (or sounds), and thus to recognize musical emotions (see also the *memory* principle mentioned in Section 12.4).

A number of functional imaging studies on music and emotion reported an involvement of the parahippocampal gyrus in networks responsive to unpleasant music, and a comparison between these studies reveals striking similarities in the coordinates of the local maxima of activity within the parahippocampal gyrus.[38] In addition, a lesion study by Gosselin *et al.* (2006) reported that patients with

---

[37]   Also note that one can buy 'fun' (for example, one can buy drugs, chocolate, sex, bungee jumps, etc.), but not happiness.

[38]   Coordinates of maxima of activation (in Talairach stereotaxic space) are: 25, 28, 21 (Blood *et al.*, 1999), -25, -26, -11/22, -26, -13 (Koelsch *et al.*, 2006a), -20 -30 -9 (in response to unpleasant stimuli; Fritz & Koelsch, 2005), -23 -29 -9 (Ter Haar *et al.*, 2007, the activation correlated with sounds of increasing dissonance).

lesions of the (left or right) parahippocampal gyrus did not rate dissonant musical excerpts as unpleasant as did control participants. Activity changes due to unpleasant music were more pronounced in the right parahippocampal gyrus in the study by Blood *et al.* (1999), and stronger in the left (though with lower statistical threshold also present in the right) parahippocampal gyrus in the study by Koelsch *et al.* (2006a).[39] The reasons for these hemispheric weightings remain to be specified. However, the consistency of the coordinates listed above suggests that the parahippocampal gyrus plays a particular role in the processing of auditory stimuli with different degrees of consonance/dissonance, and therefore presumably also with different degrees of acoustic roughness (which is evolutionarily perhaps more relevant than consonance/dissonance).[40]

12.6.5   A network comprising hippocampus, parahippocampal gyrus, and temporal poles

So far, three functional imaging studies on music and emotion have reported activity changes within the hippocampus, the parahippocampal gyrus, and the temporal poles (Baumgartner *et al.*, 2006b; Fritz & Koelsch, 2005; Koelsch *et al.*, 2006a). As mentioned above, the study by Fritz & Koelsch (2005) suggests a functional connectivity between (a central, perhaps basolateral aspect of) the amygdala, and the hippocampus, the parahippocampal gyrus, and the temporal poles (see also Figure 12.2). Thus, it appears that the memory network comprising the hippocampus, parahippocampal gyrus, and temporal poles is also involved in emotional processing. Within this network, the hippocampus is presumably involved in memory formation and generation of emotions, the parahippocampal gyrus in storing emotional memories and recognition of emotion, and the temporal poles probably in the retrieval of emotional memories.

12.6.6   Effects of music on insular and anterior cingulate cortex activity

Current theories of emotion emphasize the association between emotion and changes in physiological arousal (mainly involving changes in autonomic and endocrine activity). Changes in autonomic activity have been reported to be associated with activity changes in the anterior cingulate cortex (ACC) and the insular cortex (Critchley, 2005; Critchley *et al.*, 2000; Craig, 2009), and music studies using PET or fMRI have observed activity changes in both of these structures during music-evoked frissons (Blood & Zatorre, 2001), as well as during experiences

---

[39]   Similar results were observed by Fritz & Koelsch (2005).

[40]   Changes in acoustic roughness of vocalizations provide important cues for the emotional state of an individual. Moreover, acoustic roughness provides information about the structure of the things that were involved in the production of the sounds. For example, hearing someone walk through fine sand produces sounds with a lower degree of acoustic roughness than sounds originating from footsteps through gravel.

of fear and sadness (Baumgartner *et al.*, 2006b). Note, however, that activity changes in the ACC or insular cortex are not necessarily related to emotional processing. For example, the ACC is also involved in performance monitoring, movement-related functions, and in the perception of speech and music (e.g., Cole *et al.*, 2009; Mutschler *et al.*, 2007; Koelsch *et al.*, 2010b). It has recently been suggested (Koelsch *et al.*, 2010b) that the ACC is involved in the *synchronization of biological subsystems* (a term coined by Klaus Scherer, 2000). These systems are comprised of physiological arousal, motor expression, motivational processes (action tendencies), monitoring processes, and cognitive appraisal. The synchronization of activity of these subsystems is likely to occur as an effect of every emotional instance, and may even be indispensable for subjective emotional experiences (usually referred to as *feelings*). The ACC is in a unique position to accomplish such synchronization, due to its involvement in cognition, autonomic nervous system activity, motor activity, motivation, and monitoring.

The insula is also involved in autonomic regulation, but is in addition a visceral sensory, somatosensory, visceral motor, and motor association area (Augustine, 1996; Mutschler *et al.*, 2009). The insula has main connections with the cingulate cortex, orbitofrontal cortex (OFC), secondary somatosensory area, retroinsular area of the parietal cortex, temporal poles, superior temporal sulcus (STS), amygdala, hippocampus, and parahippocampal gyrus, as well as with the frontal operculum and lateral premotor cortex (PMC). Particularly the anterior insula (which has connections to both subcortical and cortical limbic/paralimbic structures as well as to the brain stem) appears to play a specific role in the integration of visceral and somatosensory information with autonomic activity (see also Flynn *et al.*, 1999; Mutschler *et al.*, 2009). Thus, it is likely that one important function of the anterior insula is the adjustment, or regulation, of the intensity of autonomic activity to an appropriate level according to somatosensory, visceral sensory, and autonomic information. In concert with the ACC, which may prevent emotions from going 'overboard', the insula prevents autonomic activity from going 'overboard', but also from being blunted, or from instantly ebbing away.[41]

12.7 Electrophysiological effects of music-evoked emotions

Only few EEG studies (Schmidt & Trainor, 2001; Altenmüller *et al.*, 2002; Baumgartner *et al.*, 2006a; Sammler *et al.*, 2007; Flores-Gutiérrez *et al.*, 2007) have so far investigated emotion with music (and up to now, there is a lack of studies on music and emotion using magneto-encephalography). All of these studies investigated the valence dimension (contrasting effects of pleasant with

---

[41]   For the role of the insula in interoceptive awareness see, e.g., Critchley *et al.* (2004) and Craig (2009). For the putative role of the insula in empathy see, e.g., Singer *et al.* (2009).

effects of unpleasant music). Schmidt & Trainor (2001) and Altenmüller *et al.* (2002) reported more pronounced neural activity in the left compared to the right frontal lobes in response to music with positive valence (and the opposite hemispheric weighting for music with negative valence). One of these studies (Altenmüller *et al.*, 2002) measured direct current EEG, while the other one (Schmidt & Trainor, 2001) measured oscillatory neural activity in the alpha band. However, such a lateralization was not observed in the studies by Baumgartner *et al.* (2006) and Sammler *et al.* (2007).

Instead, a study by Baumgartner *et al.* (2006) reported a bilateral increase of alpha power for happy music (combined with happy pictures) compared to sad and scary music (combined with sad and scary pictures), and a study by Sammler *et al.* (2007) did not find any differences in the alpha band (or in sub-bands of the alpha frequency range) between pleasant and unpleasant music. However, that study reported an increase in fronto-midline theta power in response to pleasant music. This increased oscillatory activity presumably originated from the dorsal anterior cingulate cortex, and presumably reflected emotional processing interlinked with attentional functions. Notably, that study also reported that the difference in theta power between conditions was significantly stronger during the second half of each excerpt (i.e., seconds 30–60) compared to the first half of each excerpt (i.e., seconds 1–30). Further research is needed to gain more insights into electrophysiological correlates of music-evoked emotions, and it appears that the most promising approach to this topic involves the investigation of oscillatory activity in different frequency bands.

12.8   Time course of emotion

The intensity, and sometimes also the quality, of an emotion (or a mix of emotions) changes over time. To date, however, little is known about the time course of emotional processes, and neural correlates of different stages of emotional episodes are not known. Intuitively, for example, it seems plausible that aversive sounds elicit immediate emotional responses (although long durations of such sounds might even increase the degree of unpleasantness), and that especially tender emotions might take a while to unfold.

One of the few psychophysiological studies that investigated the time course of emotion was conducted by Krumhansl (1997). In that study, several physiological measures (including cardiac, vascular, electrodermal, and respiratory functions) were recorded while listeners heard musical excerpts (each about 3 minutes long). Excerpts were chosen to represent one of three emotions (sadness, fear, happiness). Significant correlations were found between most of the recorded physiological responses and time (measured in one-second intervals from the beginning of the presentation of each musical excerpt). The strongest physiological effects for each emotion type tended to increase over time, suggesting that the intensity of an emotional experience is likely to increase over time during the perception of a musical excerpt. In studies measuring changes in heart rate and breathing rate to music, Orini *et al.* (2010) found that these two physiological parameters mainly change within the first 20 seconds of a musical excerpt, and then remain relatively stable (see also Lundqvist *et al.*, 2009). Recent studies have also shown physiological changes related to emotional valence and arousal as elicited by music over time (Grewe *et al.*, 2007a,b).

Activity changes over time due to emotional processing were also observed in the fMRI study by Koelsch *et al.* (2006a). In that study, the pleasant and unpleasant musical excerpts had a duration of about 1 min, and data were not only modelled for the entire excerpts, but also separately for the first 30 seconds, and for the remaining 30 seconds of each piece, in order to investigate possible differences in brain activity over time. When looking at activation differences between the first 30 seconds and the remaining 30 seconds, activity changes in the amygdala, parahippocampal gyrus, temporal poles, insula, and ventral striatum were stronger during the second 30 seconds of musical excerpts, presumably because the intensity of listeners' emotional experiences increased during the perception of both the pleasant and the unpleasant musical excerpts. Using the same stimuli in an EEG experiment, Sammler *et al.* (2007) reported that an increase in fronto-midline theta power in response to pleasant music (compared to unpleasant music) was significantly stronger during the second half of each excerpt (i.e., seconds 30–60) compared to the first half of each excerpt (i.e., seconds 1–30). Finally, the study by Salimpoor *et al.* (2011) showed that anticipation of a music-evoked frisson was associated with dopaminergic activity in a dorsal compartment of the striatum, whereas the experience of a frisson itself was associated with dopaminergic activity in the ventral striatum (presumably the nucleus accumbens).

So far, these three studies (Koelsch *et al.* 2006a; Sammler *et al.*, 2007; Salimpoor *et al.*, 2011) are the only ones that investigated the temporal dynamics of the neural correlates of emotional processing. Note that music is an ideal stimulus to investigate this issue, because music always unfolds over time. New studies of emotional processing with music could perform analyses investigating the activity of the structures involved in emotional processing over time. Information about activity changes over time of the structures implicated in emotion (such as information about how activity in one structure affects activity in another) would provide important insights into the functional significance of these structures, and into the neural substrates of human emotion.

12.9   Salutary effects of music making

Listening to music, and probably even more so music making, has beneficial effects on psychological and physiological health, and thus on the well-being of individuals. This section will enumerate several factors that appear to contribute to

Table 12.4   Factors underlying salutary effects of music making (bottom row), in relation to processes of music perception, dimensions of musical meaning, social functions, and emotion principles. *Perception* is related to auditory *feature extraction* (but includes here all other perceptive processes during music listening, and particularly during music making or dancing, such as somatosensory perception, equilibrioception, and visual perception). *Attention* is listed left of *musical expectancy formation* and *structure building* because these processes are influenced by the voluntary direction of attention (as indicated by the influence of attention on the generation of the ERAN). Long-term *memory* (procedural, implicit, semantic, episodic, and emotional memory) also occurs in the *emotion principles*, and parallels the (implicit) long-term memory operations required for *musical expectancy formation* and recognition of *symbolic* sign quality of music. *Intelligence* is related to *structure building* (thus also to the processing of *intra-musical* meaning) and *communication*. Salutary effects of *emotion* are related to *vitalization*, and *emotional musicogenic meaning*.

music perception	feature extraction	Gestalt formation	interval analysis	structure building	structural re-analysis	vitalization	premotor, immune system
musical meaning	iconic	indexical	symbolic	intra-musical	physical	emotional	personal
social functions	contact	social cognition	co-pathy	communication	coordination	cooperation	social cohesion
emotion principles	evaluation	contagion	memory	expectancy	imagination	understanding	social functions, aesthetics
salutary factors	perception	attention	long-term memory	intelligence	action	emotion	social functions

such effects (summarized in Table 12.4). Many of these factors are only indirectly linked to emotion, but it is presumed here that positive emotional valence of musical experience contributes to the salutary effects of all factors. During music listening, and particularly during music making (or dancing), these factors do usually operate in combination (for example, it is possible that all of the mentioned factors operate at the same time). The therapeutic effects of these factors are systematically used in music therapy (see also the 'heuristic working factor model for music therapy' by Thomas Hillecke *et al.*, 2005). Note, however, that the number of high-quality empirical studies in the field of music therapy is still rather limited; therefore, the notion of salutary effects of several factors listed here is still in part speculative, and the listing of these factors is meant to provide a framework for further research, rather than as a summary of previous empirical research.

(1) Perception.   Music making fosters (and supports the development of) perceptual skills. For example, Sections 5.2 and 5.3.2 reviewed studies showing that musical training modulates the decoding of acoustic features such as pitch height and frequency modulations already on the level of the brainstem and the auditory cortex, leading to increased auditory acuity of the perception of chords and phonemes. The finding that musical training has effects on basic perceptual

processes during language comprehension (e.g., Wong *et al.*, 2007) is relevant, because children with language impairment often not only suffer from productive difficulties, but also from basic perceptual difficulties (e.g., Tallal & Gaab, 2006). Therefore, it is likely that making music would have prevented such perceptual difficulties, and that music-therapeutic treatment of such perceptual difficulties can also help in the treatment of language impairment. An early treatment of language impairment is important to decrease the risk for the development of learning and reading disorders after entering school. Recent studies also showed that perception modulation, presumably on the level of the auditory cortex, can ameliorate forms of tinnitus that are due to mal-plasticity of central auditory processing structures (Okamoto *et al.*, 2010, other treatment forms include attention-related therapy of tinnitus; see also below). Note that perceptual training effects of music making are not restricted to the auditory domain, but also include proprioception, tactile perception, and multimodal integration. Potentially salutary effects of training such perceptual skills remain to be investigated. With regard to the model of music perception (Chapter 8 and first row of Table 12.4), perception is related to auditory *feature extraction* (but includes here all other perceptive processes during music listening, and particularly during music making or dancing, such as somatosensory perception, equilibrioception, and visual perception).

(2) Voluntary direction of Attention.   Music making requires voluntary direction of attention, particularly when making music together with other individuals. The training of voluntary direction of attention with music appears to have positive effects on the attentional symptoms of children with attention-deficit/hyperactivity disorder (ADHD), although the empirical evidence is still quite sparse in this area (Jackson, 2003). Moreover, focussing attention on music can distract attention from stimuli prone to evoke negative experiences; therefore, this factor probably accounts, at least in part, for anxiety-, worry-, and pain-reducing effects of music listening during medical procedures (e.g., Nelson *et al.*, 2008; Klassen *et al.*, 2008; Spintge, 2000; Koelsch *et al.*, 2011). With regard to the model of music perception, attention is listed left of *musical expectancy formation* and *structure building* because these processes are influenced by the voluntary direction of attention (as indicated by the influence of attention on the generation of the ERAN).

(3) Long-term memory.   Making music usually engages procedural, implicit, semantic, episodic, and emotional memory (in terms of encoding and retrieving musical information, or information associated with musical experiences). Such engagement might be beneficial for memory, although studies in this area remain to be conducted.[42] However, a memory factor might contribute to

---

[42]   Notably, it has been argued that simply evoking neural activity in a brain structure by activating cognitive and/or emotional processes during music listening (referred to as *cognitization*) might have beneficial effects on the recovery of stroke patients (Särkämö *et al.*, 2008, 2010).

beneficial effects of receptive music therapy on the symptoms of Alzheimer's disease (AD) (Gerdner & Swanson, 1993), although empirical evidence for such effects is still very limited.[43] Long-term *memory* (procedural, implicit, semantic, episodic, and emotional memory) also occurs in the *emotion principles* (fourth row of Table 12.4), and parallels the (implicit) long-term memory operations required for *musical expectancy formation* and recognition of *symbolic* sign quality of music.

(4) Intelligence.   Music making can involve intelligence, for example during improvisation in terms of planning and execution of hierarchically organized complex action sequences, or the invention of hierarchically organized music-syntactic structures. Intelligence involves constructive reasoning and application of logic (thus, intelligence stays in contrast to illogical, irrational thoughts).[44,45] Reasoning involves *working memory* (WM) operations, which include, during both music perception and music production, central executive function, as well as use of the phonological short-term store (and sometimes also the phonological loop). Using the tools of intelligence during music making or during listening to music might help in keeping them in good condition, although – again – empirical evidence for this notion remains to be gathered. With regard to the model of music perception, intelligence is related to *structure building* (thus also to the processing of *intra-musical* meaning) as well as to *communication* (first to third rows of Table 12.4).

(5) Action, and movement modulation.   This factor refers to the fine-tuning of (sensori-)motor skills. Engaging in *action* during music making or dance keeps muscles, joints, tendons, and peripheral nerves in good physical shape. It also re-establishes and maintains gross and fine (sensori-)motor skills in stroke patients with hemiplegia (Schneider *et al.*, 2010), and – with regard to motor processes underlying speech production – in patients with Broca's aphasia (Schlaug *et al.*, 2009; Norton *et al.*, 2009). The latter studies suggest that *Melodic Intonation Therapy*, a therapy in which patients regain their speaking by learning to sing sentences, leads to the establishment of neuronal connections in the (unlesioned) right hemisphere between auditory regions in the temporal lobe, and the right-hemispheric homologue of Broca's area. Furthermore, a musical tactus can improve gait in

---

[43]   Simmons-Stern *et al.* (2010) reported that lyrics of unfamiliar children's songs were better remembered by AD patients when they were sung, compared to when they were spoken (lyrics were spoken or sung by another individual, thus this was not a study on effects of music production). Control (healthy) older adults showed no significant difference between the two conditions.

[44]   For studies investigating effects of musical training on 'general intelligence' see Schellenberg (2006). For a study showing a stronger development of neurophysiological indices of music- and language-syntactic processing in children with musical training see Jentschke & Koelsch (2009).

[45]   Because intelligence stays in contrast to illogical, irrational thoughts, intelligence is inherently related to self-esteem and its beneficial effects for health: Irrational thoughts are a necessary condition for low self-esteem (supported, e.g., by Warren *et al.*, 1988), hampering self-respect, self-acceptance, self-assertiveness, self-responsibility, self-efficacy, integrity, and self-confidence.

neurological patients with Parkinson's disease, or apraxia (Thaut, 2003; Thaut *et al.*, 2005; Thaut & Abiru, 2010). Finally, positive effects of dancing have been reported in patients with Parkinson's disease, arthritis, and fibromyalgia (for a review see Murcia *et al.*, 2010).

(6) Emotion.   As described in Chapter 12, music-evoked emotions can modulate activity within all limbic- and paralimbic brain structures. These findings have implications for music-therapeutic approaches for the treatment of affective disorders such as depression, pathological anxiety, and post-traumatic stress disorder because these disorders are partly related to dysfunction of limbic/paralimbic structures such as amygdala, hippocampus, and orbitofrontal cortex. With regard to depression, a Cochrane Database review (Maratos *et al.*, 2008) reported that music therapy is associated with improvements in mood in patients with depression (for a study showing positive effects of music making in a group on the mood of individuals see Koelsch *et al.*, 2010a).[46] However, only a small number of studies could be included in that report, and the authors noted that the methodological quality of those studies was low. Therefore, high quality trials are required to evaluate the effectiveness of music therapy on depression. With regard to effects of music on acute, chronic or cancer pain intensity, pain relief, and analgesic requirements, another Cochrane Database review reported that listening to music reduces pain intensity levels (as well as opioid requirements; Cepeda *et al.*, 2006).[47,48] The magnitude of these beneficial effects was relatively small (compared to analgesic drugs), leading the authors to question their clinical importance.[49] Nevertheless, one example for such clinical importance was provided by Tramo *et al.* (2011), who reported that (vocal) music can ameliorate pain and stress in premature infants following heel sticks (a painful procedure applied to obtain blood samples). In that study, heart rate, and crying (due to the heel stick) decreased during a 10-minute recovery in infants exposed to music, but not in unexposed infants.

Moreover, as described in Chapter 8, music can evoke emotions with vitalizing and regenerative effects (this relates salutary effects of emotion to *vitalization*, and *emotional musicogenic meaning*; see first and third rows of Table 12.4): Emotions always have effects on the vegetative (or 'autonomic') nervous system and the hormonal (endocrine) system, thus also leading to changes within the immune

---

[46]   In that study, Koelsch *et al.* (2010) making music together in a group decreased depressive mood and fatigue, and increased vigour (as measured with the *Profile of Mood States*).

[47]   See also Hillecke *et al.* (2004).

[48]   It appears that, in clinical settings, pain reductions are often due, at least in part, to the attraction of attention by a musical stimulus. Because musical information consumes cognitive (including attentional) resources, patients listening to music are distracted from fearful and worrying thoughts, and from the perception of the noises produced by medical procedures (Koelsch *et al.*, 2011).

[49]   For example, in studies evaluating acute postoperative pain, pain intensity in patients exposed to music was 0.5 units lower on a scale ranging from zero to ten, compared to pain intensity in unexposed subjects.

system. Knowledge of the effects that music making has on these systems is still sparse, but due to the potential of music to evoke and modulate emotions, it is conceivable that music therapy can be used for the treatment of disorders related to dysfunctions and dysbalances within these systems. Such disorders do not only comprise affective disorders, but also chronic somatic disorders such as autoimmune diseases.

(7) Engaging in social functions.   This can lead to the training, restoration, and maintenance, of skills such as communication (thus also expression), coordination of actions, and cooperation. Moreover, engaging in such activities fulfils basic human needs, and can evoke emotions such as fun, joy, and happiness (related to the establishment and maintenance of social bonds). Therefore, it seems highly likely that engaging in social functions has regenerative, salutary effects on humans (including, for example, autonomic, endocrine, and immune systems effects). Although systematic research in this area is still lacking, there is some evidence that exclusion from the engagement in these social functions has deleterious effects on human health and life expectancy. For example, social isolation is related to emotional stress (including increased blood pressure and hypertension, thus posing a risk factor for cardiovascular disease), as well as to less efficacious repair and maintenance of physiological functioning, including slower wound healing and poorer sleep efficiency, which makes social isolation a risk factor for morbidity and mortality (Cacioppo & Hawkley, 2003). Engaging in musical activities, on the other hand, prevents social isolation (and thus the adverse effects of social isolation).

13

Concluding Remarks and Summary

13.1 Music and language

The previous chapters described cognitive processes (and their neural correlates) underlying the processing of music, often drawing parallels to analogous processes underlying the processing of language. In this section, I will systematically relate processes of music perception (as described by the model of music perception presented in Chapter 8) to processes of language perception as described by the neurocognitive model on auditory sentence processing by Angela Friederici (2002). Both models show several analogies, such as *Feature Extraction* (music model) and *Primary acoustic analysis* (language model), *Analysis of intervals* (music model) and *Identification of word form* (language model), *Syntactic structure building* (both models), *Re-analysis* (both models), as well as *Meaning* (music model) and *Semantic relations* (language model).

With regard to 'feature extraction', both music and speech require decoding of acoustic information. As described in Section 1.5, with regard to acoustics the terms *phoneme* and *timbre* are equivalent, because both are characterized by spectrum envelope and amplitude envelope (the two physical correlates of timbre).[1] Thus, the *Identification of phonemes* in language (see model on auditory sentence processing by Friederici, 2002) is paralleled by the identification of

---

[1] For his *Talking Piano*, Peter Ablinger converted the spectrum envelope of speech into piano pieces in a way that the original speech is still recognizable (although the sounds are produced by a piano; piano keys are played by a machine under computerized control).

Brain and Music, First Edition. Stefan Koelsch.

timbres in music. However, the segmentation of phonemic information during language perception usually requires a higher temporal resolution compared to music perception (because timbral information in music usually does not change as rapidly as phonemic information in language). This probably leads to the left-hemispheric weighting for the segmentation of phonemes during language perception, whereas segmentation of spectral information, such as melodic information of speech prosody or musical melodies, engages the right auditory cortex more strongly than the left auditory cortex (e.g., Zatorre *et al.*, 2002, Hyde *et al.*, 2008). This functional hemispheric specialization of the auditory cortex can be observed already at birth (Perani *et al.*, 2010), and presumably develops during pregnancy.[2]

Adequate processing of both music and speech also requires auditory sensory memory, and auditory scene analysis including auditory stream segregation and auditory Gestalt formation (summarized under 'Gestalt formation' in Table 13.1). This is particularly important in noisy environments (which are more common than the quiet laboratory conditions typical for experiments on language or music perception).[3]

The processes of minute *interval analysis* ('*interval analysis* 'in Table 13.1) mediate the identification of chords, e.g., whether a chord is a major or a minor chord, whether a chord is presented in root position or in inversion, etc. As mentioned in Chapter 9, such modifications alter the syntactic properties of chords. In this regard, the processes of *interval analysis* perhaps parallel those underlying the identification of word form: Both words and chords have a stem, or root, from which different versions, with different syntactic properties, emerge (i.e., inflections and inversions). That is, a chord function might parallel a lexeme, and the different versions of a chord function parallel word inflections. Note that chord inversions are not the only variations that modify syntactic properties of chords (because such properties are also determined by melodic, rhythmic, etc. information).

Likewise, syntactic structure-building up to the level of structures involving long-distance dependencies (phrase-structure/context-free grammar; see *structure building* in Table 13.1) can be found in both music and language. Processing of music- and language-syntactic information interacts at levels of morpho-syntactic processing, phrase-structure processing, and possibly word-category information (see the evidence for the interaction between ERAN and LAN, and the provisional evidence for an interaction between ERAN and ELAN described in Chapter 9).

---

[2]   Structural left-right asymmetries become recognizable by the 31st week of gestation (Chi *et al.*, 1977). The human foetus responds to sounds at 19 weeks of gestational age, and the auditory system appears to be functional by the start of the third trimester (28th week; Hepper & Shahidullah, 1994; Birnholz & Benacerraf, 1983).

[3]   For a study showing topological shifts in the activation patterns of language perception due to speech degradation (noise-band vocoding) see Obleser *et al.* (2011).

The processes of syntactic re-analysis and revision (*structural re-analysis* in Table 13.1) can also be observed in both music and language (see, e.g., *Processes of re-analysis and repair* in the model of auditory sentence processing by Friederici, 2002). That is, beyond the early interactions between music- and language-syntactic processes, it appears that cognitive and neural resources are also shared during later stages of syntactic re-analysis, integration, and revision (see also Chapter 6 and Patel, 1998). The notion of shared resources during these later stages is also referred to as *shared syntactic integration resource hypothesis* (SSIRH; Patel, 2003).

Both music and language can give rise to regenerative affective processes (*vitalization* in Table 13.1). In language, this is perhaps particularly the case for literature, for the understanding of wisdom (for example, of some philosophical or some religious texts), and for language that contributes to self-esteem.[4] So far, such emotional effects have, however, received only little attention from neuroscientists. Emotional effects of language have been investigated mainly with regard to the perception of affective prosody (e.g., Ethofer *et al.*, 2009; Wittfoth *et al.*, 2010) as well as the affective content of words (e.g., Võ *et al.*, 2009; Herbert *et al.*, 2009).

Finally, both speech and music perception involve perception-action mechanisms and premotor coding (see *premotor* in Table 13.1; premotor activations due to perception-action mediation in music are described in Chapter 11). Liberman & Mattingly (1985) proposed in their *motor theory of speech perception* that, during speech perception, speech is decoded in part by the same processes that are involved in speech production. Since then, several neuroimaging studies showed (pre)motor activations in response to the perception of speech sounds, word meanings and sentence structures (e.g., Pulvermüller & Fadiga, 2010). For example, listening to action-related sentences activates premotor areas (Tettamanti *et al.*, 2005), and perception of action words that are semantically related to different parts of the body (for example, 'lick', 'pick' and 'kick') activates premotor cortex in a somatotopic manner (Pulvermüller, 2005).

All of these processes can activate representations of meaningful concepts, in both music and speech. For example, with regard to feature extraction, a single tone or phoneme can have sign quality activating representations of meaningful concepts, syntactic-structural properties have meaning in both music and language, etc. Chapter 10 illustrated (a) that music can also convey meaningful information (and that, thus, communication of meaning is not exclusively a linguistic domain), (b) that semantic processing of musical sign qualities gives rise to N400 responses, and (c) that processing of intra-musical meaning might be reflected in the N5 (relations between musical and linguistic semantics are discussed further in the next section).

---

[4]   That is, for language that contributes to self-respect, self-acceptance, self-assertiveness, self-responsibility, self-efficacy, integrity, and self-confidence.

Corresponding to the shared processes underlying the perception of music and language, there is considerable resemblance, overlap, and interaction between the ERP components (and their neural generators) that reflect these processes. Processing of musical as well as of speech information evokes (a) FFRs originating from the auditory brainstem; (b) P1, N1, and P2 potentials originating from the auditory cortex; (c) MMN potentials originating from temporal and frontal cortical areas; (d) potentials of (early) syntactic processing that interact with each other (ERAN/LAN), and receive main contributions from the (inferior) pars opercularis (BA 44v); (e) potentials of syntactic (re)integration/re-analysis and repair (P600), and (f) N400 effects reflecting semantic processing (probably originating from posterior temporal and inferior frontal cortex).

13.2   The music-language continuum

The illustrated overlaps of the cognitive operations (and neural mechanisms) underlying music- and language-processing indicate that 'music' and 'language' are different aspects, or two poles, of a single continuous domain. I refer to this domain as the *music-language continuum*. Several *design features* (Fitch, 2006; Hockett, 1960) of 'music' and 'language' are identical within this continuum. Fitch (2006) mentioned complexity, generativity, cultural transmission, and transposability in this regard. *Complexity* means that 'musical signals (like linguistic signals) are more complex than the various innate vocalizations available in our species (groans, sobs, laughter and shouts)' (Fitch, 2006, p. 178). *Generativity* means that both 'music' and 'language' are structured according to a syntactic system (usually involving long-distance dependencies/context-free grammar).[5] *Cultural transmission* means that music, like language, is learned by experience and culturally transmitted. *Transposability* means that both 'music' and 'speech' can be produced in different keys, or with different 'starting tones', without their recognition being distorted.

Two additional design features, that could be added to this list, are *universality* (all human cultures that we know of have music as well as language), and – related to this – the human *innate learning capabilities* for the effortless acquisition of music and language: Even individuals without formal musical training show sophisticated abilities with regard to the decoding of musical information, the acquisition of knowledge about musical syntax, the processing of musical information according to that knowledge, and the understanding of music. This

---

[5]   Fitch (2006) mentions that 'a second component of linguistic 'generativity' (in the technical sense), a symmetry between listener and speaker that Hockett (1960) termed 'interchangeability', is not typically present in instrumental music. One can understand and appreciate a viola or an oboe performance despite being unable to play either instrument'. However, one should be careful here in that melodies can usually be sung. The fact that oboe and voice have different timbres is paralleled by the fact that the voice timbre of two speakers is usually not identical.

supports the notion that musicality is a natural ability of the human brain, and parallels the natural human ability to acquire language. With regard to production, many cultures do not have concepts such as 'musician' and 'non-musician', let alone 'musical' and 'unmusical' (Cross, 2008b), indicating that, at least in some cultures it is natural that everyone actively participates in music-making. Also note that musical abilities are important for the acquisition and the processing of language: Infants acquire information about word and phrase boundaries (possibly even about word meaning) in part through different types of prosodic cues such as speech melody, metre, rhythm and timbre (that is, through *musical* aspects of speech). At birth, a newborn child does not understand words such as 'milk', 'drink', or 'sleep' – therefore, the voices of other individuals are perceived musically (and already newborns have sophisticated abilities to process, group, and differentiate between sounds, e.g., Moon *et al.*, 1993; Winkler *et al.*, 2009b; Stefanics *et al.*, 2007, 2009; Háden *et al.*, 2009). The assumption of this intimate connection between music and speech is corroborated by the findings of overlapping and shared neural resources for music and language processing in both adults and children (Chapter 9). These findings suggest that the human brain, particularly at an early age, does not treat language and music as strictly separate domains, but rather treats 'language as a special case of music' (Koelsch & Siebel, 2005).

Beyond these identical design features, there are also design features that are typical for either 'music' at one end of the continuum, or 'language' at the other, but that overlap between language and music in *transitional zones*, rather than being clear-cut distinctive features for 'music' or 'language' in general. These features are *scale-organized discrete pitch*, *isochrony*, and *propositional semantics*.

Pitch information is essential for both music and speech. With regard to language, tone languages rely on a meticulous decoding of pitch information (due to tones coding lexical or grammatical meaning), and both tonal and non-tonal languages use suprasegmental variations in F0 contour (*intonation*) to code structure and meaning conveyed by speech (phrase boundaries, questions, imperatives, moods and emotions, etc.). Music often uses sets of *discrete pitches*, whereas such discrete pitches are not used for speech. However, pitches in music are often less discrete than one might think. For example, players of non-tempered instruments (e.g., string instruments) produce different pitches, depending on the key, or on whether the tone is a leading note or a target note (particularly in romantic music, where leading notes are played with a pitch that is very narrow to the target note). Thus, for example, *c sharp* and *d flat* (which are the same key on the piano) may be played with a number of different pitches throughout a piece. Moreover, in many musical styles, playing an instrument involves use of glissandos (think, e.g., of the pitch bending produced by blues guitar players). Finally, many kinds of drum music do not use any scale-organized discrete pitches at all, yet we readily refer to these kinds of music as 'music'. On the other hand, the pitch height of different pitches produced during speaking appears to be not arbitrary, but rather to follow principles of the overtone series, which is also the basis of the pitches of many musical scales (e.g., Ross *et al.*, 2007). Particularly emphatic speech (that borders on song) often uses discrete scale-like pitches (in addition to more isochronous timing of syllables). This illustrates that discrete pitches (such as piano tones) are at the musical end of the music-language continuum, and that there is a transitional zone of the use of discrete pitch, that draws into both 'music' and 'language'.

Also at the 'musical' end of the continuum is an *isochronous* tactus on which the musical signals are built in time. Though such an isochronous pulse does not appear to be a characteristic feature of spoken language, it can be found in poetry (Lerdahl, 2001a), ritualistic speech, and emphatic speech. On the other hand, not all kinds of music are based on a tactus (in particular pieces of contemporary music),[6] and many kinds of music have considerable variability in their tactus (e.g., due to expressive timing, ritardandos, and accelerandos). Thus, like discrete pitches, isochronous signals are more characteristic for the musical end of the music-language continuum, and there is a transitional zone from isochronous to non-isochronous signals in both music and speech. As mentioned above, emphatic speech borders on song, and often uses discrete scale-like pitches as well as more isochronous timing of syllables. Martin Luther King's speeches are a nice example of how it is often difficult to say whether someone is singing or speaking, and many art-forms, such as rap-music or recitatives, represent transitional zones from speech to song. Fitch (2006) noted that 'discrete time and pitch make music more acoustically predictable than language, and thus enhance acoustic integration between multiple individuals in an ensemble' (p. 179). Therefore, an isochronous pulse is a logical consequence when several individuals share the joint intention to produce sounds together.

With regard to the language-end of the music-language continuum, it appears that no musical tradition makes use of *propositional semantics*. Drum- and whistle languages can imitate language (Chapter 10), leading to music-like signals with propositional semantics.[7] In that case, however, propositional semantics is an imitative, rather than a genuine feature of 'music'. Nevertheless, on the one hand music can prime representations of quantifiers such as 'some' and 'all', and possibly also evoke at least vague associations of some modals (such as 'must' in passages conveying strong intentionality) or connectives (by establishing dependency relations between musical elements). In Western music, such capabilities of music can be used to convey narrative content of music, but clearly there is no existence of (nor necessity for) the full-blown vocabulary of propositional semantics of language. On the other hand, quantifiers, modals, or connectives are often used imprecisely in everyday language (think of the 'logical and', or the 'logical or'). Simply the existence of the two terms 'propositional' and 'non-propositional' easily leads to the illusion that there is a clear border between 'propositional' and 'non-propositional' (or that one is the opposite of the other). However, in reality

---

[6]   Think, for example, of Ferneyhough's *Etudes Transcendantales* or his *String Quartet No 3*, Ligeti's *Lux Aeterna*, or many passages of Parra's *Sirrt die Sekunde*.

[7]   See also the artificial Solresol language and, with regard to rhythm, Morse-code.

there is a transitional zone between 'propositional' and 'non-propositional', and the degrees of freedom differ between language and music with regard to the construction of propositions. That is, propositional semantics is characteristic for the language-pole of the music-language-continuum, and there is a transitional zone of propositionality drawing into language, as well as partly into music.

Fitch (2006) noted that 'lyrical music, which because it incorporates language thus automatically inherits any linguistic design features' (p. 176). Therefore, anyone interested in how listening to music with propositional semantics might feel like, just has to listen to a song with lyrics containing propositional semantics. As already mentioned in Chapter 10, the interesting phenomenon in this regard is, in my view, that humans appreciate music as a communicative medium, *even though* it has no, or only little, propositional semantics. Cross (2011) offers one interesting hypothesis in this regard: In music, due to relational goals that involve 'the formation, maintenance or restructuring of connections and affiliations between participants', true-false conditions are simply not required to be made mutually explicit (similar to the statement 'I like you', which does not require the additional statement 'and this is not a lie'). That is, in music the truth of information is either out of the question (in the case of connections and affiliations), or irrelevant (e.g., in the case of musical narratives).

Another design feature that is often taken as characteristic for language is *meaning specificity*. It is true that in all cultures that we know of, language appears to be more suitable to refer to objects of the *extra-individual world* (that is, objects that can be perceived by different individuals, and whose existence and qualities can thus be verified, or falsified, by others). However, although with a limited vocabulary, musical cultures have extra-musical sign qualities that can also convey such meaning, and, for example, the symbolic sign quality of music is, by definition, just as specific as the symbolic sign quality of words. Similarly to the terms 'propositional' and 'non-propositional', the two terms 'communication' (in the sense of conveying specific, unambiguous information with language) and 'expression' (in the sense of conveying rather unspecific, ambiguous information with music) are normally used as if there were two separate realms of conveying meaningful information (communication and expression) with a clear border between them. However, this notion is not accurate, because there is a continuous degree of specificity of meaning information, with 'expression' being located towards one end, and 'communication' towards the other.

More importantly, music can communicate states of the *intra-individual world* (that is, states that cannot be perceived by different individuals, and whose existence and qualities can thus not be falsified by others). Chapter 10 described that music can evoke sensations which, *before* they are reconfigured into words, bear greater inter-individual correspondence than the words that an individual uses to describe these sensations (I refer to this meaning quality as *a priori* musical meaning). In this sense, music has the advantage of defining a sensation without this definition being biased by the use of words. As mentioned in Chapter 10, in music we often deal with a description (not with a verification), and with certainty (not with knowledge). With regard to propositional semantics and binary (true–false) truth conditions, this also means that there is no 'true' or 'false' with regard to the inner application of a rule for the usage of a concept. In other words, although music might seem to be 'far less specific' (Slevc & Patel, 2011) than language (in terms of its semantic specificity), music can be more specific when it conveys information about sensations that are problematic to express with words (because music can operate *prior* to the reconfiguration of sensations into words). Importantly, in spoken language, affective prosody operates in part on this level, because it elicits sensational phenomena in a perceiver that bear resemblance to those that occur in the producer. This notion is supported by the observation that affective information is coded with virtually identical acoustical features in speech and music (Scherer, 1995; Juslin & Laukka, 2003).

Another design feature that seems to be considerably more typical for language than for music is *translatability* (Patel, 2008). Again, however, there is not a clear boundary between 'translatable' and 'non-translatable': There are considerable problems of accurately translating one language into another (which become obvious when using translation software), and there are several ways one might imagine that information is conveyed with different kinds of music. For example, extra-musical meaning can be conveyed in different styles, intra-musical meaning can be construed with different kinds of music, and similar musicogenic meanings might emerge in response to very different kinds of music.

Features that represent less extreme ends of the music–language continuum include performative contexts and repertoire. Music occurs more typically than language in *performative contexts*, although language also occurs in such contexts, for example in theatrical performances and traditional storytelling. The performative contexts of music and language vary considerably between cultures (Fitch, 2006). *Repertoire* is also more typical for music than for language, because songs or performances are typically repeated. This is less typical for language, except theatrical performances and traditional storytelling (I should perhaps also add lecturing), 'phatic' communication such as greetings or farewells, and ritual language such as prayers, blessings, invocations, etc. (Fitch, 2006).

The description of the design features illustrates that the notion of clear-cut dichotomies of these features, and thus of clear-cut boundaries between music and language, is too simplistic. Therefore, any clear-cut distinction between music and language (and thus also any pair of separate definitions for language and music) is likely to be inadequate, or incomplete, and a rather artificial construct.[8] Due to our language games, the meaning of 'music' and 'language' is sufficiently sharp for an adequate use in everyday language. For scientific language, however, it is more accurate to consider the transitional nature of the mentioned design features, and to distinguish a scientific use of the words 'music' and 'language' from the use of these words in everyday language. Using the term *music–language*

---

[8]   Similar to the distinction between perception and action, or motor and sensory processes, which do not occur in isolation in an individual.

continuum acknowledges both the commonalities between music and language, and the transitional nature of the design features of music and language.

13.3   Summary of the theory

This section summarizes the music–psychological theory established in this book. The essence of the theory is illustrated in Table 13.1, which contains a systematic overview of the processes, concepts, functions, and principles introduced in this book. With regard to that table, I will refer to the different areas discussed in this book as *domains*. The domains are listed in the outermost left column of Table 13.1 (i.e., I refer to *music perception, syntactic processing, musical meaning*, etc. as 'domains'). Moreover, I will refer to the seven columns listed right of the domains as *dimensions*. For example, the dimensions of the domain *music perception* are *feature extraction, Gestalt formation, interval analysis, structure building*, etc. Labels of these dimensions are provided in the first row of Table 13.1 (*opening, formation, content, combination, coordination, integration*, and *system*). These labels represent abstractions of the dimensions, and serve the purpose of communicating to which dimension a process, function, principle etc. is ascribed (for example, *feature extraction* is ascribed to the *opening dimension*).[9]

The theory is based on the model of music perception that I developed with Walter A. Siebel (Koelsch & Siebel, 2005; the model is described in Chapter 8).[10]

---

[9]   *Opening* refers to the opening of each domain in the sense that, if the domain has a strong horizontal organization, the process listed in the opening dimension is the first process, or a prerequisite for the other processes of the domain. *Opening* also refers to the opening of the system described in a domain. For example, *feature extraction* is related to perception, that is, the sensory opening for acoustic information. *Formation* refers to the formation of representations into higher-order units, usually without long-term knowledge being required. For example, during auditory *Gestalt formation*, auditory perceptual elements are formed into a representation of an auditory object based on Gestalt principles. *Content* refers to adding content from memory, e.g., to a representation of a formation. For example, a representation of a tonal hierarchy (which can be built without knowledge; see *knowledge-free structuring*) can be filled with knowledge about statistical probabilities of likely events to follow (*musical expectancy formation*). *Combination* refers to the application of intelligence, reasoning, and logic when combining elements in a way that meaning emerges (such as *structure building* when processing hierarchically organized structures). *Combination* also refers to differentiating and relating structural elements. *Coordination* refers to both coordination of (physical) movements (including correction and synchronization of movements), as well as to co-ordination of structural elements in the sense of re-ordering these elements to establish new structures. *Integration* refers to the integration of processes (often within a domain). Due to the integrative nature of emotions (typically comprising of affective, autonomic, and endocrine activity, motor expression, action tendency, and subjective feeling), the *integration* dimension has a particular relation to emotion. This parallels the particular relation of the *combination* dimension to the mind, and the relation of the *coordination* dimension to the body. *System* has a particular emphasis on the systemic effects within a domain, and/or on the overlap of one domain with another domain (for example, the *system* dimension of *music perception* overlaps with the *opening* dimension of the *action* domain).

[10]   The model, in turn, was developed in part by applying Siebel's *Noologic Theory* (Siebel *et al.*, 1990).

Table 13.1 Systematic overview of the processes and concepts introduced in this book. The outermost left column indicates the domain, the remaining columns list processes and concepts (i.e., the dimensions of each domain).

domain	opening	formation	content	combin-ation	coord-ination	integra-tion	system
music perception	feature extraction	Gestalt formation	interval analysis	structure building	structural re-analysis	vitalization	premotor, immune system
syntactic processing	element extraction	knowledge-free structuring	musical expectancy formation	structure building	structural re-analysis	syntactic integration	large-scale structuring
musical meaning	iconic	indexical	symbolic	intra-musical	physical	emotional	personal
intra-musical	build-up of structure	stability of structure	extent of structure	structural breach	post-breach structure	resolution	(large-scale) relation
action	action goal	program formation	motor commands, efference copies	differenti-ation (relating)	correction	integration	perception of action effect
social functions	contact	social cognition	co-pathy	communi-cation	coord-ination	cooper-ation	social cohesion
emotion principles	evaluation	contagion	memory	expectancy	imagina-tion	under-standing	social functions, aesthetics
salutary factors	perception	attention	long-term memory	intelligence	action	emotion	social functions

The principles underlying the stages, or dimensions, of music perception are regarded here as so fundamental for music psychology (and psychology in general), that dimensions of other domains were derived in such a way that they correspond to the dimensions of music perception.[11] That is, the theory is constructed in a way that, on an abstract level, processes, functions etc. of different domains are related within the same dimension (i.e., vertically in Table 13.1). By doing so, a coherent framework is established which integrates different (music-)psychological domains by assuming numerous shared processes and similarities.

---

[11] I follow here Siebel's Theory (which also comprises seven psychological dimensions; Siebel *et al.*, 1990). Several concepts in Table 13.1 emphasize relations to his theory, e.g., *opening*, *content*, and *communication*.

The dimensions of each domain do not necessarily stand in any causal relationship with each other. For example, with regard to the *emotion principles*, none of the principles is a sufficient condition for any of the other principles of that domain, although there are possible causal relationships (e.g., engaging in a *social function* may evoke an attachment-related emotion in the hippocampal formation, and this emotion may be *evaluated* in the orbitofrontal cortex). However, *if* a phenomenon A is a necessary condition for a phenomenon B in a row (in the sense of $\neg A \rightarrow \neg B$, or $B \rightarrow A$), then A is listed left of B. For example, *extraction of acoustic features* is a necessary condition for the other processes of *music perception* (therefore listed left of all other phenomena in that row; see first domain of Table 13.1). Note, however, that simply the appearance of a phenomenon left of another phenomenon does not require that the former is a necessary condition for the latter (nor does it require that the former is a sufficient condition for the latter). For example, the fact that *vitalization* is listed left of *premotor* processes does not mean that it is a necessary condition for premotor processes (which it is not: It is possible that, e.g., after *extraction of acoustic features* a single tone directly activates a premotor representation).

It is important to note that processes can feed back into those located left of them. For example, all processes of music perception located right of auditory *feature extraction* can feed back into, and thus modulate, processes of acoustic feature extraction. Moreover, whereas most domains show strong (horizontal) inner-domain relations (*music perception, syntactic processing, intra-musical, action*, and *social functions*), other domains show – due to their nature! – stronger vertical than horizontal relations (*musical meaning, emotion principles*, and *salutary factors*). For example, as mentioned above, the emotion principles do not stand in any necessary causal relationship to each other – therefore, they show stronger vertical relations to other domains, and only relatively weak horizontal (within-domain) relations. Also note that a domain can be nested in another domain. For example, *intra-musical* meaning as one dimension of *musical meaning* is extrapolated into an own domain with seven new dimensions.

In the following, the processes, concepts, functions, principles, and factors of all domains will be summarized, and relations between the same dimension of different domains will be illustrated. Processes and concepts related to the theory are written in oblique type (not all of these terms occur in the table, for reasons of clarity).

Music perception   As mentioned above, *extraction of acoustic features* is a necessary condition for other processes of music perception (therefore listed in the outermost left dimension). Feature extraction happens in the cochlea, the auditory brainstem, as well as the auditory cortex, and is reflected in electrophysiological indices such as frequency-following responses, auditory brainstem potentials, early- and mid-latency responses including the P1, as well as in the N1. *Gestalt formation* is a prerequisite for syntactic *structure building* (and, therefore, listed left of structure building). Gestalt formation and *grouping* are based on operations of the *auditory sensory memory*, and have been investigated using the *mismatch negativity* (grouping, auditory sensory memory, and mismatch negativity are not listed in the Table, but are all ascribed to the *formation* dimension). Operations reflected in the mismatch negativity are assumed to mainly involve neural activity in the region of the middle and posterior part of the superior temporal gyrus (STG). Syntactic *structure building* requires (in tonal music) the distinction between major and minor, or between different inversions of chords (therefore, minute *interval analysis* is listed left of structure building). Accurate analysis of intervals is assumed to involve neural activity in both the posterior and the anterior STG. Disruption of syntactic structure building (involving processing of long-distance dependencies at the phrase structure level) is assumed to be reflected in the early right anterior negativity (ERAN), which receives main contributions from BA 44v. Processes of syntactic *re-analysis and revision* follow processes of structure building, and appear to be reflected electrically in the P600, or Late Positive Component. Such processes presumably involve neural activity in the posterior superior and middle temporal cortex, the inferior parietal lobule, basal ganglia, and, due to increased working memory demands, perhaps also (dorsolateral) frontal and prefrontal regions. Music listening, and perhaps even more so music making, can have *vitalizing* effects on an individual (including activity of the autonomic nervous system, along with the conscious cognitive integration of 'musical' and 'non-musical' information). The (non-musical) physiological effects evoked by music have meaning quality for the perceiver. The interpretation of such meaning information requires conscious awareness, and therefore presumably involves multimodal association cortices such as parietal cortex in the region of BA 7. Note, however, that meaning can arise from all dimensions of music perception. The integrated activity of autonomic and endocrine activity can have regenerative effects on the *immune* system. Moreover, the late stages of music perception overlap with the early stages of action in the form of *premotor* activation. Note that not all processes listed left of premotor processes are necessary conditions for premotor processes – for example, the late stages of the perception of a single tone can activate premotor representations.

Syntactic processing   Both *music perception* and *syntactic processing* begin with the *extraction* of the basic elements that are processed in the subsequent stages. Because the *opening* dimension of syntactic processing may involve the extraction of a tactus, the term *element extraction* is used (instead of auditory feature extraction). Then, based on *knowledge-free structuring* (involving operations of the *auditory sensory memory*), a representation of in-key tones can be established, and out-of-key tones can be detected (thus giving rise to a representation of *key membership*; key membership and auditory sensory memory are not listed in the table, but ascribed to the *formation* dimension). Operations of the auditory sensory memory also serve processes of *grouping* and auditory *Gestalt formation* (auditory sensory memory, grouping, and Gestalt formation are summarized in the domain of *music perception* under *Gestalt formation*). These processes give rise to the establishment of a *hierarchy of harmonic stability* (not listed in the table, but ascribed to the *formation* dimension). Although these processes can be modulated by musical training, they are operational even without musical training (as illustrated by the term *knowledge-free structuring*). By contrast, *musical expectancy formation* is based on (implicit) knowledge of probabilities of (local) transitions of tones and chords. The establishment and application of knowledge about such statistical probabilities requires minute *interval analysis*, to determine probabilities for chord-transitions between major and minor chords, and to recognize how the position of a chord (e.g., root position, sixth, or six-four chord) influences the probabilities of chord-transitions. In addition to the disruption of syntactic structure building, violations of expectancies formed on the basis of *musical expectancy formation* presumably also evoke ERAN potentials. That is, both processes probably elicit sub-components of the ERAN. *Syntactic integration* is related to the processes of music-evoked *vitalization* due to the pleasurable experience emerging from the 'co-'operation of syntactic features (in tonal music melody, metre, rhythm, harmony, intensity, instrumentation, and texture). For example, after the closure of a cadence, and particularly when the closure resolves a previous breach of expectancy, and/or previous dissonances, the simultaneous operation of all syntactic features is perceived as particularly pleasurable and relaxing.[12] Finally, a representation of the *large-scale structuring* of a piece may emerge from the recognition of sub-structures. Large-scale structures imply the spanning of relatively long time intervals (minutes to hours), and the systemic nature of such relations across longer time-spans parallels the systemic nature of *immune* activity as an effect of autonomic and endocrine activity during music perception and production.

Musical meaning   Intra-musical meaning emerges from the processing of structural relations, and is thus related to *structure building*. The dimensions of extra-musical meaning are not grounded on such relations, and are therefore listed left of intra-musical meaning. This ordering also reflects that processing of the three extra-musical sign qualities appears to happen earlier in time than intra-musical processing (as indicated by the latencies of N400 and N5). Iconic and indexical meaning are listed left of symbolic meaning because understanding symbols communicated by another individual requires knowledge of the conventional meaning of the symbol. This parallels the knowledge of statistical probabilities underlying *musical expectancy formation*. Iconic meaning is listed left of indexical meaning because, in contrast to iconic sign quality (e.g., 'the music sounds like a bird'), indexical signs (e.g., 'the music sounds like the voice of someone being happy')

---

[12]   To illustrate this: If, after a long cadence with a decent breach of expectancy, the final resolution does not occur on a heavy beat, then it is not perceived as fully pleasurable (*even though* the succession of chords – without regard to rhythm or metre – would be absolutely correct).

never signal the presence of something that is inherent in the sign itself.[13] *Physical musicogenic meaning* emerges from movements (i.e., physical activity) in response to, and in synchrony with, music (singing, playing an instrument, dancing, clapping, conducting, head-nodding, tapping, swaying, etc.). Merely the fact that an individual shows such activity has meaning for the individual; in addition, the way in which the individual moves expresses meaning information (these movements are 'composed' by the individual). *Emotional musicogenic meaning* can emerge from emotions evoked by music. Music (like affective prosody), can evoke feeling sensations which, before they are reconfigured into words, bear greater inter-individual correspondence than the words that individuals use to describe these sensations. In other words, although music seems semantically less specific than language, music can be more specific when it conveys information about feeling sensations that are problematic to express with words because music can operate prior to the reconfiguration of feeling sensations into words. I refer to this meaning quality as *a priori musical meaning*. Emotional musicogenic meaning is related to the feeling sensations of *vitalization* during music perception. Finally, feeling sensations evoked by a particular piece of music, or music of a particular composer, can have a personal relevance, and thus meaning, for an individual in that they touch, or move, the individual more than feeling sensations evoked by other pieces of music, or music of another composer (partly due to inter-individual differences in personality, both on the side of the recipient and on the side of the producer). That is, music-evoked emotions can also be related to one's inner self, sometimes leading to the experience that one recognizes oneself in the music in a particular, personal way. I refer to this meaning quality as *personal musicogenic meaning*.

Intra-musical features   A *build-up* of a structural context is the prerequisite for the other features of this domain (therefore listed in the outermost left dimension). The build-up of structure is not identical with syntactic *structure building*, but rather a subset of structure building, as is the processing of a structural breach, of post-breach structure, and of resolution (processing of large-scale relations is a special case in this regard, because it involves memory capacities that go beyond the usual time range of short-term, or working memory). The *stability of structure* (for example, determined by the ratio of the number of chord functions of the harmonic core on the one side, and the number of chord functions not in the harmonic core on the other) relates to the *hierarchy of harmonic stability* which can be established based on *knowledge-free structuring* (all ascribed to the

---

[13]   Also note that each single tone has an iconic sign quality (although, in addition, iconic meaning often emerges from sequences of tones), whereas indexical meaning usually emerges from sequences of tones and suprasegmental features such as pitch variation, rising or falling F0 contour, sound level variation, etc. (although even a single tone can also convey emotional information). For detailed information about the acoustic cues that code emotions in music and speech see Juslin & Laukka (2003).

formation dimension). In addition to a representation of a hierarchy of stability, a representation of major–minor tonal key space gives rise to a representation of the *extent of a structure* (for example, a structure can be confined to one tonal key, or span several keys). *Structural breach* is related to the disruption of *structure building*, followed by *post-breach structure*, and ultimately by a *resolution*. These intra-musical features have different meaning qualities, and different emotional effects. In major–minor tonal music, these different features are important to communicate meaning and to evoke emotions. Note that many kinds of music (for example, the music of the Mafa people) do not make use of some of these features.[14] Intra-musical meaning emerging from *large-scale relations* relates to *large-scale structuring* as one feature of (music-)syntactic processing.

Action   Action starts with an *action goal*, and the *formation* of a motor program, which is presumably a necessary condition for the establishment of *efference copies*. With regard to the domain of *music perception*, the forward model established during *program formation* parallels the predictive forward model established based on operations of the auditory sensory memory (summarized under the concept of *Gestalt formation*). At the same time as *motor commands* are sent to the periphery, an *efference copy* is established. Particularly when actions are carried out quickly, it is likely that the efference copy contains more predictive information than the preparation of the efference copy during program formation, e.g., with regard to those sensory consequences of movements that are due to external objects (such as objects that are touched or grazed during the movement). That is, in the course of the establishment of the efference copy, predictive information about sensory consequences is probably added to the forward model. Because these consequences are based on knowledge about the nature of external objects (their weight, temperature, surface texture, etc.) this predictive information parallels the predictive information due to *musical expectancy formation*. During a goal-directed movement, sensory feedback (such as proprioceptive, tactile, and visual information), that is, information about the actual consequences of a movement, is *differentiated* from the information of the efference copy (that is, from predictive information established on the basis of the motor program). Differentiating between actual movement and predicted outcomes allows to us detect deviances between the two (likewise, sensory consequences that deviate from the predicted consequences can be detected). This parallels the processes of structure building which include the differentiation between predicted (or expected) sound and actual musical information. In the action domain, such error-detection is reflected electrically in the error-related negativity, which appears to receive main contributions from the rostral cingulate zone. The error signal can, in turn, lead to the corrective modulation of motor commands (however, movements are

---

[14]   Perhaps the principle of moving from conflict to salvation (*per aspera ad astra*) is so central in Western music because it reflects a central principle of Christian religion.

sometimes faster than the propagation of sensory information to the cortex, thus sensory feedback cannot always be used to correct movements). Processes of *correction* parallel the processes of *structural re-analysis* and revision. Corrections (as well as sensory information that deviates from the efference copy), are integrated with ongoing and subsequent movements of different effectors (such as the two hands during piano playing). Errors and their corrections can be learned (to avoid making similar mistakes in the future). Such learning is related to the actual movement, and thus different from the learning processes emerging from the *perception of action effects*.[15] Note that, once an action effect is learned, the anticipation of an action effect (which is part of the action goal) influences planning, control and execution of an action (the action–effect principle).

Social functions   Inter-individual *contact* is the prerequisite for the emergence of the other social functions (and therefore listed in the outermost left dimension). *Social cognition* parallels the interpretation of *indexical* sign qualities of music (which are related to the inner state of an individual). Moreover, social cognition is a necessary condition for the emergence of *co-pathy* (social cognition is possible without co-pathy, but not vice versa). *Communication* is a necessary condition for the *coordinated* activities that lead to *cooperation*. Moreover, coordination of actions is related to the physical activation from which *physical musicogenic meaning* emerges, as well as to the *correction* of movements during action production. Cooperation parallels the co-operated activity of syntactic features (e.g., of melody, harmony, metre, and rhythm) during *syntactic integration*. Cooperation leads to increased *social cohesion* which, in turn, is related to increased health of the individuals of a group (which relates to the effects of *music perception* on the *immune system*). Moreover, the inter-individual nature of social relations implied by social cohesion parallels *large-scale* relations, and the inter-individual liking component of social cohesion parallels *personal musicogenic meaning*. The experience of all social functions during music making may lead to spiritual experiences (in an additional dimension, not listed here).

Emotion principles   Processes of *evaluation* that elicit emotions can happen at the earliest stages of *feature extraction* (therefore listed in the outermost left dimension), for example at the level of the inferior colliculus or the thalamus (later processes of evaluation may occur at many non-cognitive and cognitive levels, and evaluative processes might happen with or without conscious awareness). Emotional *contagion* is related to *social cognition* and the processing of *indexical* sign

---

[15]   For example, an individual can learn that pressing different keys produces different tones (action effect), yet this learning is different from learning the movement itself that produces the tone. Realizing that the velocity of a keypress was too strong by hearing the produced tone (action effect), and deciding as well as perhaps memorizing this information, is a different learning process. That is, if an action goal is compared with the action effect, then the result of the action effect can be learned. A new attempt can be started, with a slightly different motor program, and the movement can be learned during the action.

quality. Moreover, it parallels processes of the formation of a motor program in the course of movement production (*program formation*). Note that contagion is a precursor of full-blown empathic responses whose social function is *co-pathy* (therefore, contagion is listed left of co-pathy). *Long-term memory* refers here to emotional, episodic, and semantic memory. A memory of an emotion can be activated by musical information, and such activation may evoke the respective emotion (the 'Darling, they are playing our tune' phenomenon). Such processes can occur even if a musical piece has not been heard before. A memory for emotions might be located in the parahippocampal cortex. Furthermore, musical information can activate an episodic memory representation which, in turn, can evoke an emotion. In this case, the musical information has (usually idiosyncratic) *symbolic* sign quality. Semantic memory refers to musical information with (more general) *symbolic* sign quality: The concept evoked by music can have an emotional valence, or connotation, which might evoke an emotional response (similar to emotions evoked by the emotional quality of words). With regard to the other domains, long-term memory is also related to the (implicit) knowledge required for *musical expectancy formation*. The emotion principle of *expectancy* is related here to the *communication* of *intra-musical* meaning (including the tension-arch), that is, to all intra-musical features that give rise to meaning, as well as to emotional responses.[16] *Imagination* refers to emotional effects of being resourceful, inventive, curious, creative, and to emotional effects of trying something out. It is related to the imaginative (and often inventive) processes of *structural re-analysis and revision*, and thus also to the processing of *post-breach structure*. *Understanding*, as well as the success of a trial, evokes feelings of reward and pleasure; this relates understanding to *vitalization*, *syntactic integration*, as well as to the pleasurable experience of the *resolution* of a structural breach (note that a musical sequence with a structural breach cannot be understood fully until it is resolved). Finally, engaging in *social functions* fulfils a basic human need, and can evoke emotions such as fun, joy, and happiness (related to the establishment and maintenance of social bonds). As mentioned above, engaging in these social functions can lead to inter-individual (*large-scale*) social relations, and has positive effects on human health (which relates to the effects of *music perception* on the *immune system*). Such regenerative effects only emerge in the absence of violence, linking social functions to experiences of beauty, and thus to *aesthetic experience*. Beyond the aesthetic experience of social functions, the beauty of musical sounds, contents, and structures can also evoke emotions.

Salutary effects   Salutary effects of *perception* include, for example, restoration of accurate hearing in children with language impairment, or restoration of normal hearing in patients with tinnitus. Perception is related to auditory *feature extraction*, but includes here all other perceptive processes during music listening, and

---

[16]   In this regard, however, empirical data are so far available only for the disruption of *structure building* (which is perceived as a violation of expectancy).

particularly during music making or dancing (such as somatosensory perception, equilibrioception, and visual perception). *Attention* refers here to the voluntary direction of attention (not to the allocation of attention by a stimulus). Salutary effects of voluntary direction of attention include, e.g., amelioration of symptoms in children with attention-deficit disorder. Attention is listed left of *musical expectancy formation* and *structure building* because these processes are influenced by the voluntary direction of attention (as indicated by the influence of attention on the generation of the ERAN). Long-term *memory* refers here to procedural, implicit, semantic, episodic, and emotional memory. Engaging memory (that is, engaging processes of memory encoding and memory retrieval) with music might have positive effects on long-term memory functions in patients with Alzheimer's disease. With regard to the other domains, long-term memory also occurs in the *emotion principles*, and parallels the (implicit) long-term memory operations required for *musical expectancy formation* and recognition of *symbolic* sign qualities of music. Engaging *intelligence* involves constructive reasoning and application of logic (thus, intelligence stays in contrast to illogical, irrational thoughts).[17] Reasoning involves *working memory* (WM) operations, which include, during both music perception and music production, central executive function, as well as use of phonological short-term store (and sometimes also the phonological loop). Using these tools of intelligence during music making or during listening to music probably helps keeping them in good condition. Intelligence is related to *structure building* (thus also to the processing of *intra-musical* meaning) and *communication*. Engaging in *action* during music making or dance keeps muscles, joints, tendons, and peripheral nerves in good physical shape. It also re-establishes and maintains (sensori-)motor skills, e.g., in patients with Parkinson's disease, apraxia, Broca's aphasia, or hemiplegia due to a cerebral stroke. Empirical investigations of salutary effects of (positive) music-evoked *emotion* are still rather sparse.[18] Such beneficial effects could originate from activity of the mesolimbic dopaminergic pathway (related to feelings of fun, pleasure and reward) and activity in the hippocampal formation (due to evocation of attachment-related emotions) in patients with pathological anxiety and depression, as well as in patients with post-traumatic stress disorder. In Table 13.1, salutary effects of *emotion* are related to *vitalization*, and *emotional musicogenic meaning*. Salutary effects of engaging in *social functions* (as well as their relations to effects of *music perception* on the *immune system*) have already been described above.

13.4   Summary of open questions

This section summarizes a few of the open questions raised in this book that give rise to further research. This summary is meant as a catalogue of those research

---

[17]   This inherently relates intelligence, for example, to self-esteem and its beneficial effects for health, because irrational thoughts are a necessary condition for low self-esteem.
[18]   However, effects of music on pain perception appear to be stable, although relatively small.

questions that are, in my view, most pressing with regard to the areas dealt with in this book. It is also supposed to provide interested students and scientists who are new to the field with possible starting points for research. Some of the questions raised here require neuroscientific methods, but many questions can also be addressed with behavioural measures.

Auditory feature extraction   It appears that FFRs are generated by brainstem neurons. However, given the importance of the auditory cortex for the high resolution of spectral information, it is likely that the auditory cortex shapes the FFR responses, and thus significantly contributes to the fidelity of FFRs (note that often the same stimulus is presented for thousands of times in one experimental session, giving the auditory cortex the opportunity to shape brainstem responses via efferent projections). How much the auditory cortex, and how much the brainstem contributes to the fidelity of the FFRs is not known. This could be tested in patients with bilateral lesion of the auditory cortex, or in studies using transcranial magnetic stimulation to inhibit activity in the auditory cortex.

Syntax   As described in Chapter 9, several different processes contribute to the syntactic processing of music. Such processes include the extraction of tactus, metre, tonal centre, establishment of a hierarchy of stability, musical expectancy formation (by applying knowledge about the probabilities for upcoming events), syntactic structure building (involving processing of long-distance dependencies), structural re-analysis and revision, syntactic integration, and processing of large-scale structures. The neural correlates of these processes have not been sufficiently discerned so far. For example, neural correlates of musical expectancy formation on the one side, and musical structure building on the other have not been disentangled. One approach would be to expose participants to stimuli that follow probabilities specified in a Markov-table (or in an n-gram model), and subsequently measure ERPs related to the formation of expectancy, as well as to the violation of such expectancies.[19] This would allow one to investigate ERPs related to predictions of local transition probabilities that are based on long-term knowledge[20] without contributions of hierarchical processing. On the other hand, an artificial phrase-structure grammar with carefully balanced local transition probabilities could test processes underlying hierarchical structure building and processing of long-distance dependencies.[21] That is, experimental factors such as requirement of long-term knowledge on the one side and knowledge-free structuring on the other, or processing of finite-state grammar on the one side and phrase-structure

---

[19]   Stimuli would have to have similar, or no, extra-musical meaning quality. Also, they should not be major–minor tonal stimuli (if Western listeners are tested).
[20]   Or based on information beyond the information available in auditory sensory and auditory short-term memory.
[21]   When discerning processing of local vs. long-distance dependencies, researchers have to be aware of the fact that this syntactic distinction usually confounds working memory requirements.

grammar on the other, as well as working memory demands, should be taken into account carefully. In this regard it is also worth noting that, up to now, no neurophysiological investigation has tested whether individuals actually perceive music cognitively according to tree-structures. One hypothesis is that both violations of local transition probabilities, as well as disruptions of hierarchical structure building elicit early anterior negativities, and that the ERAN is probably a conglomerate of such early negativities. Thus, such studies could disentangle these processes and specify sub-components of the ERAN.

Moreover, most studies in this area have investigated music-syntactic processing of harmony (and a few studies examined music-syntactic processing of melody). There is a scarcity of empirical studies investigating other syntactic aspects (metre, rhythm, intensity, instrumentation, and texture), and no study has so far addressed the interaction between these aspects. For example, how does the independent manipulation of harmonic, metric, and melodic in/correctness influence our correctness-judgements of such sequences? With regard to melodic and harmonic processing, it appears that melodic irregularities are processed earlier than irregular harmonic information (as reflected in the peak latency of the N125 elicited by irregular tones, as compared to the peak latency of the N180 elicited by irregular chords). Future studies could further specify these two components (as well as similar components evoked by other syntactic aspects), and take particular care that acoustic factors do not confound syntactic regularity (such as acoustic similarity and strength of refractoriness effects).

Other studies could investigate how syntactic processes develop during childhood, how they are influenced by attentional factors, and how they are modulated by musical training. Theoretical and ethno-musicological work could specify which of these processes are engaged for the processing of other kinds of music.

With regard to the *Syntactic Equivalence Hypothesis*, there is a lack of functional neuroimaging studies investigating the overlap, or interaction, of neural correlates of music-syntactic and language-syntactic processing. Such studies could further specify interactions between particular syntactic processes and features of both music and language. For example, are there music-syntactic processes that interact more strongly with morpho-syntactic violations than with phrase-structure violations in language? There is also a lack of studies investigating interactions between music-syntactic processing and the processing of the syntax of mathematical formulae and termini, or the processing of syntactic properties of action sequences. Particularly with regard to action sequences, it would be important to differentiate between predictive processes (such as those engaged when observing a grasping movement), and processes required for the processing of hierarchically organized action sequences.

Semantics There is only limited knowledge about the acoustical features that determine the iconic and indexical sign qualities of sounds. For example, which acoustical properties make an acoustic event sound 'warm'/'cold', 'active'/'passive', 'sweet'/'sour', 'rough'/'smooth', etc.? Moreover, no study has so far differentiated electrophysiological correlates of different sign qualities of musical information. Are N400 effects which are elicited by different extra-musical sign qualities (iconic, indexical, symbolic) identical? Studies addressing this issue could also investigate possible different cognitive processes required for the interpretation of such sign qualities. For example, whereas understanding of iconic sign quality needs the establishment of an acoustic similarity between the heard sound and an object that sounds similar, understanding of symbolic sign quality requires memory retrieval of a sound as well as of the meaning associated with that sound.

With regard to the processing of intra-musical meaning, the N5 is taken to reflect harmonic integration, and processing of intra-musical meaning. One piece of evidence supporting this assumption is the interaction between N5 and N400 reported in the study by Steinbeis & Koelsch (2008b). However, so far this is the only study reporting such an interaction. Therefore, interactions between N5 and N400 (and relations between N5 and the processing of intra-musical meaning) should be substantiated, and investigated further (e.g., with regard to whether only a specific subcomponent of the N5 interacts with the N400). Moreover, there is a lack of studies localizing the generators of the N5.[22]

With regard to semantic processing, Chapter 10 mentioned that such processing includes (a) storage of meaningful information, (b) activation of representations of meaningful information, (c) selection of representations of meaningful information, and (d) relation as well as integration of the semantic information of these representations with the previous semantic context. As for language, it remains to be specified where in the brain these processes are located, and how they are reflected electrically in ERPs.

There is also a scarcity of empirical data on the musicogenic dimensions of musical meaning, therefore future studies could empirically test the theoretical considerations that physical, emotional, and personality-related effects of music give rise to meaning. In this regard, it would be of particular interest to investigate neural correlates of (non-conceptual) meaning, and of the reconfiguration of such meaning into language.

Action   Music studies investigating action can use stimuli, and actions, that are considerably more complex, and ecologically more valid, than those investigating action in experiments using 2 or 3 buttons on a response box (a typical setting in this field). Moreover, such studies open the precious opportunity to investigate interactions between action and emotion. For example, when playing a piece, are action-related processes (such as error detection and correction) influenced by the emotional state of the player (such as 'flow experiences' during playing, or a particular mood evoked by the music)? Do emotional states influence the learning

---

[22]   Source localization should take into account that the N5 presumably receives contributions from distributed sources.

of the piece? How does playing music together with other individuals influence action-related processes?

Along these lines, studies in this area could also investigate neural mechanisms underlying neural entrainment to music. For example, alpha activity is known to entrain to isochronous external stimuli – how long does this entrainment take? Is such entrainment influenced by expertise, emotion, and/or preference? Likewise, only little is known about the neural correlates of synchronizing to an external beat, or of inter-individual synchronization during music making and music listening (and possible effects of expertise, emotion, and/or preference on such processes of synchronization are unknown). Studies investigating electrophysiological correlates of inter-individual synchronization during music making in groups can also provide important contributions to the investigation of the influence of social factors on emotional processes, as well as on cognitive processes such as action monitoring and error processing. Moreover, such studies can contribute to the investigation of neural correlates of joint action, shared intentions, and the nonverbal communication of music performers.

Emotion Due to the scarcity of studies investigating particular mechanisms, or principles, underlying the evocation of emotion with music, future studies are needed to specify neural correlates of these principles. Even if the investigation of the emotion principles is not the primary interest of a study, researchers could – in the course of planning a study on emotion – consider which emotion principle/s might be involved.

Likewise, the different neural circuits that mediate different emotions are still only poorly understood. Future studies could aim at specifying the neural signatures of different emotions, with a particular emphasis on thoroughly discriminating between different positive and different negative emotions. Such studies could also begin to approach the investigation of mixed emotions.

In Chapter 12 I endorsed the hypothesis that the hippocampal formation plays an important role for the generation of attachment-related emotions (subjectively experienced as joy, love, and happiness). Due to the large proportion of functional neuroimaging studies reporting activity changes in the hippocampal formation in association with music-evoked emotions, it seems particularly promising to investigate this hypothesis with music. However, due to the involvement of the hippocampus in memory, familiarity, novelty, and expectancy, future studies on this topic need to carefully control for such factors (to be able to pin down activity changes in the hippocampal formation to emotional processes).

With regard to EEG, only very few EEG-studies have investigated music-evoked emotions (these studies analysed oscillatory neuronal activity). Due to the potential of music to evoke emotions, and due to the advantages of EEG compared to fMRI (particularly the absence of scanner noise), future studies should further investigate electrophysiological correlates of emotion. One should bear in mind, however, that such studies are likely to capture cognitive antecedents, and cognitive effects of emotion-related activity of subcortical (limbic/paralimbic) structures, due to the fact that the main contributions to the scalp-recorded EEG are generated by (neo)cortical pyramidal neurons. One exception is the anterior cingulate cortex, which is critically involved in emotional processes (described in Chapter 12). Since most of the above-mentioned studies in this field have analysed oscillatory neuronal activity (not ERPs), the most pressing tasks are to investigate the functional significance of the observed effects with regard to emotional processes. For example, with regard to reported effects in the alpha band (Schmidt & Trainor, 2001; Baumgartner *et al.*, 2006a) or the theta band (Sammler *et al.*, 2007), it is difficult to determine whether such effects reflected rather cognitive processes (such as attention, memory, subjective feeling, or appraisal), or processes related to the modulation of peripheral-physiological activity, processes related to motor expression, to action tendencies, etc. Note that in the study by Sammler *et al.* (2007) oscillatory effects in the theta band were stronger during the second half of each excerpt (that is, in the time interval of 30 to 60 seconds). Therefore, it seems advisable to use relatively long musical stimuli (at least one minute in duration).

Along these lines, future studies could also put more emphasis on the investigation of temporal aspects of music-evoked emotions. So far, only three studies (Koelsch *et al.*, 2006a; Sammler *et al.*, 2007; Salimpoor *et al.*, 2011) have investigated the temporal dynamics of the neural correlates of emotional processing. Information about activity changes over time of the structures implicated in emotion (such as information about how activity in one structure affects activity in another) would provide important insights into the functional significance of these structures, and into the neural substrates of human emotion in general. Note that music is an extremely suitable stimulus to investigate this issue, because on the one hand certain music maintains an emotional expression over longer time intervals (such as minutes), and on the other hand many compositions purposefully move, over time, through different regions of human emotion space (for example, compositions in which the emotional expression changes between different discrete emotions, or different mixed emotions). An additional approach is to investigate neural correlates of music-evoked tension and resolution (for behavioural approaches see, e.g., Lerdahl, 2001b; Bigand *et al.*, 1996; Lerdahl & Krumhansl, 2007). Such studies, however, could take into account the different intra-musical features that give rise to meaning and emotion (such as build-up of structure, structural breach, post-breach structure, resolution; see Table 13.1). This is important for neuroimaging studies, because events such as build-up of structure and resolution might have the same tension values on the one-dimensional tension-resolution scale, but different neural correlates.

Another neural mechanism important for emotion is the *synchronization of biological subsystems* (Scherer, 2000); Scherer even defines an emotion as a synchronization of biological subsystems that comprise cognitive appraisal, physiological arousal, motor expression, motivational processes, and monitoring processes. On the basis of the different functions in which the ACC is involved, we (Koelsch *et al.*, 2010b) have previously proposed that the ACC is involved in such synchronization. Functional neuroimaging studies with different peripheral-physiological and EEG-measures (recorded in the MR scanner) could examine which brain structures are involved in such synchronization, for example by using coherence measures of peripheral-physiological and EEG data as a regressor for the modelling of functional neuroimaging data.

Finally, the fulfilment of the need to understand (i.e., of the *knowledge instinct*) is experienced as rewarding (the 'aha moment', or 'eureka moment'). Some even report that such moments give rise to frissons (reviewed in Panksepp, 1995). Future studies could investigate, e.g., whether understanding of structures at the end of musical sequences involves activity of brain structures that belong to the dopaminergic reward pathway.

Salutary effects of music making   From the perspective of neuroscience and biology, there are numerous reasons to assume that music making (sometimes even just listening to music) has beneficial effects on the psychological and physiological health of individuals. However, surprisingly few high-quality studies have actually tested, and systematically investigated, such effects. In particular, more studies that fulfil the standards of evidence-based medicine (for example, involving randomized designs with a control group) are needed to provide scientific evidence that gives rise to more systematic, more widespread, and more empirically-grounded applications of music in education and therapy.

With regard to salutary effects of voluntary direction of attention during music making, questions arise such as: Can attentional skills be trained with music? What are autonomic and endocrinological effects of focusing attention to music? Can focusing of attention with music help individuals with attention disorders? Can music help patients with Autism Spectrum Disorder to direct their attention to cues relevant for social cognition? Attention deficits often occur together with hyperactivity (e.g., in children with attention deficit/hyperactivity disorder, ADHD) – does focusing attention, e.g., while making music, help children with ADHD to calm down, concentrate, and organize behaviour? Which kinds of music are particularly suitable?

With regard to memory and intelligence, it is often surmised that use of memory and intelligence keep these functions in good shape. However, there is a lack of empirical evidence for this assumption. Long-term trials with patients with neurodegenerative diseases could compare groups of individuals that often make music (or sing or dance) with groups of individuals that participate in non-musical activities involving memory and intelligence skills (such as chess playing), and with groups of individuals undergoing interventions that do not involve either of those functions.

With regard to communicative skills, it was speculated (in Chapter 9) that children with specific language impairment (SLI) would not develop SLI if they received musical training from an early age on. This could be explored in future studies with children at risk for developing SLI, or by comparing randomly selected children with regard to (a) their musical training and (b) language impairment (such studies have to carefully match socio-economic status between groups).

With regard to salutary effects of emotion, several psychopathological disorders are associated with functional and structural anomaly of the hippocampal formation (such as depression, anxiety disorders, and post-traumatic stress disorder, PTSD). Due to the capability of music to evoke activity changes in the hippocampus (Chapter 12), it is conceivable that music therapy with such patient groups has positive effects on the up-regulation of neurogenesis in the hippocampus. Studies investigating this issue could measure (music-evoked) hippocampal activity and hippocampal volume in patients before and after a period with a musical intervention. Such studies could also obtain immune parameters, due to the association between depression and various immune parameters.[23] The rationale of such studies would be that engaging in the social functions during music making (or dancing) together in groups leads to the emergence of attachment-related emotions generated in the hippocampus, thus leading to functional and structural changes within the hippocampal formation. One of the great powers of music is that it can not only be fun, but that it can also make people happy. This power needs to be explored in the future, to provide more systematic, more widespread, and more theoretically-grounded applications of music in education and therapy.

---

[23] Particularly interferones and cytokines such as several interleukins, or tumor necrosis factor (TNF) α.

References

Achaibou, A., Pourtois, G., Schwartz, S. & Vuilleumier, P. (2008). Simultaneous recording of EEG and facial muscle reactions during spontaneous emotional mimicry, *Neuropsychologia* 46(4): 1104–1113.

Alain, C., Woods, D. L. & Knight, R. T. (1998). A distributed cortical network for auditory sensory memory in humans, *Brain Research* 812: 23–37.

Alho, K., Sainio, K., Sajaniemi, N., Reinikainen, K. & Näätänen, R. (1990). Event-related brain potential of human newborns to pitch change of an acoustic stimulus, *Electroencephalography and Clinical Neurophysiology/Evoked Potentials Section* 77(2): 151–155.

Alho, K., Tervaniemi, M., Huotilainen, M., *et al.* (1996). Processing of complex sounds in the human auditory cortex as revealed by magnetic brain responses, *Psychophysiology* 33(4): 369–375.

Allen, R., Hill, E. & Heaton, P. (2009). 'Hath charms to soothe...': An exploratory study of how high-functioning adults with ASD experience music, *Autism* 13(1): 21–41.

Alperson, P. (1994). *What is Music? An Introduction to the Philosophy of Music*, Pennsylvania State University Press.

Altenmüller, E., Schürmann, K., Lim, V. & Parlitz, D. (2002). Hits to the left, flops to the right: Different emotions during listening to music are reflected in cortical lateralisation patterns, *Neuropsychologia* 40(13): 2242–2256.

Amiez, C. & Petrides, M. (2009). Anatomical organization of the eye fields in the human and non-human primate frontal cortex, *Progress in Neurobiology* 89(2): 220–230.

Amunts, K., Lenzen, M., Friederici, A. D., *et al.* (2010). Broca's region: Novel organizational principles and multiple receptor mapping, *PLoS Biology* 8(9): e1000489.

Apel, W. (1970). *Harvard Dictionary of Music*, Cambridge, MA: MIT Press.

Aramaki, M., Marie, C., Kronland-Martinet, R., Ystad, S. & Besson, M. (2010). Sound categorization and conceptual priming for nonlinguistic and linguistic sounds, *Journal of Cognitive Neuroscience* 22: 2555–2569.

Armstrong, T. & Detweiler-Bedell, B. (2008). Beauty as an emotion: The exhilarating prospect of mastering a challenging world, *Review of General Psychology* 12(4): 305–329.

Attneave, F. & Olson, K. (1971). Pitch as a medium: A new approach to psychophysical scaling, *American Journal of Psychology* 84: 147–166.

Augustine, J. R. (1996). Circuitry and functional aspects of the insular lobe in primates including humans, *Brain Research Reviews* 22: 229–244.

Bach, P., Gunter, T. C., Knoblich, G., Prinz, W. & Friederici, A. (2009). N400-like negativities in action perception reflect the activation of two components of an action representation, *Social Neuroscience* 4(3): 212–232.

Bahlmann, J., Rodriguez-Fornells, A., Rotte, M. & Münte, T. (2007). An fMRI study of canonical and noncanonical word order in German, *Human Brain Mapping* 28(10): 940–949.

Bahlmann, J., Schubotz, R. I., Mueller, J., Koester, D. & Friederici, A. D. (2009). Neural circuits of hierarchical visuo-spatial sequence processing, *Brain Research* 1298: 161–170.

Ball, T., Rahm, B., Eickhoff, S., *et al.* (2007). Response properties of human amygdala subregions: Evidence based on functional MRI combined with probabilistic anatomical maps, *PLoS One* 2(3): e307.

Bangert, M. & Altenmüller, E. O. (2003). Mapping perception to action in piano practice: A longitudinal DC-EEG study, *BMC Neuroscience* 4(1): 26–39.

Bangert, M., Peschel, T., Schlaug, G., *et al.* (2006). Shared networks for auditory and motor processing in professional pianists: Evidence from fMRI conjunction, *Neuroimage* 30(3): 917–926.

Barbas, H., Ghashghaei, H., Dombrowski, S. M. & Rempel-Clower, N. L. (1999). Medial prefrontal cortices are unified by common connections with superior temporal cortices and distinguished by input from memory-related areas in the rhesus monkey, *Journal of Comparative Neurology* 410(3): 343–367.

Barrett, S. E. & Rugg, M. D. (1990). Event-related potentials and the semantic matching of pictures, *Brain and Cognition* 14: 201–212.

Baumeister, R. & Leary, M. (1995). The need to belong: Desire for interpersonal attachments as a fundamental human motivation, *Psychological Bulletin* 117(3): 497–497.

Baumgartner, T., Esslen, M. & Jäncke, L. (2006a). From emotion perception to emotion experience: Emotions evoked by pictures and classical music, *International Journal of Psychophysiology* 60(1): 34–43.

Baumgartner, T., Lutz, K., Schmidt, C. & Jäncke, L. (2006b). The emotional power of music: How music enhances the feeling of affective pictures, *Brain Research* 1075(1): 151–164.

Beck, R., Cesario, T., Yousefi, A. & Enamoto, H. (2000). Choral singing, performance perception, and immune system changes in salivary immunoglobulin A and cortisol, *Music Perception* 18: 87–106.

Belin, P., Fecteau, S. & Bédard, C. (2004). Thinking the voice: Neural correlates of voice perception, *Trends in Cognitive Sciences* 8(3): 129–135.

Bendor, D. & Wang, X. (2005). The neuronal representation of pitch in primate auditory cortex, *Nature* 436(7054): 1161–1165.

Bentin, S. & Deouell, L. (2000). Structural encoding and identification in face processing: ERP evidence for separate mechanisms, *Cognitive Neuropsychology* 17(1): 35–55.

Berent, I. & Perfetti, C. A. (1993). An on-line method in studying music parsing, *Cognition* 46: 203–222.

Berridge, K., Robinson, T. & Aldridge, J. (2009). Dissecting components of reward: Liking, wanting, and learning, *Current Opinion in Pharmacology* 9(1): 65–73.

Besson, M. & Faita, F. (1995). An event-related potential (ERP) study of musical expectancy: Comparison of musicians with nonmusicians, *Journal of Experimental Psychology: Human Perception and Performance* 21(6): 1278–1296.

Besson, M., Faita, F., Peretz, I., Bonnel, A. M. & Requin, J. (1998). Singing in the brain: Independence of lyrics and tunes, *Psychological Science* 9(6): 494–498.

Besson, M., Frey, A. & Aramaki, M. (2011). Is the distinction between intra- and extra-musical meaning implemented in the brain? (Comment), *Physics of Life Reviews* 8(2): 112–113.

Besson, M. & Macar, F. (1986). Visual and auditory event-related potentials elicited by linguistic and non-linguistic incongruities, *Neuroscience Letters* 63(2): 109–114.

Besson, M. & Macar, F. (1987). An event-related potential analysis of incongruity in music and other non-linguistic contexts, *Psychophysiology* 24: 14–25.

Bharucha, J. (1984). Anchoring effects in music: The resolution of dissonance, *Cognitive Psychology* 16: 485–518.

Bharucha, J. & Krumhansl, C. (1983). The representation of harmonic structure in music: Hierarchies of stability as a function of context, *Cognition* 13: 63–102.

Bharucha, J. & Stoeckig, K. (1986). Reaction time and musical expectancy: Priming of chords, *Journal of Experimental Psychology: Human Perception and Performance* 12: 403–410.

Bharucha, J. & Stoeckig, K. (1987). Priming of chords: Spreading activation or overlapping frequency spectra? *Perception and Psychophysics* 41(6): 519–524.

Bigand, E., Madurell, F., Tillmann, B. & Pineau, M. (1999). Effect of global structure and temporal organization on chord processing, *Journal of Experimental Psychology: Human Perception and Performance* 25(1): 184–197.

Bigand, E., Parncutt, R. & Lerdahl, J. (1996). Perception of musical tension in short chord sequences: The influence of harmonic function, sensory dissonance, horinzontal motion, and musical training, *Perception and Psychophysics* 58(1): 125–141.

Bigand, E. & Pineau, M. (1997). Global context effects on musical expectancy, *Perception and Psychophysics* 59(7): 1098–1107.

Bigand, E., Poulin, B., Tillmann, B., Madurell, F. & D'Adamo, D. A. (2003). Sensory versus cognitive components in harmonic priming, *Journal of Experimental Psychology: Human Perception and Performance* 29(1): 159–171.

Bigand, E. & Poulin-Charronnat, B. (2006). Are we 'experienced listeners'? A review of the musical capacities that do not depend on formal musical training, *Cognition* 100(1): 100–130.

Bigand, E., Tillmann, B., Poulin, B., D'Adamo, D. & Madurell, F. (2001). The effect of harmonic context on phoneme monitoring in vocal music, *Cognition* 81(1): B11–B20.

Birnholz, J. & Benacerraf, B. (1983). The development of human fetal hearing, *Science* 222(4623): 516–518.

Björklund, A. & Dunnett, S. (2007). Dopamine neuron systems in the brain: An update, *Trends in Neurosciences* 30(5): 194–202.

Bleaney, B. I. & Bleaney, B. (1976). *Electricity and Magnetism* 3rd ed. Oxford: Oxford University Press.

Block, N. (2005). Two neural correlates of consciousness, *Trends in Cognitive Sciences* 9(2): 46–52.

Blood, A. J., Zatorre, R., Bermudez, P. & Evans, A. C. (1999). Emotional responses to pleasant and unpleasant music correlate with activity in paralimbic brain regions, *Nature Neuroscience* 2(4): 382–387.

Blood, A. & Zatorre, R. (2001). Intensely pleasurable responses to music correlate with activity in brain regions implicated in reward and emotion, *Proceedings of the National Academy of Sciences* 98(20): 11818–11823.

Boehm, S. & Paller, K. (2006). Do I know you? Insights into memory for faces from brain potentials, *Clinical EEG and Neuroscience* 37(4): 322.

Bonnel, A. M., Faita, F., Peretz, I. & Besson, M. (2001). Divided attention between lyrics and tunes of operatic songs: Evidence for independent processing, *Perception and Psychophysics* 63(7): 1201–1213.

Bornkessel-Schlesewsky, I. & Schlesewsky, M. (2008). An alternative perspective on semantic P600 effects in language comprehension, *Brain Research Reviews* 59(1): 55–73.

Botvinick, M., Braver, T., Barch, D., Carter, C. & Cohen, J. (2001). Conflict monitoring and cognitive control, *Psychological Review* 108(3): 624–652.

Brandão, M., Tomaz, C., Leão Borges, P., Coimbra, N. & Bagri, A. (1988). Defense reaction induced by microinjections of bicuculline into the inferior colliculus, *Physiology and Behavior* 44(3): 361–365.

Brattico, E., Tervaniemi, M., Naatanen, R. & Peretz, I. (2006). Musical scale properties are automatically processed in the human auditory cortex, *Brain Research* 1117(1): 162–174.

Bregman, A. (1994). *Auditory Scene Analysis: The Perceptual Organization of Sound*, The MIT Press.

Bremner, J. (1999). Does stress damage the brain? *Biological Psychiatry* 45(7): 797–805.

Broadbent, D. (1957). A mechanical model for human attention and immediate memory, *Psychological Review* 64(3): 205–215.

Broadbent, D. E. (1958). *Perception and Communication*, New York, Pergamon Press.

Brown, H., Butler, D. & Jones, M. (1994). Musical and temporal influences on key discovery, *Music Perception* 11: 371–407.

Brown, S., Martinez, M. & Parsons, L. (2004). Passive music listening spontaneously engages limbic and paralimbic systems, *NeuroReport* 15(13): 2033–2037.

Brown, S. & Volgsten, U. (2006). *Music and manipulation: On the social uses and social control of music*, Berghahn Books, Oxford.

Budd, M. (1996). *Values of Art*, Penguin Books.

Burns, D. (2001). The effect of the Bonny method of guided imagery and music on the mood and life quality of cancer patients, *Journal of Music Therapy* 38(1): 51–65.

Buxton, R. (2002). *Introduction to Functional Magnetic Resonance Imaging: Principles and Techniques*, Cambridge University Press.

Cacioppo, J. & Hawkley, L. (2003). Social isolation and health, with an emphasis on underlying mechanisms, *Perspectives in Biology and Medicine* 46(3): S39–S52.

Callan, D., Tsytsarev, V., Hanakawa, T., *et al.* (2006). Song and speech: Brain regions involved with perception and covert production, *Neuroimage* 31(3): 1327–1342.

Caplin, W. (2004). The classical cadence: Conceptions and misconceptions, *Journal of American Musicological Society* 57(1): 51–118.

Carlyon, R. (2004). How the brain separates sounds, *Trends in Cognitive Sciences* 8(10): 465–471.

Carral, V., Huotilainen, M., Ruusuvirta, T., *et al.* (2005). A kind of auditory 'primitive intelligence' already present at birth, *European Journal of Neuroscience* 21(11): 3201–3204.

Carrión, R. & Bly, B. (2008). The effects of learning on event-related potential correlates of musical expectancy, *Psychophysiology* 45(5): 759–775.

Celesia, G. & Puletti, F. (1971). Auditory input to the human cortex during states of drowsiness and surgical anesthesia, *Electroencephalography and Clinical Neurophysiology* 31(6): 603–609.

Cepeda, M., Carr, D., Lau, J. & Alvarez, H. (2006). Music for pain relief, *Cochrane Database of Systematic Reviews (Online)* 2: CD 004843.

Ceponiene, R., Kushnerenko, E., Fellman, V., *et al.* (2002). Event-related potential features indexing central auditory discrimination by newborns, *Cognitive Brain Research* 13(1): 101–114.

Cheour, M., Ceponiene, R., Leppanen, P., *et al.* (2002a). The auditory sensory memory trace decays rapidly in newborns, *Scandinavian Journal of Psychology* 43(1): 33–39.

Cheour, M., Kushnerenko, E., Ceponiene, R., Fellman, V. & Naatanen, R. (2002b). Electric brain responses obtained from newborn infants to changes in duration in complex harmonic tones, *Developmental Neuropsychology* 22(2): 471–479.

Cheour, M., Leppänen, H. T. & Kraus, N. (2000). Mismatch negativity (MMN) as a tool for investigating auditory discrimination and sensory memory in infants and children, *Clinical Neurophysiology* 111(1): 4–16.

Chi, J., Dooling, E. & Gilles, F. (1977). Left-right asymmetries of the temporal speech areas of the human fetus, *Archives of Neurology* 34(6): 346.

Chwilla, D., Brown, C. & Hagoort, P. (1995). The N400 as a function of the level of processing, *Psychophysiology* 32(3): 274–285.

Clynes, M. (1969). Dynamics of vertex evoked potentials: The R-M brain function, *in* E. Donchin and D. Lindsley (eds), *Average Evoked Potentials: Methods, Results and Evaluations*, Washington: U.S. Government Printing Office, pp. 363–374.

Cole, M., Yeung, N., Freiwald, W. & Botvinick, M. (2009). Cingulate cortex: Diverging data from humans and monkeys, *Trends in Neurosciences* 32(11): 566–574.

Cook, N. (1987). The perception of large-scale tonal closure, *Music Perception* 5: 197–205.

Cook, N. (1992). *Music, Imagination, and Culture*, Oxford University Press.

Cook, P. (ed.) (2001). *Music, Cognition, and Computerized Sound. An Introduction to Psychoacoustics*, Cambridge, MA: MIT Press.

Conard N., Malina M. & Münzel S. (2009). New flutes document the earliest musical tradition in south western Germany, *Nature* 460(7256): 737–740.

Coulson, S., King, J. & Kutas, M. (1998). Expect the unexpected: Event-related brain response to morphosyntactic violations, *Language and Cognitive Processes* 13: 21–58.

Courchesne, E., Hillyard, S. & Galambos, R. (1975). Stimulus novelty, task relevance and the visual evoked potential in man, *Electroencephalography and Clinical Neurophysiology* 39(4): 131–143.

Craig, A. (2009). How do you feel–now? The anterior insula and human awareness, *Nat Rev Neurosci* 10: 59–70.

Crapse, T. & Sommer, M. (2008). Corollary discharge circuits in the primate brain, *Current Opinion in Neurobiology* 18(6): 552–557.

Critchley, H. (2005). Neural mechanisms of autonomic, affective, and cognitive integration, *The Journal of Comparative Neurology* 493(1): 154–166.

Critchley, H., Corfield, D., Chandler, M., Mathias, C. & Dolan, R. (2000). Cerebral correlates of autonomic cardiovascular arousal: A functional neuroimaging investigation in humans, *The Journal of Physiology* 523(1): 259–270.

Critchley, H., Wiens, S., Rotshtein, P., Öhman, A. & Dolan, R. (2004). Neural systems supporting interoceptive awareness, *Nature Neuroscience* 7(2): 189–195.

Cross, I. (2008a). The evolutionary nature of musical meaning, *Musicae Scientiae* 13: 179–200.

Cross, I. (2008b). Musicality and the human capacity for culture, *Musicae Scientiae* 12(1 suppl): 147–167.

Cross, I. (2011). The meanings of musical meanings (Comment), *Physics of Life Reviews* 8(2): 116–119.

Cross, I. & Morley, I. (2008). The evolution of music: Theories, definitions and the nature of the evidence, *in* S. Malloch and C. Trevarthen (eds), *Communicative Musicality: Exploring the Basis of Human Companionship*, Oxford: Oxford University Press, pp. 61–82.

Cumming, N. (1994). Metaphor in Roger Scruton's aesthetics of music, *in* A. Pople (ed.), *Theory, Analysis and Meaning in Music*, Cambridge: Cambridge University Press, pp. 3–28.

Cummings, A., Ceponiene, R., Koyama, A., *et al.* (2006). Auditory semantic networks for words and natural sounds, *Brain Research* 1115(1): 92–107.

Cycowicz, Y. M. & Friedman, D. (1998). Effect of sound familiarity on the event-related potentials elicited by novel environmental sounds, *Brain and Cognition* 36(1): 30–51.

Dahlhaus, C. (ed.) (1980). *Neues Handbuch der Musikwissenschaft*, Wiesbaden: Laaber.

Dahlhaus, C. & Eggebrecht, H. (eds) (1979). *Brockhaus-Riemann-Musiklexikon*, Vol. 2, Wiesbaden: Laaber.

Dalgleish, T. (2004). The emotional brain, *Nature Reviews Neuroscience* 5(7): 583–589.

Daltrozzo, J. & Schön, D. (2009a). Is conceptual processing in music automatic? An electrophysiological approach, *Brain Research* 1270: 88–94.

Daltrozzo, J. & Schön, D. (2009b). Conceptual processing in music as revealed by N400 effects on words and musical targets, *Journal of Cognitive Neuroscience* 21(10): 1882–1892.

Darwin, C. (1872). *The expression of emotion in man and animals*, London: Murray.

Darwin, C. (1874). *The Descent of Man*, London: Murray.

Darwin, C. (1997). Auditory grouping, *Trends in Cognitive Sciences* 1(9): 327–333.

Darwin, C. (2008). Listening to speech in the presence of other sounds, *Philosophical Transactions of the Royal Society B: Biological Sciences* 363(1493): 1011–1021.

Davies, J. (1978). *The Psychology of Music*, Stanford University Press.

Davies, S. (1994). *Musical Meaning and Expression*, Cornell University Press.

Davis, M. & Whalen, P. (2001). The amygdala: Vigilance and emotion, *Molecular Psychiatry* 6(1): 13–34.

De Sanctis, P., Ritter, W., Molholm, S., Kelly, S. & Foxe, J. (2008). Auditory scene analysis: The interaction of stimulation rate and frequency separation on pre-attentive grouping, *European Journal of Neuroscience* 27(5): 1271–1276.

Debruille, J., Pineda, J. & Renault, B. (1996). N400-like potentials elicited by faces and knowledge inhibition, *Cognitive Brain Research* 4(2): 133–144.

Deouell, L. (2007). The frontal generator of the mismatch negativity revisited, *Journal of Psychophysiology* 21(3/4): 188–203.

Desmurget, M. & Grafton, S. (2000). Forward modeling allows feedback control for fast reaching movements, *Trends in Cognitive Sciences* 4(11): 423–431.

Desmurget, M. & Sirigu, A. (2009). A parietal-premotor network for movement intention and motor awareness, *Trends in Cognitive Sciences* 13(10): 411–419.

Deutsch, D. (ed.) (1982). *The Psychology of Music*, New York: Academic Press.

Deutsch, J. & Deutsch, D. (1963). Attention: Some theoretical considerations, *Psychological Review* 70(1): 80–90.

Di Pietro, M., Laganaro, M., Leemann, B. & Schnider, A. (2004). Receptive amusia: Temporal auditory processing deficit in a professional musician following a left temporo-parietal lesion, *Neuropsychologia* 42(7): 868–877.

Dick, F., Lee, H. L., Nusbaum, H. & Price, C. J. (2011). Auditory-motor expertise alters "speech selectivity" in professional musicians and actors, *Cerebral Cortex* 21(4): 938–948.

Diedrichsen, J., Shadmehr, R. & Ivry, R. (2010). The coordination of movement: Optimal feedback control and beyond, *Trends in Cognitive Sciences* 14(1): 31–39.

Donchin, E. & Coles, M. (1988). Is the P300 component a manifestation of context updating? *Behavioral and Brain Sciences* 11: 357–374.

Donchin, E. & Coles, M. G. H. (1998). Context updating and the P300, *Behavioral and Brain Sciences* 21(1): 152–154.

Donchin, E., Ritter, W. & McCallum, W. C. (1978). Cognitive psychology: The endogenous components of the ERP, *in* E. Callaway, P. Tueting and S. H. Koslow (eds), *Event-Related Brain Potentials in Man*, New York: Academic Press, pp. 349–411.

Donkers, F., Nieuwenhuis, S. & van Boxtel, G. (2005). Mediofrontal negativities in the absence of responding, *Cognitive Brain Research* 25(3): 777–787.

Draganova, R., Eswaran, H., Murphy, P., *et al.* (2005). Sound frequency change detection in fetuses and newborns, a magnetoencephalographic study, *Neuroimage* 28(2): 354–361.

Drevets, W., Price, J., Bardgett, M., *et al.* (2002). Glucose metabolism in the amygdala in depression: Relationship to diagnostic subtype and plasma cortisol levels, *Pharmacology Biochemistry and Behavior* 71(3): 431–447.

Drost, U., Rieger, M., Brass, M., Gunter, T. C. & Prinz, W. (2005a). Action-effect coupling in pianists, *Psychological Research* 69(4): 233–241.

Drost, U., Rieger, M., Brass, M., Gunter, T. C. & Prinz, W. (2005b). When hearing turns into playing: Movement induction by auditory stimuli in pianists, *The Quarterly Journal of Experimental Psychology Section A* 58(8): 1376–1389.

Duncan, J. (1980). The locus of interference in the perception of simultaneous stimuli, *Psychological Review* 87(3): 272–300.

Eggebrecht, H. H. (ed.) (1967). *Riemann Musik Lexikon (Sachteil)*, Wiesbaden: Schott's Söhne.

Eggebrecht, H. H. (ed.) (1972). *Handwörterbuch der Musikalischen Terminologie*, Wiesbaden: Steiner.

Ekman, P. (1999). Basic emotions, *in* T. Dalgleish and M. Power (eds), *Handbook of Cognition and Emotion*, Wiley Online Library, pp. 45–60.

Elbert, T. (1998). Neuromagnetism, *in* W. Andrä and H. Nowak (eds), *Magnetism in Medicine*, Berlin: Wiley-VCH.

Eldar, E., Ganor, O., Admon, R., Bleich, A. & Hendler, T. (2007). Feeling the real world: Limbic response to music depends on related content, *Cerebral Cortex* 17(12): 2828–2840.

Ellis, C. J. (1965). Pre-instrumental scales, *Ethnomusicology* 9: 126–144.

Ericsson, K. & Lehmann, A. (1996). Expert and exceptional performance: Evidence of maximal adaption to task constraints, *Annual Review of Psychology* 47(1): 273–305.

Escher, J., Hoehmann, U., Anthenien, L., *et al.* (1993). Music during gastroscopy, *Schweizerische Medizinische Wochenschrift* 123(26): 1354–1358.

Ethofer, T., Kreifelts, B., Wiethoff, S., *et al.* (2009). Differential influences of emotion, task, and novelty on brain regions underlying the processing of speech melody, *Journal of Cognitive Neuroscience* 21(7): 1255–1268.

Evarts, E. (1974). Precentral and postcentral cortical activity in association with visually triggered movement, *Journal of Neurophysiology* 37(2): 373–381.

Evers, S. & Suhr, B. (2000). Changes of the neurotransmitter serotonin but not of hormones during short time music perception, *European Archives of Psychiatry and Clinical Neuroscience* 250(3): 144–147.

Falkenstein, M., Hohnsbein, J., Hoormann, J. & Blanke, L. (1990). Effects of errors in choice reaction tasks on the ERP under focused and divided attention, *in* C. Brunia, A. Gaillard and A. Kok (eds), *Psychophysiological Brain Research*, Tilburg: Tilburg University Press, pp. 192–195.

Faro, S. & Mohamed, F. (2006). *Functional MRI: Basic Principles and Clinical Applications*, Springer Verlag.

Fazio, P., Cantagallo, A., Craighero, L., *et al.* (2009). Encoding of human action in Broca's area, *Brain* 132(7): 1980–1988.

Fechner, G. T. (1873). *Einige Ideen zur Schöpfungs- und Entwicklungsgeschichte der Organismen*, Leipzig: Breitkopf und Härtel.

Fedorenko, E., Patel, A., Casasanto, D., Winawer, J. & Gibson, E. (2009). Structural integration in language and music: Evidence for a shared system, *Memory and Cognition* 37(1): 1–19.

Ferdinand, N., Mecklinger, A. & Kray, J. (2008). Error and deviance processing in implicit and explicit sequence learning, *Journal of Cognitive Neuroscience* 20(4): 629–642.

Fernald, A. (1989). Intonation and communicative intent in mothers' speech to infants: Is the melody the message? *Child Development* 60: 1497–1510.

Fettiplace, R. & Hackney, C. (2006). The sensory and motor roles of auditory hair cells, *Nature Reviews Neuroscience* 7(1): 19–29.

Fiebach, C. & Schubotz, R. I. (2006). Dynamic anticipatory processing of hierarchical sequential events: A common role for Broca's area and ventral premotor cortex across domains? *Cortex* 42(4): 499–502.

Finney, S. (1997). Auditory feedback and musical keyboard performance, *Music Perception* 15(2): 153–174.

Finney, S. & Palmer, C. (2003). Auditory feedback and memory for music performance: Sound evidence for an encoding effect, *Memory and Cognition* 31(1): 51–64.

Fischler, I., Bloom, P. A., Childers, D. G., Roucos, S. E. & Perry, N. W. (1983). Brain potentials related to stages of sentence verification, *Psychophysiology* 20: 400–409.

Fischler, I., Childers, D. G., Achariyapaopan & Perry, N. W. (1985). Brain potentials during sentence verification: Automatic aspects of comprehension, *Biological Psychology* 21: 83–106.

Fishman, Y., Reser, D., Arezzo, J. & Steinschneider, M. (2001). Neural correlates of auditory stream segregation in primary auditory cortex of the awake monkey, *Hearing Research* 151(1-2): 167–187.

Fitch, W. (2005). The evolution of music in comparative perspective, *Annals of the New York Academy of Sciences* 1060(1): 29–49.

Fitch, W. (2006). The biology and evolution of music: A comparative perspective, *Cognition* 100(1): 173–215.

Fitch, W. & Gingras, B. (2011). Multiple varieties of musical meaning (Comment), *Physics of Life Reviews* 8(2): 108–109.

Fitch, W. & Hauser, M. (2004). Computational constraints on syntactic processing in a nonhuman primate, *Science* 303(5656): 377–380.

Flores-Gutiérrez, E., Díaz, J., Barrios, F., *et al.* (2007). Metabolic and electric brain patterns during pleasant and unpleasant emotions induced by music masterpieces, *International Journal of Psychophysiology* 65(1): 69–84.

Flynn, F., Benson, D. & Ardila, A. (1999). Anatomy of the insula functional and clinical correlates, *Aphasiology* 13(1): 55–78.

Fodor, J., Mann, V. & Samuel, A. (1991). Panel discussion: The modularity of speech and language, *in Modularity and the Motor Theory of Speech Perception: Proceedings of a Conference to Honor Alvin M. Liberman*, Lawrence Erlbaum, p. 359.

Folstein, J. & Van Petten, C. (2008). Influence of cognitive control and mismatch on the N2 component of the ERP: A review, *Psychophysiology* 45(1): 152–170.

Frey, A., Marie, C., Prod'Homme, L., Timsit-Berthier, M., Schön, D. & Besson, M. (2009). Temporal semiotic units as minimal meaningful units in music? An electrophysiological approach, *Music Perception* 23: 247–256.

Friederici, A. D. (1982). Syntactic and semantic processes in aphasic deficits: The availability of prepositions, *Brain and Language* 15(2): 249–258.

Friederici, A. D. (1999). *Language Comprehension: A Biological Perspective*, Springer.

Friederici, A. D. (2002). Towards a neural basis of auditory sentence processing, *Trends in Cognitive Sciences* 6(2): 78–84.

Friederici, A. D. (2004). Event-related brain potential studies in language, *Current Neurology and Neuroscience Reports* 4(6): 466–470.

Friederici, A. D. (2005). Neurophysiological markers of early language acquisition: From syllables to sentences, *Trends in Cognitive Sciences* 9(10): 481–488.

Friederici, A. D. (2009). Pathways to language: Fiber tracts in the human brain, *Trends in Cognitive Sciences* 13(4): 175–181.

Friederici, A. D., Bahlmann, J., Heim, S., Schubotz, R. I. & Anwander, A. (2006). The brain differentiates human and non-human grammars: Functional localization and structural connectivity, *Proceedings of the National Academy of Sciences* 103(7): 2458–2463.

Friederici, A. D., Friedrich, M. & Weber, C. (2002a). Neural manifestation of cognitive and precognitive mismatch detection in early infancy, *NeuroReport* 13(10): 1251–1254.

Friederici, A. D., Hahne, A. & Mecklinger, A. (1996). Temporal structure of syntactic parsing: Early and late event-related brain potential effects elicited by syntactic anomalies, *Journal of Experimental Psychology: Learning, Memory, and Cognition* 22: 1219–1248.

Friederici, A. D., Pfeifer, E. & Hahne, A. (1993). Event-related brain potentials during natural speech processing: Effects of semantic, morphological and syntactic violations, *Cognitive Brain Research* 1: 183–192.

Friederici, A. D., Steinhauer, K. & Pfeifer, E. (2002b). Brain signatures of artificial language processing: Evidence challenging the critical period hypothesis, *Proceedings of the National Academy of Sciences* 99(1): 529–534.

Friederici, A. D., Wang, Y., Herrmann, C. S., Maess, B. & Oertel, U. (2000). Localisation of early syntactic processes in frontal and temporal cortical areas: An MEG study, *Human Brain Mapping* 11: 1–11.

Friederici, A. D. & Wartenburger, I. (2010). Language and brain, *Wiley Interdisciplinary Reviews: Cognitive Science* 1(2): 150–159.

Friederici, A. D. & Weissenborn, J. (2007). Mapping sentence form onto meaning: The syntax-semantic interface, *Brain Research* 1146: 50–58.

Friedrich, M. & Friederici, A. D. (2006). Early N400 development and later language acquisition, *Psychophysiology* 43(1): 1–12.

Friedrich, R. & Friederici, A. D. (2009). Mathematical logic in the human brain: Syntax, *PLoS ONE* 4(5): e5599.

Friston, K., Buechel, C., Fink, G., Morris, J., Rolls, E. & Dolan, R. (1997). Psychophysiological and modulatory interactions in neuroimaging, *Neuroimage* 6(3): 218–229.

Friston, K., Harrison, L. & Penny, W. (2003). Dynamic causal modelling, *Neuroimage* 19(4): 1273–1302.

Fritz, T., Jentschke, S., Gosselin, N., *et al.* (2009). Universal recognition of three basic emotions in music, *Current Biology* 19(7): 573–576.

Fritz, T. & Koelsch, S. (2005). Initial response to pleasant and unpleasant music: An fMRI study, *Neuroimage* 26(16): 2005.

Fritz, T. & Koelsch, S. (2008). The role of semantic association and emotional contagion for the induction of emotion with music (Comment), *Behavioral and Brain Sciences* 31(05): 579–580.

Fujioka, T., Trainor, L., Ross, B., Kakigi, R. & Pantev, C. (2004). Musical training enhances automatic encoding of melodic contour and interval structure, *Journal of Cognitive Neuroscience* 16(6): 1010–1021.

Fujioka, T., Trainor, L., Ross, B., Kakigi, R. & Pantev, C. (2005). Automatic encoding of polyphonic melodies in musicians and nonmusicians, *Journal of Cognitive Neuroscience* 17(10): 1578–1592.

Gaab, N., Gabrieli, J. & Glover, G. (2007). Assessing the influence of scanner background noise on auditory processing. I. An fMRI study comparing three experimental designs with varying degrees of scanner noise, *Human Brain Mapping* 28(8): 703–720.

Gaab, N., Gaser, C., Zaehle, T., Jäncke, L. & Schlaug, G. (2003). Functional anatomy of pitch memory – an fMRI study with sparse temporal sampling, *Neuroimage* 19(4): 1417–1426.

Gabrielson, A. & Juslin, P. (2003). Emotional expression in music, *in* Davidson, R.J. (ed.), *Handbook of Affective Sciences*, New York: Oxford University Press, pp. 503–534.

Galaburda, A. & Sanides, F. (1980). Cytoarchitectonic organization of the human auditory cortex, *The Journal of Comparative Neurology* 190: 597–610.

Galbraith, G., Threadgill, M., Hemsley, J., *et al.* (2000). Putative measure of peripheral and brainstem frequency-following in humans, *Neuroscience Letters* 292(2): 123–127.

Garcia-Cairasco, N. (2002). A critical review on the participation of inferior colliculus in acoustic-motor and acoustic-limbic networks involved in the expression of acute and kindled audiogenic seizures, *Hearing Research* 168(1-2): 208–222.

Garrido, M., Kilner, J., Stephan, K. & Friston, K. (2009). The mismatch negativity: A review of underlying mechanisms, *Clinical Neurophysiology* 120(3): 453–463.

Gebauer, G. (in press). Wie können wir über emotionen sprechen? *in* G. Gebauer, M. Holodinski, S. Koelsch, C. Moll and C. von Schewe (eds),

Emotion und Sprache, Weilerswist-Metternich: Velbrück.

Geisler, C. (1998). *From Sound to Synapse: Physiology of the Mammalian Ear*, Oxford University Press, USA.

Gerdner, L. & Swanson, E. (1993). Effects of individualized music on confused and agitated elderly patients, *Archives of Psychiatric Nursing* 7(5): 284–291.

Gerra, G., Zaimovic, A., Franchini, D., *et al.* (1998). Neuroendocrine responses of healthy volunteers to techno-music: Relationships with personality traits and emotional state, *International Journal of Psychophysiology* 28(1): 99–111.

Giard, M., Perrin, F. & Pernier, J. (1990). Brain generators implicated in processing of auditory stimulus deviance. A topographic ERP study, *Psychophysiology* 27: 627–640.

Goebel, R., Roebroeck, A., Kim, D. & Formisano, E. (2003). Investigating directed cortical interactions in time-resolved fMRI data using vector autoregressive modeling and Granger causality mapping, *Magnetic Resonance Imaging* 21(10): 1251–1261.

Goerlich, K., Witteman, J., Aleman, A. & Martens, S. (2011). Hearing feelings: Affective categorization of music and speech in alexithymia, an ERP study, *PloS One* 6(5): e19501.

Goldstein, A. (1980). Thrills in response to music and other stimuli, *Physiological Psychology* 8: 126–129.

Gomes, H., Molholm, S., Ritter, W., *et al.* (2000). Mismatch negativity in children and adults, and effects of an attended task, *Psychophysiology* 37(06): 807–816.

Gordon, R., Schön, D., Magne, C., Astésano, C. & Besson, M. (2010). Words and melody are intertwined in perception of sung words: EEG and behavioral evidence, *PLoS-one* 5(3): e9889.

Gosselin, N., Peretz, I., Noulhiane, M., *et al.* (2005). Impaired recognition of scary music following unilateral temporal lobe excision, *Brain* 128(3): 628–640.

Gosselin, N., Samson, S., Adolphs, R., *et al.* (2006). Emotional responses to unpleasant music correlates with damage to the parahippocampal cortex, *Brain* 129(10): 2585–2592.

Grahn, J. (2009). The role of the basal ganglia in beat perception, *Annals of the New York Academy of Sciences* 1169(1): 35–45.

Grahn, J. & Brett, M. (2007). Rhythm and beat perception in motor areas of the brain, *Journal of Cognitive Neuroscience* 19(5): 893–906.

Grahn, J. & Rowe, J. (2009). Feeling the beat: Premotor and striatal interactions in musicians and nonmusicians during beat perception, *The Journal of Neuroscience* 29(23): 7540–7548.

Green, A., Bærentsen, K., Stødkilde-Jørgensen, H., *et al.* (2008). Music in minor activates limbic structures: A relationship with dissonance? *Neuroreport* 19(7): 711–715.

Grewe, O., Nagel, F., Kopiez, R. & Altenmüller, E. (2007a). Emotions over time: Synchronicity and development of subjective, physiological, and facial affective reactions of music, *Emotion* 7(4): 774–788.

Grewe, O., Nagel, F., Kopiez, R. & Altenmüller, E. (2007b). Listening to music as a re-creative process: Physiological, psychological, and psychoacoustical correlates of chills and strong emotions, *Music Perception* 24(3): 297–314.

Grieser-Painter, J. & Koelsch, S. (2011). Can out-of-context musical sounds convey meaning? An ERP study on the processing of meaning in music, *Psychophysiology* 48(5): 645–655.

Griffiths, T. & Warren, J. (2002). The planum temporale as a computational hub, *Trends in Neurosciences* 25(7): 348–353.

Griffiths, T. & Warren, J. (2004). What is an auditory object? *Nature Reviews Neuroscience* 5(11): 887–892.

Griffiths, T., Warren, J., Dean, J. & Howard, D. (2004). "When the feeling's gone": A selective loss of musical emotion, *Journal of Neurobiology, Neurosurgery and Psychiatry* 75(2): 341–345.

Grigor, J., Van Toller, S., Behan, J. & Richardson, A. (1999). The effect of odour priming on long latency visual evoked potentials of matching and mismatching objects, *Chemical Senses* 24(2): 137–144.

Grimm, S. & Schroger, E. (2005). Pre-attentive and attentive processing of temporal and frequency characteristics within long sounds, *Cognitive Brain Research* 25(3): 711–721.

Grodzinsky, Y. & Friederici, A. D. (2006). Neuroimaging of syntax and syntactic processing, *Current Opinion in Neurobiology* 16(2): 240–246.

Groussard, M., Viader, F., Hubert, V., *et al.* (2010). Musical and verbal semantic memory: Two distinct neural networks? *Neuroimage* 49(3): 2764–2773.

Gunter, T. C. & Bach, P. (2004). Communicating hands: ERPs elicited by meaningful symbolic hand postures, *Neuroscience Letters* 372(1-2): 52–56.

Gunter, T. C. & Friederici, A. D. (1999). Concerning the automaticity of syntactic processing, *Psychophysiology* 36: 126–137.

Gunter, T. C., Friederici, A. D. & Schriefers, H. (2000). Syntactic gender and semantic expectancy: ERPs reveal early autonomy and late interaction, *Journal of Cognitive Neuroscience* 12(4): 556–568.

Gunter, T. C., Stowe, L. & Mulder, G. (1997). When syntax meets semantics, *Psychophysiology* 34: 660–676.

Gutschalk, A., Oxenham, A., Micheyl, C., Wilson, E. & Melcher, J. (2007). Human cortical activity during streaming without spectral cues suggests a general neural substrate for auditory stream segregation, *Journal of Neuroscience* 27(48): 13074–13081.

Hackett, T. A. & Kaas, J. (2004). Auditory cortex in primates: Functional subdivisions and processing streams, *in* M. S. Gazzaniga (ed.), *The Cognitive Neurosciences*, Cambridge, MA: MIT Press, pp. 215–232.

Hackley, S. (1993). An evaluation of the automaticity of sensory processing using event-related potentials and brain-stem reflexes, *Psychophysiology* 30(5): 415–428.

Háden, G., Stefanics, G., Vestergaard, M., *et al.* (2009). Timbre-independent extraction of pitch in newborn infants, *Psychophysiology* 46(1): 69–74.

Hagoort, P., Brown, C. & Groothusen, J. (1993). The syntactic positive shift (SPS) as an ERP measure of syntactic processing, *Language and Cognitive Processes* 8(4): 439–483.

Hahne, A. (1999). *Charakteristika Syntaktischer und Semantischer Prozesse Bei der Auditiven Sprachverarbeitung*, Leipzig: MPI Series.

Hajcak, G., Holroyd, C., Moser, J. & Simons, R. (2005). Brain potentials associated with expected and unexpected good and bad outcomes, *Psychophysiology* 42(2): 161–170.

Hajcak, G., McDonald, N. & Simons, R. (2003). To err is autonomic: Error-related brain potentials, ANS activity, and post-error compensatory behavior, *Psychophysiology* 40(6): 895–903.

Hajcak, G., Moser, J., Holroyd, C. & Simons, R. (2007). It's worse than you thought: The feedback negativity and violations of reward prediction in gambling tasks, *Psychophysiology* 44(6): 905–912.

Hald, L., Steenbeek-Planting, E. & Hagoort, P. (2007). The interaction of discourse context and world knowledge in online sentence comprehension. Evidence from the N400, *Brain Research* 1146: 210–218.

Hall, D., Haggard, M., Akeroyd, M., *et al.* (1999). "Sparse" temporal sampling in auditory fMRI, *Human Brain Mapping* 7(3): 213–223.

Hall, J. (1979). Auditory brainstem frequency following responses to waveform envelope periodicity, *Science* 205: 1297–1299.

Hämäläinen, M., Hari, R., Ilmoniemi, R. J., Knuutila, J. & Lounasmaa, O. (1993). Magnetoencephalography–theory, instrumentation, and applications to noninvasive studies of the working human brain, *Reviews of Modern Physics* 65(2): 413–497.

Hamm, J., Johnson, B. & Kirk, I. (2002). Comparison of the N300 and N400 ERPs to picture stimuli in congruent and incongruent contexts, *Clinical Neurophysiology* 113(8): 1339–1350.

Hanslick, E. (1854). *Vom Musikalisch-Schönen: Ein Beitrag zur Revision der Ästhetik der Tonkunst*, Leipzig: Weigel.

Harlow, H. (1958). The nature of love, *American Psychologist* 13: 673–685.

Haslinger, B., Erhard, P., Altenmüller, E., *et al.* (2005). Transmodal sensorimotor networks during action observation in professional pianists, *Journal of Cognitive Neuroscience* 17(2): 282–293.

Hasting, A. & Kotz, S. (2008). Speeding up syntax: On the relative timing and automaticity of local phrase structure and morphosyntactic processing as reflected in event-related brain potentials, *Journal of Cognitive Neuroscience* 20(7): 1207–1219.

Hasting, A., Kotz, S. & Friederici, A. D. (2007). Setting the stage for automatic syntax processing:

The mismatch negativity as an indicator of syntactic priming, *Journal of Cognitive Neuroscience* 19(3): 386–400.

Hatfield, E., Cacioppo, J. & Rapson, R. (1993). Emotional contagion, *Current Directions in Psychological Science* 2(3): 96–100.

Hatfield, E., Rapson, R. & Le, Y. (2009). Emotional contagion and empathy, *in* J. Decety and W. Ickes (eds), *The Social Neuroscience of Empathy*, Cambridge: MIT Press, pp. 19–30.

Hatsopoulos, N., Xu, Q. & Amit, Y. (2007). Encoding of movement fragments in the motor cortex, *The Journal of Neuroscience* 27(19): 5105–5114.

Haueisen, J. & Knösche, T. (2001). Involuntary motor activity in pianists evoked by music perception, *Journal of Cognitive Neuroscience* 13(6): 786–792.

Hauptmann, M. (1873). *Die Natur der Harmonik und der Metrik: Zur Theorie der Musik*, Leipzig: Breitkopf und Härtel.

Hayashi, R., Imaizumi, S., Mori, K., *et al.* (2001). Elicitation of N400m in sentence comprehension due to lexical prosody incongruity, *NeuroReport* 12(8): 1753–1756.

Haynes, J. & Rees, G. (2005). Predicting the orientation of invisible stimuli from activity in human primary visual cortex, *Nature Neuroscience* 8(5): 686–691.

Heim, S. & Alter, K. (2006). Prosodic pitch accents in language comprehension and production: ERP data and acoustic analyses, *Acta Neurobiologiae Experimentalis* 66(1): 55–68.

Heim, S., Eickhoff, S., Opitz, B. & Friederici, A. D. (2006). BA 44 in Broca's area supports syntactic gender decisions in language production, *NeuroReport* 17(11): 1097–1101.

Heinke, W., Kenntner, R., Gunter, T., *et al.* (2004). Differential effects of increasing propofol sedation on frontal and temporal cortices: An ERP study, *Anesthesiology* 100: 617–625.

Heinke, W. & Koelsch, S. (2005). The effects of anesthetics on brain activity and cognitive function, *Current Opinion in Anesthesiology* 18(6): 625–631.

Hepper, P. & Shahidullah, S. (1994). The development of fetal hearing, *Fetal and Maternal Medicine Review* 6(03): 167–179.

Herbert, C., Ethofer, T., Anders, S., *et al.* (2009). Amygdala activation during reading of emotional adjectives - an advantage for pleasant content, *Social Cognitive and Affective Neuroscience* 4(1): 35–49.

Herholz, S., Lappe, C., Knief, A. & Pantev, C. (2008). Neural basis of music imagery and the effect of musical expertise, *European Journal of Neuroscience* 28(11): 2352–2360.

Herrmann, B., Maess, B., Hasting, A. & Friederici, A. D. (2009). Localization of the syntactic mismatch negativity in the temporal cortex: An MEG study, *Neuroimage* 48(3): 590–600.

Herrojo-Ruiz, M., Jabusch, H. & Altenmüller, E. (2009a). Detecting wrong notes in advance: Neuronal correlates of error monitoring in pianists, *Cerebral Cortex* 19(11): 2625–2639.

Herrojo-Ruiz, M., Koelsch, S. & Bhattacharya, J. (2009b). Decrease in early right alpha band phase synchronization and late gamma band oscillations in processing syntax in music, *Human Brain Mapping* 30(4): 1207–1225.

Herrojo-Ruiz, M., Strübing, F., Jabusch, H. C. & Altenmüller, E. (2010). EEG oscillatory patterns are associated with error prediction during music performance and are altered in musician's dystonia, *Neuroimage* 55: 1791–1803.

Hevner, K. (1936). Experimental studies of the elements of expression in music, *American Journal of Psychology* 48: 246–268.

Hickok, G., Buchsbaum, B., Humphries, C. & Muftuler, T. (2003). Auditory-motor interaction revealed by fMRI: Speech, music, and working memory in are Spt, *Journal of Cognitive Neuroscience* 15(5): 673–682.

Hickok, G. & Poeppel, D. (2007). The cortical organization of speech processing, *Nature Reviews Neuroscience* 8(5): 393–402.

Hillecke, T., Nickel, A. & Bolay, H. (2005). Scientific perspectives on music therapy, *Annals of the New York Academy of Sciences* 1060 (The Neurosciences and Music II: From Perception to Performance): 271–282.

Hillecke, T., Wormit, A., Bardenheuer, H. & Bolay, H. (2004). Schmerz, *Musik-, Tanz und Kunsttherapie* 15(2): 92–94.

Hockett, C. F. (1960). Logical considerations in the study of animal communication, *in* W. Lanyon and W. Tavolga (eds), *Animal Sounds and Communication*, Washington, DC: American Institute of Biological Sciences, pp. 392–430.

Holcomb, P. J. & McPherson, W. B. (1994). Event-related brain potentials reflect semantic priming in an object decision task, *Brain and Cognition* 24: 259–276.

Holcomb, P. J. & Neville, H. J. (1990). Semantic priming in visual and auditory lexical decision: A between modality comparison, *Language and Cognitive Processes* 5: 281–312.

Holdefer, R. & Miller, L. (2002). Primary motor cortical neurons encode functional muscle synergies, *Experimental Brain Research* 146(2): 233–243.

Holle, H. & Gunter, T. C. (2007). The role of iconic gestures in speech disambiguation: ERP evidence, *Journal of Cognitive Neuroscience* 19(7): 1175–1192.

Holle, H., Gunter, T. C., Rüschemeyer, S., Hennenlotter, A. & Iacoboni, M. (2008). Neural correlates of the processing of co-speech gestures, *Neuroimage* 39(4): 2010–2024.

Hommel, B., Müsseler, J., Aschersleben, G. & Prinz, W. (2001). The theory of event coding (TEC): A framework for perception and action planning, *Behavioral and Brain Sciences* 24(05): 849–878.

Hoormann, J., Falkenstein, M., Hohnsbein, J. & Blanke, L. (1992). The human frequency-following response (FFR): Normal variability and relation to the click-evoked brainstem response, *Hearing Research* 59(2): 179–188.

Hoover, J. & Strick, P. (1999). The organization of cerebellar and basal ganglia outputs to primary motor cortex as revealed by retrograde transneuronal transport of herpes simplex virus type 1, *The Journal of Neuroscience* 19(4): 1446–1463.

House, J., Landis, K. & Umberson, D. (1988). Social relationships and health, *Science* 241(4865): 540–545.

Hucklebridge, F., Lambert, S., Clow, A., *et al.* (2000). Modulation of secretory immunoglobulin A in saliva: Response to manipulation of mood, *Biological Psychology* 53(1): 25–35.

Huotilainen, M., Kujala, A., Hotakainen, M., *et al.* (2005). Short-term memory functions of the human fetus recorded with magnetoencephalography, *NeuroReport* 16(1): 81–84.

Huron, D. (2001). Is music an evolutionary adaptation? *in* R. J. Zatorre & I. Peretz (eds), *The Biological Foundations of Music*, Vol. 930, Annals of the New York Academy of Sciences, pp. 43–61.

Huron, D. (2006). *Sweet Anticipation: Music and the Psychology of Expectation*, The MIT Press.

Huron, D. & Parncutt, R. (1993). An improved model of tonality perception incorporating pitch salience and echoic memory, *Psychomusicology* 12: 154–171.

Hyde, K., Peretz, I. & Zatorre, R. (2008). Evidence for the role of the right auditory cortex in fine pitch resolution, *Neuropsychologia* 46(2): 632–639.

Indefrey, P. (2004). Hirnaktivierungen bei syntaktischer sprachverarbeitung: Eine meta-analyse, *in* H. Mueller and S. Rickheit (eds), *Neurokognition der Sprache*, Vol. 1, Tübingen: Stauffenburg Verlag, pp. 31–50.

Indefrey, P. & Levelt, W. (2004). The spatial and temporal signatures of word production components, *Cognition* 92(1-2): 101–144.

Istók, E., Brattico, E., Jacobsen, T., *et al.* (2009). Aesthetic responses to music: A questionnaire study, *Musicae Scientiae* 13(2): 183–206.

Jackson, N. (2003). A survey of music therapy methods and their role in the treatment of early elementary school children with ADHD, *Journal of Music Therapy* 40(4): 302–323.

Jacobsen, T., Schröger, E., Horenkamp, T. & Winkler, I. (2003). Mismatch negativity to pitch change: Varied stimulus proportions in controlling effects of neural refractoriness on human auditory event-related brain potentials, *Neuroscience Letters* 344(2): 79–82.

Jacobsen, T., Schubotz, R. I., Höfel, L. & Cramon, D. Y. (2006). Brain correlates of aesthetic judgment of beauty, *Neuroimage* 29(1): 276–285.

Jakobson, R. (1960). Closing statement: Linguistics and poetics, *in* T. A. Sebeok (ed), *Style in Language*, New York: Wiley.

James, C., Britz, J., Vuilleumier, P., Hauert, C. & Michel, C. (2008). Early neuronal responses in right limbic structures mediate harmony incongruity processing in musical experts, *Neuroimage* 42(4): 1597–1608.

James, W. (1890). *The Principles of Psychology*, New York: Holt.

Janata, P. (1995). ERP measures assay the degree of expectancy violation of harmonic contexts in music, *Journal of Cognitive Neuroscience* 7(2): 153–164.

Janata, P. (2001). Brain electrical activity evoked by mental formation of auditory expectations and images, *Brain Topography* 13(3): 169–193.

Janata, P. (2007). Navigating tonal space, *Tonal Theory for the Digital Age (Computing in Musicology)* 15: 39–50.

Janata, P. (2009). The neural architecture of music-evoked autobiographical memories, *Cerebral Cortex* 19(11): 2579–2594.

Janata, P., Birk, J., Van Horn, J., *et al.* (2002a). The cortical topography of tonal structures underlying Western music, *Science* 298(5601): 2167–2170.

Janata, P., Tillmann, B. & Bharucha, J. (2002b). Listening to polyphonic music recruits domain-general attention and working memory circuits, *Cognitive, Affective, & Behavioral Neuroscience* 2(2): 121–140.

Jasper, H. H. (1958). The ten-twenty electrode system of the international federation, *Electroencephalography and Clinical Neuropsychology* 10: 371–375.

Jemel, B., George, N., Olivares, E., Fiori, N. & Renault, B. (1999). Event-related potentials to structural familiar face incongruity processing, *Psychophysiology* 36(4): 437–452.

Jentschke, S. (2007). Neural correlates of processing syntax in music and language – Influences of development, musical training, and language impairment, PhD thesis, University of Leipzig.

Jentschke, S. & Koelsch, S. (2009). Musical training modulates the development of syntax processing in children, *Neuroimage* 47(2): 735–744.

Jentschke, S., Koelsch, S., Sallat, S. & Friederici, A. D. (2008). Children with specific language impairment also show impairment of music-syntactic processing, *Journal of Cognitive Neuroscience* 20(11): 1940–1951.

Johnson, K., Nicol, T. & Kraus, N. (2005). Brain stem response to speech: A biological marker of auditory processing, *Ear and Hearing* 26(5): 424–434.

Johnson, K., Nicol, T., Zecker, S. & Kraus, N. (2008). Developmental plasticity in the human auditory brainstem, *Journal of Neuroscience* 28(15): 4000–4007.

Johnsrude, I., Penhune, V. & Zatorre, R. (2000). Functional specificity in the right human auditory cortex for perceiving pitch direction, *Brain* 123(1): 155–163.

Jonaitis, E. & Saffran, J. (2009). Learning harmony: The role of serial statistics, *Cognitive Science* 33(5): 951–968.

Jones, M. (1981). Music as stimulus for psychological motion: Part I. some determinants of expectancies, *Psychomusicology* 1: 34–51.

Jones, M. (1982). Music as stimulus for psychological motion: Part II. an expectancy model, *Psychomusicology* 2: 1–13.

Jürgens, U. (2002). Neural pathways underlying vocal control, *Neuroscience and Biobehavioral Reviews* 26: 235–258.

Juslin, P. & Laukka, P. (2003). Communication of emotions in vocal expression and music performance: Different channels, same code? *Psychological Bulletin* 129(5): 770–814.

Juslin, P., Liljeström, S., Laukka, P., Västfjäll, D. & Lundqvist, L. (2011). Emotional reactions to music in a nationally representative sample of Swedish adults, *Musicae Scientiae* 15(2): 174–207.

Juslin, P. & Västfjäll, D. (2008). Emotional responses to music: The need to consider underlying mechanisms, *Behavioral and Brain Sciences* 31(05): 559–575.

Kaan, E., Harris, A., Gibson, E. & Holcomb, P. (2000). The P600 as an index of syntactic integration difficulty, *Language and Cognitive Processes* 15(2): 159–201.

Kaas, J. & Hackett, T. (2000). Subdivisions of auditory cortex and processing streams in primates, *Proceedings of the National Academy of Sciences of the United States of America* 97(22): 11793–11799.

Kahneman, D. (1973). *Attention and Effort*, Prentice Hall.

Kandel, E. R., Schwartz, J. H. & Jessell, T. M. (2000). *Principles of Neural Science*, Connecticut: Appleton and Lange.

Kanske, P. & Kotz, S. (2007). Concreteness in emotional words: ERP evidence from a hemifield study, *Brain Research* 1148: 138–148.

Karbusický, V. (1986). *Grundriß der musikalischen Semantik*, Wissenschaftliche Buchgesellschaft, Darmstadt.

Karg-Elert, S. (1931). *Polaristische Klang-und Tonalitätslehre (Harmonologik)*, FEC Leuckart.

Katahira, K., Abla, D., Masuda, S. & Okanoya, K. (2008). Feedback-based error monitoring processes during musical performance: An ERP study, *Neuroscience Research* 61(1): 120–128.

Kellenbach, M., Wijers, A. & Mulder, G. (2000). Visual semantic features are activated during the processing of concrete words: Event-related potential evidence for perceptual semantic priming, *Cognitive Brain Research* 10(1-2): 67–75.

Khalfa, S., Guye, M., Peretz, I., *et al.* (2008). Evidence of lateralized anteromedial temporal structures involvement in musical emotion processing, *Neuropsychologia* 46(10): 2485–2493.

Khalfa, S., Isabelle, P., Jean-Pierre, B. & Manon, R. (2002). Event-related skin conductance responses to musical emotions in humans, *Neuroscience Letters* 328(2): 145–149.

Khalfa, S., Schon, D., Anton, J. & Liégeois-Chauvel, C. (2005). Brain regions involved in the recognition of happiness and sadness in music, *Neuroreport* 16(18): 1981–1984.

Kilpatrick, L. & Cahill, L. (2003). Amygdala modulation of parahippocampal and frontal regions during emotionally influenced memory storage, *Neuroimage* 20(4): 2091–2099.

Kimble, D., Rogers, L. & Hendrickson, C. (1967). Hippocampal lesions disrupt maternal, not sexual, behavior in the albino rat, *Journal of Comparative and Physiological Psychology* 63(3): 401–407.

Kirschner, S. & Tomasello, M. (2009). Joint drumming: Social context facilitates synchronization in preschool children, *Journal of Experimental Child Psychology* 102(3): 299–314.

Kivy, P. (1991). *Music alone: Philosophical reflections on the purely musical experience*, Cornell University Press.

Kivy, P. (1999). Feeling the musical emotions, *British Journal of Aesthetics* 39(1): 1–13.

Klassen, J., Liang, Y., Tjosvold, L., Klassen, T. & Hartling, L. (2008). Music for pain and anxiety in children undergoing medical procedures: A systematic review of randomized controlled trials, *Ambulatory Pediatrics* 8(2): 117–128.

Kluender, R. & Kutas, M. (1993). Subjacency as a processing phenomenon, *Language and Cognitive Processes* 8: 573–633.

Knight, R. T. (1990). Electrophysiology in behavioral neurology, *in* M. Marsel (ed.), *Principles of Behavioral Neurology*, Phiadelphia: F.A. Davis Co., pp. 327–346.

Knösche, T., Neuhaus, C., Haueisen, J., *et al.* (2005). Perception of phrase structure in music, *Human Brain Mapping* 24(4): 259–273.

Knösche, T. R. (1997). Solutions of the Neuroelectromagnetic Inverse Problem, PhD thesis, University of Enschede, Netherlands.

Koch, M., Lingenhöhl, K. & Pilz, P. (1992). Loss of the acoustic startle response following neurotoxic lesions of the caudal pontine reticular formation: Possible role of giant neurons, *Neuroscience* 49(3): 617–625.

Koechlin, E. & Jubault, T. (2006). Broca's area and the hierarchical organization of human behavior, *Neuron* 50(6): 963–974.

Koelsch, S. (2005). Neural substrates of processing syntax and semantics in music, *Current Opinion in Neurobiology* 15(2): 207–212.

Koelsch, S. (2006). Significance of Broca's area and ventral premotor cortex for music-syntactic processing, *Cortex* 42(4): 518–520.

Koelsch, S. (2009). Music-syntactic processing and auditory memory: Similarities and differences between ERAN and MMN, *Psychophysiology* 46(1): 179–190.

Koelsch, S. (2010). Towards a neural basis of music-evoked emotions, *Trends in Cognitive Sciences* 14(3): 131–137.

Koelsch, S. & Friederici, A. D. (2003). Towards the neural basis of processing structure in music: Comparative results of different neurophysiological investigation methods (EEG, MEG, fMRI), *Annals of the New York Academy of Sciences* 999: 15–27.

Koelsch, S., Fritz, T., Cramon, D. Y., Müller, K. & Friederici, A. D. (2006a). Investigating emotion with music: An fMRI study, *Human Brain Mapping* 27(3): 239–250.

Koelsch, S., Fritz, T. & Schlaug, G. (2008a). Amygdala activity can be modulated by unexpected chord functions during music listening, *NeuroReport* 19(18): 1815–1819.

Koelsch, S., Fritz, T., Schulze, K., Alsop, D. & Schlaug, G. (2005a). Adults and children processing music: An fMRI study, *Neuroimage* 25(4): 1068–1076.

Koelsch, S., Fuermetz, J., Sack, U., *et al.* (2011). Effects of music listening on cortisol levels and propofol consumption during spinal anesthesia, *Frontiers in Psychology* 2(58): 1–9.

Koelsch, S., Grossmann, T., Gunter, T. C., Hahne, A., Schroger, E. & Friederici, A. D. (2003a). Children processing music: Electric brain responses reveal musical competence and gender differences, *Journal of Cognitive Neuroscience* 15(5): 683–693.

Koelsch, S., Gunter, T. C., Friederici, A. D. & Schröger, E. (2000). Brain Indices of Music Processing: "Non-musicians" are musical, *Journal of Cognitive Neuroscience* 12(3): 520–541.

Koelsch, S., Gunter, T. C., Schröger, E., *et al.* (2001). Differentiating ERAN and MMN: An ERP-study, *NeuroReport* 12(7): 1385–1389.

Koelsch, S., Gunter, T. C., von Cramon, D. Y., *et al.* (2002a). Bach speaks: A cortical "language-network" serves the processing of music, *Neuroimage* 17: 956–966.

Koelsch, S., Gunter, T. C., Wittfoth, M. & Sammler, D. (2005b). Interaction between syntax processing in language and in music: An ERP study, *Journal of Cognitive Neuroscience* 17(10): 1565–1577.

Koelsch, S., Gunter, T., Schroger, E. & Friederici, A. D. (2003b). Processing tonal modulations: An ERP study, *Journal of Cognitive Neuroscience* 15(8): 1149–1159.

Koelsch, S., Heinke, W., Sammler, D. & Olthoff, D. (2006b). Auditory processing during deep propofol sedation and recovery from unconsciousness, *Clinical Neurophysiology* 117(8): 1746–1759.

Koelsch, S. & Jentschke, S. (2008). Short-term effects of processing musical syntax: An ERP study, *Brain Research* 1212: 55–62.

Koelsch, S. & Jentschke, S. (2010). Differences in electric brain responses to melodies and chords, *Journal of Cognitive Neuroscience* 22(10): 2251–2262.

Koelsch, S., Jentschke, S., Sammler, D. & Mietchen, D. (2007a). Untangling syntactic and sensory processing: An ERP study of music perception, *Psychophysiology* 44(3): 476–490.

Koelsch, S., Kasper, E., Sammler, D., *et al.* (2004a). Music, language, and meaning: Brain signatures of semantic processing, *Nature Neuroscience* 7(3): 302–307.

Koelsch, S., Kilches, S., Steinbeis, N. & Schelinski, S. (2008a). Effects of unexpected chords and of performer's expression on brain responses and electrodermal activity, *PLoS One* 3(7): e2631.

Koelsch, S., Maess, B., Grossmann, T. & Friederici, A. D. (2002b). Sex difference in music-syntactic processing, *NeuroReport* 14: 709–712.

Koelsch, S., Maess, B., Grossmann, T. & Friederici, A. D. (2003c). Electric brain responses reveal gender differences in music processing, *NeuroReport* 14(5): 709–713.

Koelsch, S. & Mulder, J. (2002). Electric brain responses to inappropriate harmonies during listening to expressive music, *Clinical Neurophysiology* 113(6): 862–869.

Koelsch, S., Offermanns, K. & Franzke, P. (2010a). Music in the treatment of affective disorders: An exploratory investigation of a new method for music-therapeutic research, *Music Perception* 27(4): 307–316.

Koelsch, S., Remppis, A., Sammler, D., *et al.* (2007b). A cardiac signature of emotionality, *European Journal of Neuroscience* 26(11): 3328–3338.

Koelsch, S. & Sammler, D. (2008). Cognitive components of regularity processing in the auditory domain, *PLoS ONE* 3(7): e2650.

Koelsch, S., Schmidt, B. H. & Kansok, J. (2002d). Influences of musical expertise on the ERAN: An ERP-study, *Psychophysiology* 39: 657–663.

Koelsch, S., Schmidt, B. & Kansok, J. (2002c). Effects of musical expertise on the early right anterior negativity: An event-related brain potential study, *Psychophysiology* 39(05): 657–663.

Koelsch, S., Schröger, E. & Gunter, T. C. (2002e). Music matters: Preattentive musicality of the human brain, *Psychophysiology* 39: 1–11.

Koelsch, S., Schröger, E. & Tervaniemi, M. (1999). Superior pre-attentive auditory processing in musicians, *NeuroReport* 10(6): 1309–1313.

Koelsch, S., Schulze, K., Sammler, D., *et al.* (2009). Functional architecture of verbal and tonal working memory: An FMRI study, *Human Brain Mapping* 30(3): 859–873.

Koelsch, S. & Siebel, W. (2005). Towards a neural basis of music perception, *Trends in Cognitive Sciences* 9(12): 578–584.

Koelsch, S., Siebel, W. A. & Fritz, T. (2010b). Functional neuroimaging, *in* P. Juslin and J. Sloboda (eds), *Handbook of Music and Emotion: Theory, Research, Applications, 2nd ed.*, Oxford: Oxford University Press, pp. 313–346.

Koelsch, S. & Stegemann, T. (in press). The brain and positive biological effects in healthy and clinical populations, *in* R. MacDonald, D. Kreutz and L. Mitchell (eds), *Music, Health and Well-Being*, Oxford: Oxford University Press.

Koelsch, S., Wittfoth, M., Wolf, A., Müller, J. & Hahne, A. (2004b). Music perception in cochlear implant users: An ERP study, *Clinical Neurophysiology* 115(4): 966–972.

Kohler, E., Keysers, C., Umilta, M., *et al.* (2002). Hearing sounds, understanding actions: Action representation in mirror neurons, *Science* 297(5582): 846–848.

Konečni, V. (2003). Review of music and emotion: Theory and research, *Music Perception* 20(3): 332–341.

Koopman, C. & Davies, S. (2001). Musical meaning in a broader perspective, *Journal of Aesthetics and Art Criticism* 59(3): 261–273.

Kopp, B. & Wolff, M. (2000). Brain mechanisms of selective learning: Event-related potentials provide evidence for error-driven learning in humans, *Biological Psychology* 51(2-3): 223–246.

Korzyukov, O., Winkler, I., Gumenyuk, V. & Alho, K. (2003). Processing abstract auditory features in the human auditory cortex, *Neuroimage* 20(4): 2245–2258.

Kotz, S., Schwartze, M. & Schmidt-Kassow, M. (2009). Non-motor basal ganglia functions: A review and proposal for a model of sensory predictability in auditory language perception, *Cortex* 45(8): 982–990.

Kraus, N. & Nicol, T. (2005). Brainstem origins for cortical "what" and "where" pathways in the auditory system, *Trends in Neurosciences* 28(4): 176–181.

Kreutz, G., Bongard, S., Rohrmann, S., Hodapp, V. & Grebe, D. (2004). Effects of choir singing or listening on secretory immunoglobulin A, cortisol, and emotional state, *Journal of Behavioral Medicine* 27(6): 623–635.

Kriegstein, K., Kleinschmidt, A., Sterzer, P. & Giraud, A. (2005). Interaction of face and voice areas during speaker recognition, *Journal of Cognitive Neuroscience* 17(3): 367–376.

Krumhansl, C. L. (1979). The Psychological representation of musical pitch in a tonal context, *Cognitive Psychology* 11: 346–374.

Krumhansl, C. L. (1990). *Cognitive Foundations of Musical Pitch*, Oxford University Press, USA.

Krumhansl, C. L. (1996). A perceptual analysis of Mozart's Piano Sonata K. 282: Segmentation, tension, and musical ideas, *Music Perception* 13: 401–432.

Krumhansl, C. L. (1997). An exploratory study of musical emotions and psychophysiology, *Canadian Journal of Experimental Psychology* 51(4): 336–353.

Krumhansl, C. L., Bharucha, J. & Castellano, M. (1982a). Key distance effects on perceived harmonic structure in music, *Perception and Psychophysics* 32(2): 96–108.

Krumhansl, C. L., Bharucha, J. & Kessler, E. (1982b). Perceived harmonic structure of chords in three related musical keys, *Journal of Experimental Psychology: Human Perception and Performance* 8(1): 24–36.

Krumhansl, C. L. & Cuddy, L. (2010). A theory of tonal hierarchies in music, *Music Perception* 51–87.

Krumhansl, C. & Kessler, E. (1982a). Tracing the dynamic changes in perceived tonal organization in a spatial representation of musical keys, *Psychological Review* 89(4): 334–368.

Krumhansl, C. L. & Kessler, E. (1982b). Tracing the dynamic changes in perceived tonal organization in a spatial representation of musical keys, *Psychological Review* 89(4): 334–368.

Krumhansl, C. L. & Shepard, R. (1979). Quantification of the hierarchy of tonal functions within a diatonic context, *Experimental Psychology: Human Perception and Performance* 5(4): 579–594.

Kujala, A., Huotilainen, M., Hotakainen, M., *et al.* (2004). Speech-sound discrimination in neonates as measured with MEG, *NeuroReport* 15(13): 2089–2092.

Kushnerenko, E., Winkler, I., Horvath, J., *et al.* (2007). Processing acoustic change and novelty in newborn infants, *European Journal of Neuroscience* 26(1): 265–274.

Kutas, M. & Federmeier, K. (2000). Electrophysiology reveals semantic memory use in language comprehension, *Trends in Cognitive Sciences* 4(12): 463–470.

Kutas, M. & Hillyard, S. (1980). Reading senseless sentences: Brain potentials reflect semantic incongruity, *Science* 207: 203–205.

Kutas, M. & Hillyard, S. (1983). Event-related brain potentials to grammatical errors and semantic anomalies, *Memory and Cognition* 11(5): 539–550.

Kutas, M., Lindamond, T. & Hillyard, S. (1984). Word expectancy and event-related brain potentials during sentence processing, *in* S. Kornblum and J. Requin (eds), *Preparatory States and Processes*, New Jersey: Erlbaum, pp. 217–238.

Kuwada, S., Yin, T. & Wickesberg, R. (1979). Response of cat inferior colliculus neurons to binaural beat stimuli: Possible mechanisms for sound localization, *Science* 206(4418): 586–588.

La Vaque, T. (1999). The History of EEG Hans Berger, *Journal of Neurotherapy* 3(2): 1–9.

Lahav, A., Saltzman, E. & Schlaug, G. (2007). Action representation of sound: Audiomotor recognition network while listening to newly acquired actions, *Journal of Neuroscience* 27(2): 308–314.

Lamont, A. & Cross, I. (1994). Children's cognitive representations of musical pitch, *Music Perception* 12: 27–55.

Langer, S. (1942). *Philosophy in a New Key*, Cambridge, MA: Harvard University Press.

Langer, S. (1953). *Feeling and Form*, New York: Scribner's.

Langner, G., Albert, M. & Briede, T. (2002). Temporal and spatial coding of periodicity information in the inferior colliculus of awake chinchilla (Chinchilla laniger), *Hearing Research* 168(1-2): 110–130.

Lau, E., Phillips, C. & Poeppel, D. (2008). A cortical network for semantics: (De)constructing the N400, *Nature Reviews Neuroscience* 9(12): 920–933.

Lauritzen, M. (2008). On the neural basis of fMRI signals, *Clinical Neurophysiology* 119: 729–730.

Leardi, S., Pietroletti, R., Angeloni, G., *et al.* (2007). Randomized clinical trial examining the effect of music therapy in stress response to day surgery, *British Journal of Surgery* 94(8): 943–947.

LeDoux, J. (2000). Emotion circuits in the brain, *Ann Rev Neurosci* 23: 155–184.

LeDoux, J. (2007). The amygdala, *Current Biology* 17(20): R868–R874.

LeDoux, J., Farb, C. & Ruggiero, D. (1990). Topographic organization of neurons in the acoustic thalamus that project to the amygdala, *Journal of Neuroscience* 10(4): 1043–1054.

Leino, S., Brattico, E., Tervaniemi, M. & Vuust, P. (2007). Representation of harmony rules in the human brain: Further evidence from event-related potentials, *Brain Research* 1142: 169–177.

Leman, M. (2000). An auditory model of the role of short-term memory in probe-tone ratings, *Music Perception* 481–509.

Lerdahl, F. (2001a). The sounds of poetry viewed as music, *in* R. J. Zatorre and I. Peretz (eds), *The Biological Foundations of Music*, Vol. 930, New York: The New York Academy of Sciences, 337–354.

Lerdahl, F. (2001b). *Tonal Pitch Space*, Oxford University Press.

Lerdahl, F. (2009). Genesis and architecture of the GTTM project, *Music Perception* pp. 187–194.

Lerdahl, F. & Jackendoff, R. (1999). *A Generative Theory of Music*, Cambridge: MIT.

Lerdahl, F. & Krumhansl, C. (2007). Modeling tonal tension, *Music Perception* 24(4): 329–366.

Lerner, Y., Papo, D., Zhdanov, A., Belozersky, L. & Hendler, T. (2009). Eyes wide shut: Amygdala mediates eyes-closed effect on emotional experience with music, *PLoS One* 4(7): e6230.

Leuthold, H. & Jentzsch, I. (2002). Spatiotemporal source localisation reveals involvement of medial premotor areas in movement reprogramming, *Experimental Brain Research* 144(2): 178–188.

Levinson, J. (1990). *Music and Negative Emotion*, Ithaca, NY: Cornell University Press.

Levinson, J. (2004). Musical chills and other delights of music, *in* J. Davidson (ed.), *The Music Practitioner: Research For the Music Performer, Teacher, and Listener*, Ashgate Pub Ltd, pp. 335–352.

Levitt, P. & Moore, R. (1979). Origin and organization of brainstem catecholamine innervation in the rat, *The Journal of Comparative Neurology* 186(4): 505–528.

Li, W. & Yang, Y. (2009). Perception of prosodic hierarchical boundaries in Mandarin Chinese sentences, *Neuroscience* 158(4): 1416–1425.

Liberman, A. & Mattingly, I. (1985). The motor theory of speech perception revised, *Cognition* 21(1): 1–36.

Liebenthal, E., Ellingson, M., Spanaki, M., *et al.* (2003). Simultaneous ERP and fMRI of the auditory cortex in a passive oddball paradigm, *Neuroimage* 19(4): 1395–1404.

Liegeois-Chauvel, C., Peretz, I., Babaie, M., Laguitton, V. & Chauvel, P. (1998). Contribution of different cortical areas in the temporal lobes to music processing, *Brain* 121(10): 1853–1867.

Logan, G. (1992). Attention and preattention in theories of automaticity, *The American Journal of Psychology* 105(2): 317–339.

Lohmann, G., Erfurth, K., Müller, K. & Turner, R. (2012). Critical comments on dynamic causal modelling, *Neuroimage* 59(3): 2322–2329.

Lohmann, G., Margulies, D., Horstmann, A., *et al.* (2010). Eigenvector centrality mapping for analyzing connectivity patterns in fMRI data of the human brain, *PLoS-one* 5(4): e10232.

Lotze, H. (1852). *Medicinische Psychologie Oder Physiologie der Seele*, Leipzig: Weidmann.

Loui, P., Grent-'t Jong, T., Torpey, D. & Woldorff, M. (2005). Effects of attention on the neural processing of harmonic syntax in Western music, *Cognitive Brain Research* 25(3): 678–687.

Loui, P., Wu, E., Wessel, D. & Knight, R. (2009). A generalized mechanism for perception of pitch patterns, *The Journal of Neuroscience* 29(2): 454–459.

Lundqvist, L., Carlsson, F., Hilmersson, P. & Juslin, P. (2009). Emotional responses to music: Experience, expression, and physiology, *Psychology of Music* 37(1): 61–90.

MacLean, P. (1990). *The Triune Brain in Evolution: Role in Paleocerebral Functions*, New York: Plenum Press.

Maess, B., Jacobsen, T., Schröger, E. & Friederici, A. D. (2007). Localizing pre-attentive auditory memory-based comparison: Magnetic mismatch negativity to pitch change, *Neuroimage* 37(2): 561–571.

Maess, B., Koelsch, S., Gunter, T. C. & Friederici, A. D. (2001). Musical syntax is processed in the area of Broca: An MEG-study, *Nature Neuroscience* 4(5): 540–545.

Magne, C., Schön, D. & Besson, M. (2006). Musician children detect pitch violations in both music and language better than nonmusician children: Behavioral and electrophysiological approaches, *Journal of Cognitive Neuroscience* 18(2): 199–211.

Maidhof, C. & Koelsch, S. (2011). Effects of selective attention on syntax processing in music and language, *Journal of Cognitive Neuroscience* (in press).

Maidhof, C., Rieger, M., Prinz, W. & Koelsch, S. (2009). Nobody is perfect: ERP effects prior to performance errors in musicians indicate fast monitoring processes, *PLoS One* 4(4): e5032.

Maidhof, C., Vavatzanidis, N., Prinz, W., Rieger, M. & Koelsch, S. (2010). Processing expectancy violations during music performance and perception: An ERP study, *Journal of Cognitive Neuroscience* 22(10): 2401–2413.

Maratos, A., Gold, C., Wang, X. & Crawford, M. (2008). Music therapy for depression, *Cochrane Database of Systematic Reviews* 1(4): 1–16.

Marteniuk, R., MacKenzie, C. & Baba, D. (1984). Bimanual movement control: Information processing and interaction effects, *The Quarterly Journal of Experimental Psychology A: Human Experimental Psychology* 36(2): 335–365.

Matthews, B., Chang, C., De May, M., Engstrom, J. & Miller, B. (2009). Pleasurable emotional response to music: A case of neurodegenerative generalized auditory agnosia, *Neurocase* 15(3): 248–259.

Mattout, J., Phillips, C., Penny, W., Rugg, M. & Friston, K. (2006). MEG source localization under multiple constraints: An extended Bayesian framework, *Neuroimage* 30(3): 753–767.

Maurer, U., Bucher, K., Brem, S. & Brandeis, D. (2003). Development of the automatic mismatch response: From frontal positivity in kindergarten children to the mismatch negativity, *Clinical Neurophysiology* 114(5): 808–817.

McAlpine, D., Jiang, D., Shackleton, T. & Palmer, A. (2000). Responses of neurons in the inferior colliculus to dynamic interaural phase cues: Evidence for a mechanism of binaural adaptation, *Journal of Neurophysiology* 83(3): 1356–1365.

McCarthy, G. & Donchin, E. (1981). A metric for thought: A comparison of P300 latency and reaction time, *Science* 211(4477): 77–80.

McCraty, R., Atkinson, M., Rein, G. & Watkins, A. (1996). Music enhances the effect of positive emotional states on salivary IgA, *Stress Medicine* 167–175.

McKinney, C., Antoni, M., Kumar, M., Tims, F. & McCabe, P. (1997). Effects of guided imagery and music (GIM) therapy on mood and cortisol in healthy adults, *Health Psychology* 16(4): 390–400.

McKinnon, R. & Osterhout, L. (1996). Constraints on movement phenomena in sentence processing: Evidence from event-related brain potentials, *Language and Cognitive Processes* 11: 495–523.

McIntosh, A. & Gonzalez-Lima, F. (1994). Structural equation modeling and its application to network analysis in functional brain imaging, *Human Brain Mapping* 2(1-2): 2–22.

McPherson, W. & Holcomb, P. (1999). An electrophysiological investigation of semantic priming with pictures of real objects, *Psychophysiology* 36: 53–65.

Mecklinger, A. (1998). On the modularity of recognition memory for object and spatial location–topographic ERP analysis, *Neuropsychologia* 36(5): 441–460.

Mecklinger, A., Schriefers, H., Steinhauer, K. & Friederici, A. D. (1995). Processing relative clauses varying on syntactic and semantic dimensions: An analysis with event-related potentials, *Memory and Cognition* 23: 477–494.

Mega, M. S., Cummings, J. L., Salloway, S. & Malloy, P. (1997). The limbic system: An anatomic, phylogenetic, and clinical perspective, *in* S. Salloway, P. Malloy and J. L. Cummings (eds), *The Neuropsychiatry of Limbic and Subcortical Disorders*, American Psychiatric Press, pp. 3–18.

Menning, H., Roberts, L. & Pantev, C. (2000). Plastic changes in the auditory cortex induced by intensive frequency discrimination training, *Neuroreport* 11(4): 817–822.

Menon, V. & Levitin, D. (2005). The rewards of music listening: Response and physiological connectivity of the mesolimbic system, *Neuroimage* 28(1): 175–184.

Messner, B., Jipson, A., Becker, P. & Byers, B. (2007). The hardest hate: A sociological analysis of country hate music, *Popular Music and Society* 30(4): 513–531.

Meyer, L. (1956). *Emotion and Meaning in Music*, Chicago: University of Chicago Press.

Meyer, M., Alter, K., Friederici, A. D., Lohmann, G. & von Cramon, D. Y. (2002). FMRI reveals brain regions mediating slow prosodic modulations in spoken sentences, *Human Brain Mapping* 17(2): 73–88.

Meyer, M., Steinhauer, K., Alter, K., Friederici, A. D. & von Cramon, D. Y. (2004). Brain activity varies with modulation of dynamic pitch variances in sentence melody, *Brain and Language* 89: 277–289.

Miall, R. & Wolpert, D. (1996). Forward models for physiological motor control, *Neural Networks* 9(8): 1265–1279.

Middleton, F. & Strick, P. (2000). Basal ganglia and cerebellar loops: Motor and cognitive circuits, *Brain Research Reviews* 31(2-3): 236–250.

Miltner, W., Braun, C. & Coles, M. (1997). Event-related brain potentials following incorrect feedback in a time-estimation task: Evidence for a "generic" neural system for error detection, *Journal of Cognitive Neuroscience* 9(6): 788–798.

Miluk-Kolasa, B., Obminski, Z., Stupnicki, R. & Golec, L. (1994). Effects of music treatment on salivary cortisol in patients exposed to pre-surgical stress, *Experimental and Clinical Endocrinology and Diabetes* 102(2): 118–120.

Miranda, R. & Ullman, M. (2007). Double dissociation between rules and memory in music: An event-related potential study, *Neuroimage* 38(2): 331–345.

Mithen, S. (2006). *The Singing Neanderthals: The Origins of Music, Language, Mind, and Body*, Harvard University Press.

Mitterschiffthaler, M. T., Fu, C. H., Dalton, J. A., Andrew, C. M. & Williams, S. C. (2007). A functional MRI study of happy and sad affective states evoked by classical music, *Human Brain Mapping* 28: 1150–1162.

Mizuno, T. & Sugishita, M. (2007). Neural correlates underlying perception of tonality-related emotional contents, *Neuroreport* 18(16): 1651–1655.

Molholm, S., Martinez, A., Ritter, W., Javitt, D. & Foxe, J. (2005). The neural circuitry of pre-attentive auditory change-detection: An fMRI study of pitch and duration mismatch negativity generators, *Cerebral Cortex* 15(5): 545–551.

Moon, C., Cooper, R. & Fifer, W. (1993). Two-day-olds prefer their native language, *Infant Behavior and Development* 16(4): 495–500.

Moore, B. (2008). *An Introduction to the Psychology of Hearing*, 5 ed, Bingley, UK: Emerald.

Moors, A. & De Houwer, J. (2006). Automaticity: A theoretical and conceptual analysis, *Psychological Bulletin* 132(2): 297–326.

Moors, A. & Kuppens, P. (2008). Distinguishing between two types of musical emotions and reconsidering the role of appraisal, *Behavioral and Brain Sciences* 31: 588–589.

Moreno, S. & Besson, M. (2006). Musical training and language-related brain electrical activity in children, *Psychophysiology* 43(3): 287–291.

Morosan, P., Rademacher, J., Palomero-Gallagher, N. & Zilles, K. (2005). Anatomical organization of the human auditory cortex: Cytoarchitecture and transmitter receptors, *in* P. Heil, E. König & E. Budinger (eds), *Auditory Cortex: Towards a synthesis of Human and Animal Research*. Mahwah, NJ: Lawrence Erlbaum.

Morosan, P., Rademacher, J., Schleicher, A., *et al.* (2001). Human primary auditory cortex: Cytoarchitectonic subdivisions and mapping into a spatial reference system, *Neuroimage* 13(4): 684–701.

Moscovitch, M., Nadel, L., Winocur, G., Gilboa, A. & Rosenbaum, R. (2006). The cognitive neuroscience of remote episodic, semantic and spatial memory, *Current Opinion in Neurobiology* 16(2): 179–190.

Müller, M., Höfel, L., Brattico, E. & Jacobsen, T. (2010). Aesthetic judgments of music in experts and laypersons – An ERP study, *International Journal of Psychophysiology* 76(1): 40–51.

Münte, T. F., Matzke, M. & Johannes, S. (1997). Brain activity associated with incongruities in words and pseudowords, *Journal of Cognitive Neuroscience* 9: 318–329.

Murcia, C., Kreutz, G., Clift, S. & Bongard, S. (2010). Shall we dance? an exploration of the perceived benefits of dancing on well-being, *Arts and Health* 2(2): 149–163.

Murray, E. (2007). The amygdala, reward and emotion, *Trends in Cognitive Sciences* 11(11): 489–497.

Musacchia, G., Sams, M., Skoe, E. & Kraus, N. (2007). Musicians have enhanced subcortical auditory and audiovisual processing of speech and music, *Proceedings of the National Academy of Sciences* 104(40): 15894–15898.

Mutschler, I., Schulze-Bonhage, A., Glauche, V., *et al.* (2007). A rapid sound-action association effect in human insular cortex, *PLoS One* 2(2): e259.

Mutschler, I., Wieckhorst, B., Kowalevski, S., *et al.* (2009). Functional organization of the human anterior insular cortex, *Neuroscience Letters* 457(2): 66–70.

Näätänen, R. (1990). The role of attention in auditory information processing as revealed by event-related potentials and other brain measures of cognitive function, *Behavioral and Brain Sciences* 13: 201–288.

Näätänen, R. (1992). *Attention and Brain Function*, Hillsdale, NJ: Erlbaum.

Näätänen, R., Astikainen, P., Ruusuvirta, T. & Huotilainen, M. (2010). Automatic auditory intelligence: An expression of the sensory-cognitive core of cognitive processes, *Brain Research Reviews* 64(1): 123–136.

Näätänen, R. & Gaillard, A. (1983). The N2 deflection of ERP and the orienting reflex, *in* A. Gaillard and W. Ritter (eds), *EEG Correlates of Information*

Processing: Theoretical Issues, Amsterdam: North Holland, pp. 119–141.

Näätänen, R., Jacobsen, T. & Winkler, I. (2005). Memory-based or afferent processes in mismatch negativity (MMN): A review of the evidence, *Psychophysiology* 42(1): 25–32.

Näätänen, R., Lehtokoski, A., Lennes, M., *et al.* (1997). Language-specific phoneme representations revealed by magnetic brain responses, *Nature* 385(6615): 432–434.

Näätänen, R., Lehtokoski, A., Lennes, M., *et al.* (1997). Language-specific phoneme representations revealed by electric and magnetic brain responses, *European Heart Journal* 385: 432–434.

Näätänen, R., Paavilainen, P., Alho, K., Reinikainen, K. & Sams, M. (1987). The mismatch negativity to intensity changes in an auditory stimulus sequence, *Electroencephalography and Clinical Neurophysiology* 40: 125–131.

Näätänen, R., Paavilainen, P. & Reinikainen, K. (1989). Do event-related potentials to infrequent decrements in duration of auditory stimuli demonstrate a memory trace in man? *Neuroscience Letters* 107: 347–352.

Näätänen, R., Paavilainen, P., Rinne, T. & Alho, K. (2007). The mismatch negativity (MMN) in basic research of central auditory processing: A review, *Clinical Neurophysiology* 118(12): 2544–2590.

Näätänen, R. & Picton, T. (1987). The N1 wave of the human electric and magnetic response to sound: A review and an analysis of the component structure, *Psychophysiology* 24: 375–425.

Näätänen, R., Simpson, M. & Loveless, N. (1982). Stimulus deviance and evoked potentials, *Biological Psychology* 14: 53–98.

Nachev, P., Kennard, C. & Husain, M. (2008). Functional role of the supplementary and pre-supplementary motor areas, *Nat Rev Neurosci* 9(11): 856–869.

Nadel, L. (2008). Hippocampus and context revisited, *in* S. Mizumori (ed.), *Hippocampal Place Fields: Relevance to Learning and Memory*, New York: Oxford University Press, pp. 3–15.

Nan, Y., Knösche, T. & Friederici, A. D. (2006). The perception of musical phrase structure: A cross-cultural ERP study, *Brain Research* 1094(1): 179–191.

Nattiez, J. (1990). *Music and Discourse: Toward a Semiology of Music*, Princeton University Press.

Nelken, I. (2004). Processing of complex stimuli and natural scenes in the auditory cortex, *Current Opinion in Neurobiology* 14(4): 474–480.

Nelson, A., Hartl, W., Jauch, K., *et al.* (2008). The impact of music on hypermetabolism in critical illness, *Current Opinion in Clinical Nutrition and Metabolic Care* 11(6): 790–794.

Neuert, V., Verhey, J. & Winter, I. (2005). Temporal representation of the delay of iterated rippled noise in the dorsal cochlear nucleus, *Journal of Neurophysiology* 93(5): 2766–2776.

Neuhaus, C., Knösche, T. & Friederici, A. D. (2006). Effects of musical expertise and boundary markers on phrase perception in music, *Journal of Cognitive Neuroscience* 18(3): 472–493.

Neville, H., Nicol, J., Barss, A., Forster, K. & Garrett, M. (1991). Syntactically based sentence processing classes: Evidence from event-related brain potentials, *Journal of Cognitive Neuroscience* 3: 151–165.

Nicola, S. (2007). The nucleus accumbens as part of a basal ganglia action selection circuit, *Psychopharmacology* 191(3): 521–550.

Niedermeyer, E. & Da Silva, F. (2005). *Electroencephalography: Basic Principles, Clinical Applications, and Related Fields*, Lippincott Williams and Wilkins.

Nieuwenhuis, S., Holroyd, C., Mol, N. & Coles, M. (2004). Reinforcement-related brain potentials from medial frontal cortex: Origins and functional significance, *Neuroscience and Biobehavioral Reviews* 28(4): 441–448.

Nieuwenhuis, S., Ridderinkhof, K., Blom, J., Band, G. & Kok, A. (2001). Error-related brain potentials are differentially related to awareness of response errors: Evidence from an antisaccade task, *Psychophysiology* 38(5): 752–760.

Nieuwenhuys, R., Voogd, J. & Huijzen, C. V. (2008). *The Human Central Nervous System*, Berlin: Springer.

Nieuwland, M. & Kuperberg, G. (2008). When the truth is not too hard to handle, *Psychological Science* 19(12): 1213–1218.

Nilsson, U. (2009). The effect of music intervention in stress response to cardiac surgery in a randomized clinical trial, *Heart and Lung: The Journal of Acute and Critical Care* 38(3): 201–207.

Nilsson, U., Unosson, M. & Rawal, N. (2005). Stress reduction and analgesia in patients exposed to calming music postoperatively: A randomized controlled trial, *European Journal of Anaesthesiology* 22(02): 96–102.

Nittono, H. (2006). Voluntary stimulus production enhances deviance processing in the brain, *International Journal of Psychophysiology* 59(1): 15–21.

Nobre, A., Coull, J., Frith, C. & Mesulam, M. (1999). Orbitofrontal cortex is activated during breaches of expectation in tasks of visual attention, *Nature Neuroscience* 2: 11–12.

Norton, A., Zipse, L., Marchina, S. & Schlaug, G. (2009). Melodic intonation therapy, *Annals of the New York Academy of Sciences* 1169(1): 431–436.

Nouvian, R., Beutner, D., Parsons, T. & Moser, T. (2006). Structure and function of the hair cell ribbon synapse, *Journal of Membrane Biology* 209(2): 153–165.

Novitski, N., Huotilainen, M., Tervaniemi, M., Naatanen, R. & Fellman, V. (2007). Neonatal frequency discrimination in 250-4000 Hz range: Electrophysiological evidence, *Clinical Neurophysiology* 118(2): 412–419.

Noy, P. (1993). How music conveys emotion, *in* S. Feder, R. Karmel and G. Pollock (eds), *Psychoanalytic Explorations in Music*, International Universities Press, pp. 125–149.

Nunez, P. L. (1981). *Electric Fields of the Brain*, Oxford: Oxford University Press.

Obermeier, C., Holle, H. & Gunter, T. C. (2011). What iconic gesture fragments reveal about gesture–speech integration: When synchrony is lost, memory can help, *Journal of Cognitive Neuroscience* 23(7): 1648–1663.

Obleser, J. & Eisner, F. (2009). Pre-lexical abstraction of speech in the auditory cortex, *Trends in Cognitive Sciences* 13(1): 14–19.

Obleser, J., Meyer, L. & Friederici, A. D. (2011). Dynamic assignment of neural resources in auditory comprehension of complex sentences, *Neuroimage* 56(4): 2310–2320.

Okamoto, H., Stracke, H., Stoll, W. & Pantev, C. (2010). Listening to tailor-made notched music reduces tinnitus loudness and tinnitus-related auditory cortex activity, *Proceedings of the National Academy of Sciences* 107(3): 1207–1210.

Oliveira, F., McDonald, J. & Goodman, D. (2007). Performance monitoring in the anterior cingulate is not all error related: Expectancy deviation and the representation of action-outcome associations, *Journal of Cognitive Neuroscience* 19(12): 1994–2004.

Öngür, D. & Price, J. L. (2000). The organization of networks within the orbital and medial prefrontal cortex of rats, monkeys and humans, *Cerebral Cortex* 10(3): 206–219.

Opitz, B. & Friederici, A. D. (2007). Neural basis of processing sequential and hierarchical syntactic structures, *Human Brain Mapping* 28(7): 585–592.

Opitz, B. & Kotz, S. (2011). Ventral Premotor Cortex Lesions disrupt learning of sequential grammatical structures, *Cortex* (in press).

Opitz, B., Mecklinger, A., Friederici, A. D. & von Cramon, D. Y. (1999a). The functional neuroanatomy of novelty processing: Integrating ERP and fMRI results, *Cerebral Cortex* 9(4): 379–391.

Opitz, B., Mecklinger, A., von Cramon, D. Y. & Kruggel, F. (1999b). Combining electrophysiological and hemodynamic measures of the auditory oddball, *Psychophysiology* 36(1): 142–147.

Opitz, B., Rinne, T., Mecklinger, A., von Cramon, D. Y. & Schröger, E. (2002). Differential contribution of frontal and temporal cortices to auditory change detection: fMRI and ERP results, *Neuroimage* 15: 167–174.

Orgs, G., Lange, K., Dombrowski, J. & Heil, M. (2006). Conceptual priming for environmental sounds and words: An ERP study, *Brain and Cognition* 62(3): 267–272.

Orgs, G., Lange, K., Dombrowski, J. & Heil, M. (2007). Is conceptual priming for environmental sounds obligatory? *International Journal of Psychophysiology* 65(2): 162–166.

Orini, M., Bailón, R., Enk, R., *et al.* (2010). A method for continuously assessing the autonomic response to music-induced emotions through HRV analysis, *Medical and Biological Engineering and Computing* 48(5): 423–433.

Osterhout, L. (1999). A superficial resemblance does not necessarily mean you are part of the family: Counterarguments to Coulson, King and Kutas (1998) in the P600/SPS-P300 debate, *Language and Cognitive Processes* 14(1): 1–14.

Osterhout, L. & Holcomb, P. (1995). ERPs and language comprehension, *in* M. Rugg and M. Coles (eds), *Electrophysiology of Mind. Event-Related Potentials and Cognition*, Oxford: Oxford University Press, pp. 192–208.

Osterhout, L. & Holcomb, P. J. (1992). Event-related potentials and syntactic anomaly, *Journal of Memory and Language* 31: 785–804.

Osterhout, L. & Holcomb, P. J. (1993). Event-related potentials and syntactic anomaly: Evidence of anomaly-detection during the perception of continuous speech, *Language and Cognitive Processes* 8: 413–437.

Osterhout, L., Holcomb, P. J. & Swinney, D. (1994). Brain potentials elicited by garden-path sentences: Evidence of the application of verb information during parsing, *Journal of Experimental Psychology: Learning, Memory, and Cognition* 20: 786–803.

Osterhout, L. & Mobley, L. A. (1995). Event-related brain potentials elicited by failure to agree, *Journal of Memory and Language* 34: 739–773.

Overy, K. & Molnar-Szakacs, I. (2009). Being together in time: Musical experience and the mirror neuron system, *Music Perception* 26(5): 489–504.

Owings, D. & Morton, E. (1998). *Animal Vocal Communication: A New Approach*, Cambridge University Press.

Paavilainen, P., Arajarvi, P. & Takegata, R. (2007). Preattentive detection of nonsalient contingencies between auditory features, *Neuroreport* 18(2): 159–163.

Paavilainen, P., Degerman, A., Takegata, R. & Winkler, I. (2003). Spectral and temporal stimulus characteristics in the processing of abstract auditory features, *Neuroreport* 14(5): 715–718.

Paavilainen, P., Jaramillo, M. & Näätänen, R. (1998). Binaural information can converge in abstract memory traces, *Psychophysiology* 35(5): 483–487.

Paavilainen, P., Karlsson, M. L., Reinikainen, K. & Näätänen, R. (1989). The mismatch negativity to change in spatial location of an auditory stimulus,

Electroencephalography and Clinical Neurophysiology 73: 129–141.

Paavilainen, P., Simola, J., Jaramillo, M., Naatanen, R. & Winkler, I. (2001). Preattentive extraction of abstract feature conjunctions from auditory stimulation as reflected by the mismatch negativity (MMN), *Psychophysiology* 38(02): 359–365.

Paller, K. A., McCarthy, G. & Wood, C. C. (1992). Event-related potentials elicited by deviant endings to melodies, *Psychophysiology* 29(2): 202–206.

Palmer, C. (1997). Music performance, *Annual Review of Psychology* 48: 115–38.

Panksepp, J. (1995). The emotional sources of "chills" induced by music, *Music Perception* 13: 171–207.

Panksepp, J. (1998). *Affective Neuroscience: The Foundations of Human and Animal Emotions*, Oxford University Press, USA.

Panksepp, J. & Bernatzky, G. (2002). Emotional sounds and the brain: The neuro-affective foundations of musical appreciation, *Behavioural Processes* 60(2): 133–155.

Pannekamp, A., Toepel, U., Alter, K., Hahne, A. & Friederici, A. (2005). Prosody-driven sentence processing: An event-related brain potential study, *Journal of Cognitive Neuroscience* 17(3): 407–421.

Pantev, C., Roberts, L. E., Schulz, M., Engelien, A. & Ross, B. (2001). Timbre-specific enhancement of auditory cortical representation in musicians, *NeuroReport* 12(1): 169–174.

Papez, J. (1937). A proposed mechanism of emotion, *Archives of Neurology and Psychiatry* 38(4): 725–743.

Papoušek, H. (1996). Musicality in infancy research, *in* J. Sloboda and I. Deliege (eds), *Musical Beginnings*, Oxford: Oxford University Press, pp. 37–55.

Parncutt, R. (1989). *Harmony: A Psychoacoustical Approach*, Berlin: Springer.

Parncutt, R. (2006). Commentary on Keith Mashinter's 'Calculating sensory dissonance: Some discrepancies arising from the models of Kameoka and Kuriyagawa, and Hutchinson and Knopoff', *Empirical Musicology Review* 1: 1–5.

Patel, A. (1998). Syntactic processing in language and music: Different cognitive operations, similar neural resources? *Music Perception* 16(1): 27–42.

Patel, A. (2003). Language, music, syntax and the brain, *Nature Neuroscience* 6(7): 674–681.

Patel, A. (2006). Musical rhythm, linguistic rhythm, and human evolution, *Music Perception* 24(1): 99–103.

Patel, A. (2008). *Music, Language, and the Brain*, Oxford University Press.

Patel, A. & Balaban, E. (2001). Human pitch perception is reflected in the timing of stimulus-related cortical activity, *Nature Neuroscience* 4(8): 839–844.

Patel, A., Gibson, E., Ratner, J., Besson, M. & Holcomb, P. (1998). Processing syntactic relations in language and music: An event-related potential study, *Journal of Cognitive Neuroscience* 10(6): 717–733.

Patel, A., Iversen, J., Bregman, M. & Schulz, I. (2009). Experimental evidence for synchronization to a musical beat in a nonhuman animal, *Current Biology* 19(10): 827–830.

Patel, A., Iversen, J., Wassenaar, M. & Hagoort, P. (2008). Musical syntactic processing in agrammatic Broca's aphasia, *Aphasiology* 22(7): 776–789.

Patel, S. & Azzam, P. (2005). Characterization of N200 and P300: Selected studies of the event-related potential, *International Journal of Medical Sciences* 2(4): 147–154.

Patterson, R. & Moore, B. (1986). Auditory filters and excitation patterns as representations of frequency resolution, *in* B. Moore (ed.), *Frequency Selectivity in Hearing*, London: Academic, pp. 123–177.

Patterson, R., Uppenkamp, S., Johnsrude, I. & Griffiths, T. (2002). The processing of temporal pitch and melody information in auditory cortex, *Neuron* 36(4): 767–776.

Peirce, C. (1931/1958). *The Collected Papers of Charles Sanders Peirce*, Cambridge, MA: Harvard University Press.

Perani, D., Saccuman, M., Scifo, P., *et al.* (2010). Functional specializations for music processing in the human newborn brain, *Proceedings of the National Academy of Sciences* 107(10): 4758–4763.

Peretz, I., Brattico, E., Järvenpää, M. & Tervaniemi, M. (2009). The amusic brain: In tune, out of key, and unaware, *Brain* 132(5): 1277–1286.

Peretz, I. & Coltheart, M. (2003). Modularity of music processing, *Nature Neuroscience* 6(7): 688–691.

Peretz, I. & Zatorre, R. (2005). Brain organization for music processing, *Annual Reviews in Psychology* 56: 89–114.

Perlovsky, L. (2007). Neural dynamic logic of consciousness: The knowledge instinct, *Neurodynamics of Cognition and Consciousness* 2: 73–108.

Perruchet, P. & Vinter, A. (1998). PARSER: A model for word segmentation, *Journal of Memory and Language* 39(2): 246–263.

Petkov, C., Kayser, C., Augath, M. & Logothetis, N. (2006). Functional imaging reveals numerous fields in the monkey auditory cortex, *PLoS Biol* 4(7): e215.

Petkov, C., Kayser, C., Steudel, T., *et al.* (2008). A voice region in the monkey brain, *Nature Neuroscience* 11(3): 367–374.

Pfordresher, P. (2003). Auditory feedback in music performance: Evidence for a dissociation of sequencing and timing, *Journal of Experimental Psychology* 29(4): 949–964.

Pfordresher, P. (2005). Auditory feedback in music performance: The role of melodic structure and musical skill, *Journal of Experimental Psychology* 31(6): 1331–1345.

Pfordresher, P. (2006). Coordination of perception and action in music performance, *Advances in Cognitive Psychology* 2(2): 183–198.

Phelps, M. (2006). *PET: Physics, Instrumentation, and Scanners*, Springer Verlag.

Pickles, J. (2008). *An Introduction to the Physiology of Hearing*, 3rd ed, Bingley, UK: Emerald.

Picton, T., Durieux-Smith, A. & Moran, L. (1994). Recording auditory brainstem responses from infants, *International Journal of Pediatric Otorhinolaryngology* 28(2-3): 93–110.

Picton, T. W. (1980). The use of human event-related potentials in psychology, *in* O. Martin and P. Venables (eds), *Techniques in Psychophysiology*, New York: Wiley, pp. 357–395.

Plack, C. (2005). *The Sense of Hearing*, New York: Lawrence Erlbaum.

Platel, H., Baron, J., Desgranges, B., Bernard, F. & Eustache, F. (2003). Semantic and episodic memory of music are subserved by distinct neural networks, *Neuroimage* 20: 244–256.

Plomp, R. & Levelt, W. (1965). Tonal consonance and critical bandwidth, *The Journal of the Acoustical Society of America* 38: 548–560.

Plomp, R. & Steeneken, H. J. M. (1968). Interference between two simple tones, *The Journal of the Acoustical Society of America* 43: 883–884.

Porter, R. & Lewis, M. (1975). Relationship of neuronal discharges in the precentral gyrus of monkeys to the performance of arm movements, *Brain Research* 98(1): 21–36.

Poulet, J. & Hedwig, B. (2007). New insights into corollary discharges mediated by identified neural pathways, *Trends in Neurosciences* 30(1): 14–21.

Poulin-Charronnat, B., Bigand, E. & Koelsch, S. (2006). Processing of musical syntax tonic versus subdominant: An event-related potential study, *Journal of Cognitive Neuroscience* 18(9): 1545–1554.

Poulin-Charronnat, B., Bigand, E., Madurell, F. & Peereman, R. (2005). Musical structure modulates semantic priming in vocal music, *Cognition* 94(3): B67–B78.

Price, J. (2005). Free will versus survival: Brain systems that underlie intrinsic constraints on behavior, *The Journal of Comparative Neurology* 493(1): 132–139.

Prinz, W. (1990). A common coding approach to perception and action, *in* O. Neumann and W. Prinz (eds), *Relationships Between Perception and Action*, Springer, pp. 167–201.

Prinz, W. (2005). An ideomotor approach to imitation, *in* S. Hurley and N. Chater (eds), *Perspectives on Imitation: Mechanisms of imitation and imitation in animals*, Cambridge, MA: MIT Press, pp. 141–156.

Pritchard, W. (1981). Psychophysiology of P300: A Review, *Psychological Bulletin* 89: 506–540.

Pulvermüller, F. (2005). Brain mechanisms linking language and action, *Nature Reviews Neuroscience* 6(7): 576–582.

Pulvermüller, F. & Fadiga, L. (2010). Active perception: Sensorimotor circuits as a cortical basis for language, *Nature Reviews Neuroscience* 11(5): 351–360.

Pulvermüller, F. & Shtyrov, Y. (2006). Language outside the focus of attention: The mismatch negativity as a tool for studying higher cognitive processes, *Progress in Neurobiology* 79(1): 49–71.

Pulvermüller, F., Shtyrov, Y., Hasting, A. & Carlyon, R. (2008). Syntax as a reflex: Neurophysiological evidence for early automaticity of grammatical processing, *Brain and Language* 104(3): 244–253.

Purwins, H., Blankertz, B. & Obermayer, K. (2007). Toroidal models in tonal theory and pitch-class analysis, *Computing in Musicology* 15: 73–98.

Quiroga Murcia, C., Bongard, S. & Kreutz, G. (2009). Emotional and neurohumoral responses to dancing tango argentino, *Music and Medicine* 1(1): 14–21.

Rameau, J.-P. (1722). *Traité de l'Harmonie Reduite á ses Principes Naturels*, Paris.

Rammsayer, T. & Altenmüller, E. (2006). Temporal information processing in musicians and nonmusicians, *Music Perception* 24(1): 37–48.

Rauschecker, J. & Scott, S. (2009). Maps and streams in the auditory cortex: Nonhuman primates illuminate human speech processing, *Nature Neuroscience* 12(6): 718–724.

Regel, S., Gunter, T. C. & Friederici, A. D. (2011). Isn't it ironic? an electrophysiological exploration of figurative language processing, *Journal of Cognitive Neuroscience* 23(2): 277–293.

Regnault, P., Bigand, E. & Besson, M. (2001). Different brain mechanisms mediate sensitivity to sensory consonance and harmonic context: Evidence from auditory event-related brain potentials, *Journal of Cognitive Neuroscience* 13(2): 241–255.

Reich, U. (2011). The meanings of semantics (Comment), *Physics of Life Reviews* 8(2): 120–121.

Ridderinkhof, K., Ullsperger, M., Crone, E. & Nieuwenhuis, S. (2004). The role of the medial frontal cortex in cognitive control, *Science* 306(5695): 443.

Riemann, H. (1877/1971). *Musikalische Syntaxis: Grundriss einer harmonischen Satzbildungslehre*, Niederwalluf: Sändig.

Rilling, J., Gutman, D., Zeh, T., *et al.* (2002). A neural basis for social cooperation, *Neuron* 35(2): 395–405.

Rinne, T., Antila, S. & Winkler, I. (2001). Mismatch negativity is unaffected by top-down predictive information, *NeuroReport* 12(10): 2209–2213.

Rinne, T., Degerman, A. & Alho, K. (2005). Superior temporal and inferior frontal cortices are activated by infrequent sound duration decrements: An fMRI study, *Neuroimage* 26(1): 66–72.

Ritter, W. & Ruchkin, D. S. (1992). A review of event-related potential components discovered in the context of studying P3, *Annual Report of the New York Academy of Science* 658: 1–32.

Rizzolatti, G. & Craighero, L. (2004). The mirror-neuron system, *Annual Reviews of Neuroscience* 27: 169–192.

Rizzolatti, G. & Sinigaglia, C. (2010). The functional role of the parieto-frontal mirror circuit: Interpretations and misinterpretations, *Nat Rev Neurosci* 11(4): 264–274.

Rohrmeier, M. (2005). Towards modelling movement in music: Analysing properties and dynamic aspects of pc set sequences in Bach's chorales, Master's thesis, University of Cambridge.

Rohrmeier, M. (2007). A generative grammar approach to diatonic harmonic structure, *in Proceedings of the 4th Sound and Music Computing Conference*, pp. 97–100, Lefkada, Greece.

Rohrmeier, M. (2011). Towards a generative syntax of tonal harmony, *Journal of Mathematics and Music* 5(1): 35–53.

Rohrmeier, M. & Cross, I. (2008). Statistical properties of tonal harmony in Bach's chorales, *in Proc 10th Intl Conf on Music Perception and Cognition*, Hokkaido Univeristy, Sapporo (Japan).

Rolls, E. & Grabenhorst, F. (2008). The orbitofrontal cortex and beyond: From affect to decision-making, *Progress in Neurobiology* 86(3): 216–244.

Rösler, F., Friederici, A. D., Pütz, P. & Hahne, A. (1993). Event-related brain potentials while encountering semantic and syntactic constraint violations, *Journal of Cognitive Neuroscience* 5: 345–362.

Ross, B., Borgmann, C., Draganova, R., Roberts, L. & Pantev, C. (2000). A high-precision magnetoencephalographic study of human auditory steady-state responses to amplitude-modulated tones, *The Journal of the Acoustical Society of America* 108: 679–691.

Ross, D., Choi, J. & Purves, D. (2007). Musical intervals in speech, *Proceedings of the National Academy of Sciences* 104(23): 9852–9857.

Rowe, M. (1981). The brainstem auditory evoked response in neurological disease: A review, *Ear and Hearing* 2(1): 41–51.

Rozin, P. & Schiller, D. (1980). The nature and acquisition of a preference for chili pepper by humans, *Motivation and Emotion* 4(1): 77–101.

Rugg, M. & Coles, M. (1995). *Electrophysiology of Mind. Event-Related Brain Potentials and Cognition*, Oxford: Oxford University Press.

Rugg, M. & Yonelinas, A. (2003). Human recognition memory: A cognitive neuroscience perspective, *Trends in Cognitive Sciences* 7(7): 313–319.

Rüschemeyer, S., Fiebach, C., Kempe, V. & Friederici, A. (2005). Processing lexical semantic and syntactic information in first and second language: fMRI evidence from German and Russian, *Human Brain Mapping* 25(2): 266–286.

Rüsseler, J., Altenmüller, E., Nager, W., Kohlmetz, C. & Münte, T. (2001). Event-related brain potentials to sound omissions differ in musicians and non-musicians, *Neuroscience Letters* 308(1): 33–36.

Ruusuvirta, T., Huotilainen, M., Fellman, V. & Naatanen, R. (2004). Newborn human brain identifies repeated auditory feature conjunctions of low sequential probability, *European Journal of Neuroscience* 20(10): 2819–2821.

Ruusuvirta, T., Huotilainen, M., Fellman, V., *et al.* (2003). The newborn human brain binds sound features together, *NeuroReport* 14(16): 2117–2119.

Saarinen, J., Paavilainen, P., Schröger, E., Tervaniemi, M. & Näätänen, R. (1992). Representation of abstract attributes of auditory stimuli in the human brain, *Neuroreport* 3(12): 1149–1151.

Salimpoor, V., Benovoy, M., Larcher, K., Dagher, A. & Zatorre, R. (2011). Anatomically distinct dopamine release during anticipation and experience of peak emotion to music, *Nature Neuroscience* 14(2): 257–262.

Sambeth, A., Huotilainen, M., Kushnerenko, E., Fellman, V. & Pihko, E. (2006). Newborns discriminate novel from harmonic sounds: A study using magnetoencephalography, *Clinical Neurophysiology* 117(3): 496–503.

Sammler, D. (2008). The Neuroanatomical Overlap of Syntax Processing in Music and Language Evidence from Lesion and Intracranial ERP Studies. PhD thesis, University of Leipzig.

Sammler, D., Grigutsch, M., Fritz, T. & Koelsch, S. (2007). Music and emotion: Electrophysiological correlates of the processing of pleasant and unpleasant music, *Psychophysiology* 44(2): 293–304.

Sammler, D., Koelsch, S. & Friederici, A. D. (2011). Are left fronto-temporal brain areas a prerequisite for normal music-syntactic processing? *Cortex* 47: 659–673.

Sams, M., Paavilainen, P., Alho, K. & Näätänen, R. (1985). Auditory frequency discrimination and event-related potentials, *Electroencephalography and Clinical Neurophysiology* 62: 437–448.

Särkämö, T., Pihko, E., Laitinen, S., *et al.* (2010). Music and speech listening enhance the recovery of early sensory processing after stroke, *Journal of Cognitive Neuroscience* 22(12): 2716–2727.

Särkämö, T., Tervaniemi, M., Laitinen, S., *et al.* (2008). Music listening enhances cognitive recovery and mood after middle cerebral artery stroke, *Brain* 131(3): 866–876.

Schellenberg, E. (2006). Long-term positive associations between music lessons and IQ, *Journal of Educational Psychology* 98(2): 457–468.

Schellenberg, E., Bigand, E., Poulin-Charronnat, B., Garnier, C. & Stevens, C. (2005). Children's implicit knowledge of harmony in Western music, *Developmental Science* 8(6): 551–566.

Schenker, H. (1956). *Neue musikalische Theorien und Phantasien: Der Freie Satz*, 2nd ed, Wien.

Scherer, K. (2001). Appraisal considered as a process of multilevel sequential checking, *in* K. Scherer, A. Schorr and T. Johnstone (eds), *Appraisal Processes in Emotion: Theory, Methods, Research*, NY: Oxford University Press, pp. 120–144.

Scherer, K. (2004). Which emotions can be evoked by music? What are the underlying mechanisms? And how can we measure them? *Journal of New Music Research* 33: 239–251.

Scherer, K. R. (1995). Expression of emotion in voice and music, *Journal of Voice* 9(3): 235–248.

Scherer, K. R. (2000). Emotions as episodes of sub-system synchronization driven by nonlinear appraisal processes, *in* M. Lewis and I. Granic (eds), *Emotion, Development, and Self-organization: Dynamic Systems Approaches to Emotional Development*, Cambridge University Press, pp. 70–99.

Scherer, K. & Zentner, M. (2001). Emotional effects of music: Production rules, *in* P. Juslin and J. Sloboda (eds), *Music and Emotion: Theory and Research*, Oxford: Oxford University Press, pp. 361–392.

Scherer, K. & Zentner, M. (2008). Music evoked emotions are different – more often aesthetic than utilitarian (Comment), *Behavioral and Brain Sciences* 31(05): 595–596.

Scherg, M. (1990). Fundamentals of dipole source potential analysis, *in* M. Grandori (ed.), *Auditory Evoked Magnetic Fields and Electric Potentials. Advances in Audiology*, Basel: Karger, pp. 40–69.

Scherg, M. & Picton, T. W. (1991). Separation and identification of event-related potential components by brain electric source analysis, *in* C. Brunia, G. Mulder and M. Verbaten (eds), *Event-Related Brain Research (Electroencephalography and Clinical Neurophysiology, Suppl.42)*, Amsterdam: Elsevier, pp. 24–37.

Scherg, M., Vajsar, J. & Picton, T. W. (1989). A source analysis of the late human auditory evoked potentials, *Journal of Cognitive Neuroscience* 1: 336–355.

Schirmer, A. & Kotz, S. (2006). Beyond the right hemisphere: Brain mechanisms mediating vocal emotional processing, *Trends in Cognitive Sciences* 10(1): 24–30.

Schlaug, G., Marchina, S. & Norton, A. (2009). Evidence for plasticity in white-matter tracts of patients with chronic broca's aphasia undergoing intense intonation-based speech therapy, *Annals of the New York Academy of Sciences* 1169(1): 385–394.

Schmidt-Kassow, M. & Kotz, S. (2009). Event-related brain potentials suggest a late interaction of meter and syntax in the P600, *Journal of Cognitive Neuroscience* 21(9): 1693–1708.

Schmidt, L. & Trainor, L. (2001). Frontal brain electrical activity (EEG) distinguishes valence and intensity of musical emotions, *Cognition and Emotion* 15(4): 487–500.

Schneider, N., Schedlowski, M., Schürmeyer, T. & Becker, H. (2001). Stress reduction through music in patients undergoing cerebral angiography, *Neuroradiology* 43(6): 472–476.

Schneider, S., Münte, T., Rodriguez-Fornells, A., Sailer, M. & Altenmüller, E. (2010). Music-supported training is more efficient than functional motor training for recovery of fine motor skills in stroke patients, *Music Perception* 27(4): 271–280.

Schneider, W. & Shiffrin, R. (1977). Controlled and automatic human information processing: I. Detection, search, and attention, *Psychological Review* 84(1): 1–66.

Schoenberg, A. (1978). *Theory of Harmony*, University of California Press.

Schon, D. & Besson, M. (2005). Visually induced auditory expectancy in music reading: A behavioral and electrophysiological study, *Journal of Cognitive Neuroscience* 17(4): 694–705.

Schön, D., Magne, C. & Besson, M. (2004). The music of speech: Music training facilitates pitch processing in both music and language, *Psychophysiology* 41(3): 341–349.

Schön, D., Ystad, S., Kronland-Martinet, R. & Besson, M. (2010). The evocative power of sounds: Conceptual priming between words and nonverbal sounds, *Journal of Cognitive Neuroscience* 22(5): 1026–1035.

Schönberg, A. (1969). *Structural Functions of Harmony*, rev. edn, New York: Norton.

Schönwiesner, M., Novitski, N., Pakarinen, S., *et al.* (2007). Heschl's gyrus, posterior superior temporal gyrus, and mid-ventrolateral prefrontal cortex have different roles in the detection of acoustic changes, *Journal of Neurophysiology* 97(3): 2075–2082.

Schouten, J., Ritsma, R. & Cardozo, B. (1962). Pitch of the residue, *Journal of the Acoustical Society of America* 34(9, Pt. II): 1418–1424.

Schröger, E. (2007). Mismatch negativity: A microphone into auditory memory, *Journal of Psychophysiology* 21(3–4): 138–146.

Schröger, E., Bendixen, A., Trujillo-Barreto, N. & Roeber, U. (2007). Processing of abstract rule violations in audition, *PLoS One* 2(11): e1131.

Schröger, E. & Wolff, C. (1998). Attentional orienting and reorienting is indicated by human event-related brain potentials, *NeuroReport* 9(15): 3355–3358.

Schubotz, R. I. (2007). Prediction of external events with our motor system: Towards a new framework, *Trends in Cognitive Sciences* 11(5): 211–218.

Schulze, K., Mueller, K. & Koelsch, S. (2011a). Neural correlates of strategy use during auditory working memory in musicians and non-musicians, *European Journal of Neuroscience* 33: 189–196.

Schulze, K., Zysset, S., Mueller, K., Friederici, A. D. & Koelsch, S. (2011b). Neuroarchitecture of verbal and tonal working memory in nonmusicians and musicians, *Human Brain Mapping* 32(5): 771–783.

Schwartze, M., Keller, P., Patel, A. & Kotz, S. (2010). The impact of basal ganglia lesions on sensorimotor synchronization; spontaneous motor tempo; and the detection of tempo changes, *Behavioural Brain Research* 216(2): 685–691.

Schwarzbauer, C., Davis, M., Rodd, J. & Johnsrude, I. (2006). Interleaved silent steady state (ISSS) imaging: A new sparse imaging method applied to auditory fMRI, *Neuroimage* 29(3): 774–782.

Scott, S. (2005). Auditory processing – speech, space and auditory objects, *Current Opinion in Neurobiology* 15(2): 197–201.

Scruton, R. (1983). *The Aesthetic Understanding: Essays in the Philosophy of Art and Culture*, Routledge Kegan and Paul.

Scruton, R. (1999). *The Aesthetics of Music*, Oxford University Press, USA.

Seifert, U. (2011). Signification and significance: Music, brain, and culture (Comment), *Physics of Life Reviews* 8(2): 122–124.

Servan-Schreiber, E. & Anderson, J. (1990). Learning artificial grammars with competitive chunking, *Journal of Experimental Psychology: Learning, Memory, and Cognition* 16(4): 592–608.

Sharbrough, F. (1991). American electroencephalographic society guidelines for standard electrode postion nomenclature, *Journal of Clinical Neurophysiology* 8: 200–202.

Shepard, R. (1982a). Geometrical approximations to the structure of musical pitch, *Psychological Review* 89: 305–333.

Shepard, R. N. (1982b). Structural representations of musical pitch, *in* D. Deutsch (ed.), *Psychology of Music*, New York: Academic Press, pp. 343–390.

Shepard, R. N. (1999). Pitch perception and measurement, *in* P. R. Cook (ed.), *Music, Cognition, and Computerized Sound. An Introduction to Psychoacoustics*, Cambridge, MA: MIT Press, pp. 149–165.

Shibasaki, H. (2008). Human brain mapping: Hemodynamic response and electrophysiology, *Clinical Neurophysiology* 119(4): 731–743.

Shiffrin, R. & Schneider, W. (1977). Controlled and automatic human information processing: II. Perceptual learning, automatic attending and a general theory, *Psychological Review* 84(2): 127–190.

Shinn-Cunningham, B. (2008). Object-based auditory and visual attention, *Trends in Cognitive Sciences* 12(5): 182–186.

Siebel, W. A. (2009). Thalamic balance can be misunderstood as happiness, *Interdis – Journal for Interdisciplinary Research* 3: 48–50.

Siebel, W. A., Winkler, T. & Seitz-Bernhard, B. (1990). *Noosomatik I: Theoretische Grundlegung*, Langwedel: Glaser u. Wohlschlegel.

Siebel, W. & Winkler, T. (1996). *Noosomatik V: Noologie, Neurologie, Kardiologie,* 2nd ed. Wiesbaden: Glaser.

Simmons-Stern, N., Budson, A. & Ally, B. (2010). Music as a memory enhancer in patients with alzheimer's disease, *Neuropsychologia* 48(10): 3164–3167.

Singer, T., Critchley, H. & Preuschoff, K. (2009). A common role of insula in feelings, empathy and uncertainty, *Trends in Cognitive Sciences* 13(8): 334–340.

Singer, T. & Lamm, C. (2009). The social neuroscience of empathy, *Annals of the New York Academy of Sciences* 1156(1): 81–96.

Skoe, E. & Kraus, N. (2010). Auditory brain stem response to complex sounds: A tutorial, *Ear and Hearing* 31(3): 302–324.

Slevc, L. & Patel, A. (2011). Meaning in music and language: Three key differences (Comment), *Physics of Life Reviews* 8(2): 110–111.

Slevc, L., Rosenberg, J. & Patel, A. (2009). Making psycholinguistics musical: Self-paced reading time evidence for shared processing of linguistic and musical syntax, *Psychonomic Bulletin and Review* 16(2): 374–381.

Sloboda, J. (2000). Individual differences in music performance, *Trends in Cognitive Sciences* 4(10): 397–403.

Sloboda, J. A. (1991). Music structure and emotional response: Some empirical findings, *Psychology of Music* 19: 110–120.

Smith, J., Marsh, J. & Brown, W. (1975). Farfield recorded frequency-following responses: Evidence for the locus of brainstem sources, *Electroencephalography and Clinical Neurophysiology* 39(5): 465–472.

Snyder, J. & Alain, C. (2007). Toward a neurophysiological theory of auditory stream segregation, *Psychological Bulletin* 133(5): 780–799.

Song, J., Skoe, E., Wong, P. & Kraus, N. (2008). Plasticity in the adult human auditory brainstem following short-term linguistic training, *Journal of Cognitive Neuroscience* 20(10): 1892–1902.

Spencer, K. M., Dien, J. & Donchin, E. (1999). A componential analysis of the ERP elicited by novel events using a dense electrode array, *Psychophysiology* 36(3): 409–414.

Spijkers, W., Heuer, H., Kleinsorge, T. & van der Loo, H. (1997). Preparation of bimanual movements with same and different amplitudes: Specification interference as revealed by reaction time, *Acta Psychologica* 96(3): 207–227.

Spintge, R. (2000). Music and anesthesia in pain therapy, *Anästhesiologie, Intensivmedizin, Notfallmedizin, Schmerztherapie* 35(4): 254–261.

Squires, N. K., Squires, K. C. & Hillyard, S. A. (1975). Two varieties of long-latency positive waves evoked by unpredictable auditory stimuli in man, *Electroencephalography and Clinical Neurophysiology* 38(4): 387–440.

Steel, K. & Kros, C. (2001). A genetic approach to understanding auditory function, *Nature Genetics* 27(2): 143–149.

Stefanics, G., Haden, G., Huotilainen, M., *et al.* (2007). Auditory temporal grouping in newborn infants, *Psychophysiology* 44(5): 697–702.

Stefanics, G., Háden, G., Sziller, I., *et al.* (2009). Newborn infants process pitch intervals, *Clinical Neurophysiology* 120(2): 304–308.

Stein, M., Koverola, C., Hanna, C., Torchia, M. & McClarty, B. (1997). Hippocampal volume in women victimized by childhood sexual abuse, *Psychological Medicine* 27(04): 951–959.

Stein, M., Simmons, A., Feinstein, J. & Paulus, M. (2007). Increased amygdala and insula activation during emotion processing in anxiety-prone subjects, *American Journal of Psychiatry* 164(2): 318–327.

Steinbeis, N. (2008). *Investigating the Meaning of Music using EEG and fMRI*, Leipzig: Risse.

Steinbeis, N. & Koelsch, S. (2008a). Comparing the processing of music and language meaning using EEG and FMRI provides evidence for similar and distinct neural representations, *PLoS One* 3(5): e2226.

Steinbeis, N. & Koelsch, S. (2008b). Shared neural resources between music and language indicate semantic processing of musical tension-resolution patterns, *Cerebral Cortex* 18(5): 1169–1178.

Steinbeis, N. & Koelsch, S. (2008c). Understanding the intentions behind man-made products elicits neural activity in areas dedicated to mental state attribution, *Cerebral Cortex* 19(3): 619–623.

Steinbeis, N. & Koelsch, S. (2011). Affective priming effects of musical sounds on the processing of word meaning, *Journal of Cognitive Neuroscience* 23: 604–621.

Steinbeis, N., Koelsch, S. & Sloboda, J. (2006). The role of harmonic expectancy violations in musical emotions: Evidence from subjective, physiological, and neural responses, *Journal of Cognitive Neuroscience* 18(8): 1380–1393.

Steinhauer, K., Alter, K. & Friederici, A. D. (1999). Brain potentials indicate immediate use of prosodic cues in natural speech processing, *Nature Neuroscience* 2(2): 191–196.

Stern, T. (1957). Drum and whistle "languages": An analysis of speech surrogates, *American Anthropologist* 59(3): 487–506.

Stevens, S. S., Volkmann, J. & Newman, E. B. (1937). A scale of measurement of the psychological magnitude of pitch, *Journal of the Acoustical Society of America* 35: 2346–2353.

Stewart, L., Von Kriegstein, K., Warren, J. & Griffiths, T. (2006). Music and the brain: Disorders of musical listening, *Brain* 129(10): 2533–2553.

Steyvers, M., Etoh, S., Sauner, D., *et al.* (2003). High-frequency transcranial magnetic stimulation of the supplementary motor area reduces bimanual coupling during anti-phase but not in-phase movements, *Experimental Brain Research* 151(3): 309–317.

Sturt, P., Pickering, M. & Crocker, M. (1999). Structural change and reanalysis difficulty in language comprehension, *Journal of Memory and Language* 40(1): 136–150.

Sussman, E. (2007). A new view on the MMN and attention debate: The role of context in processing auditory events, *Journal of Psychophysiology* 21(3): 164–175.

Sussman, E., Kujala, T., Halmetoja, J., *et al.* (2004). Automatic and controlled processing of acoustic and phonetic contrasts, *Hearing Research* 190(1–2): 128–140.

Suzuki, M., Okamura, N., Kawachi, Y., *et al.* (2008). Discrete cortical regions associated with the musical beauty of major and minor chords, *Cognitive, Affective and Behavioral Neuroscience* 8(2): 126–131.

Swinnen, S. & Wenderoth, N. (2004). Two hands, one brain: Cognitive neuroscience of bimanual skill, *Trends in Cognitive Sciences* 8(1): 18–25.

Tallal, P. & Gaab, N. (2006). Dynamic auditory processing, musical experience and language development, *Trends in Neurosciences* 29(7): 382–390.

Ter Haar, S., Mietchen, D., Fritz, T. & Koelsch, S. (2007). Auditory perception of acoustic roughness and frequency sweeps, *in Evolution of Emotional Communication: From Sounds in Nonhuman Mammals to Speech and Music in Man*, Hannover: Germany.

Terhardt, E. (1974). On the perception of periodic sound fluctuations (roughness), *Acustica* 30(4): 201–213.

Terhardt, E. (1976). Ein psychoakustisch begründetes Konzept der musikalischen Konsonanz, *Acustica* 36: 121–137.

Terhardt, E. (1978). Psychoacoustic evaluation of musical sounds, *Attention, Perception, & Psychophysics* 23(6): 483–492.

Terhardt, E. (1984). The concept of musical consonance: A link between music and psychoacoustics, *Music Perception* 1(3): 276–295.

Tervaniemi, M. (2009). Musicians – same or different? *Annals of the New York Academy of Sciences* 1169(The Neurosciences and Music III Disorders and Plasticity): 151–156.

Tervaniemi, M. & Huotilainen, M. (2003). The promises of change-related brain potentials in cognitive neuroscience of music, *Annals of the New York Academy of Sciences* 999 (The Neurosciences and Music): 29–39.

Tervaniemi, M., Ilvonen, T., Karma, K., Alho, K. & Näätänen, R. (1997a). The musical brain: Brain waves reveal the neurophysiological basis of musicality in human subjects, *Neuroscience Letters* 226(1): 1–4.

Tervaniemi, M., Just, V., Koelsch, S., Widmann, A. & Schröger, E. (2005). Pitch discrimination accuracy in musicians vs nonmusicians: An event-related potential and behavioral study, *Experimental Brain Research* 161(1): 1–10.

Tervaniemi, M., Kruck, S., De Baene, W., *et al.* (2009). Top-down modulation of auditory processing: Effects of sound context, musical expertise and attentional focus, *European Journal of Neuroscience* 30(8): 1636–1642.

Tervaniemi, M., Kujala, A., Alho, K., *et al.* (1999). Functional specialization of the human auditory cortex in processing phonetic and musical sounds: A magnetoencephalographic (MEG) study, *Neuroimage* 9(3): 330–336.

Tervaniemi, M., Medvedev, S., Alho, K., Pakhomov, S., Roudas, M., von Zuijen, T. & Näätänen, R. (2000). Lateralized automatic auditory processing of phonetic versus musical information: A PET study, *Human Brain Mapping* 10(2): 74–79.

Tervaniemi, M., Rytkönen, M., Schröger, E., Ilmoniemi, R. & Näätänen, R. (2001). Superior formation of cortical memory traces for melodic patterns in musicians, *Learning and Memory* 8(5): 295–300.

Tervaniemi, M., Castaneda, A., Knoll, M. & Uther, M. (2006a). Sound processing in amateur musicians and nonmusicians: Event-related potential and behavioral indices, *Neuroreport* 17(11): 1225–1228.

Tervaniemi, M., Szameitat, A., Kruck, S., *et al.* (2006b). From air oscillations to music and speech: Functional magnetic resonance imaging evidence for fine-tuned neural networks in audition, *Journal of Neuroscience* 26(34): 8647–8652.

Tervaniemi, M., Winkler, I. & Näätänen, R. (1997a). Pre-attentive categorization of sounds by timbre as revealed by event-related potentials, *NeuroReport* 8(11): 2571–2574.

Tettamanti, M., Buccino, G., Saccuman, M., *et al.* (2005). Listening to action-related sentences activates fronto-parietal motor circuits, *Journal of Cognitive Neuroscience* 17(2): 273–281.

Tettamanti, M. & Weniger, D. (2006). Broca's area: A supramodal hierarchical processor? *Cortex* 42(4): 491–494.

Thach, W. (1978). Correlation of neural discharge with pattern and force of muscular activity, joint position, and direction of intended next movement in motor cortex and cerebellum, *Journal of Neurophysiology* 41(3): 654–676.

Thaut, M. (2003). Neural basis of rhythmic timing networks in the human brain, *Annals of the New York Academy of Sciences* 999(1): 364–373.

Thaut, M. & Abiru, M. (2010). Rhythmic auditory stimulation in rehabilitation of movement disorders: A review of current research, *Music Perception* 27(4): 263–269.

Thaut, M., Peterson, D. & McIntosh, G. (2005). Temporal Entrainment of Cognitive Functions, *Annals of the New York Academy of Sciences* 1060(1): 243–254.

Tillmann, B. (2005). Implicit investigations of tonal knowledge in nonmusician listeners, *Annals of the New York Academy of Sciences* 1060(1): 100–110.

Tillmann, B. (2009). Music cognition: Learning, perception, expectations, *Computer Music Modeling and Retrieval. Sense of Sounds*: 4th International Symposium, CMMR 2007, Copenhagen, Denmark, pp. 11–33.

Tillmann, B., Bharucha, J. & Bigand, E. (2000). Implicit learning of tonality: A self-organized approach, *Psychological Review* 107(4): 885–913.

Tillmann, B., Janata, P. & Bharucha, J. (2003). Activation of the inferior frontal cortex in musical priming, *Cognitive Brain Research* 16(2): 145–161.

Tillmann, B., Koelsch, S., Escoffier, N., *et al.* (2006). Cognitive priming in sung and instrumental music: Activation of inferior frontal cortex, *Neuroimage* 31(4): 1771–1782.

Toiviainen, P. & Krumhansl, C. (2003). Measuring and modeling real-time responses to music: The dynamics of tonality induction, *Perception* 32(6): 741–766.

Tomasello, M., Carpenter, M., Call, J., Behne, T. & Moll, H. (2005). Understanding and sharing intentions: The origins of cultural cognition, *Behavioral and Brain Sciences* 28(05): 675–691.

Tomic, S. & Janata, P. (2008). Beyond the beat: Modeling metric structure in music and performance, *The Journal of the Acoustical Society of America* 124: 4024–4041.

Townsend, D. & Bever, T. (2001). *Sentence comprehension: The integration of habits and rules*, Cambridge, MA: MIT Press.

Trainor, L. & Trehub, S. (1994). Key membership and implied harmony in Western tonal music: Developmental perspectives, *Perception and Psychophysics* 56(2): 125–132.

Tramo, M. J., Cariani, P. A., Delgutte, B. & Braida, L. D. (2001). Neurobiological foundations for the theory of harmony in western tonal music, *in* R. J. Zatorre and I. Peretz (eds), *The Biological Foundations of Music*, Vol. 930, New York: The New York Academy of Sciences.

Tramo, M., Lense, M., Van Ness, C., *et al.* (2011). Effects of music on physiological and behavioral indices of acute pain and stress in premature infants, *Music and Medicine* 3(2): 72–83.

Tramo, M., Shah, G. & Braida, L. (2002). Functional role of auditory cortex in frequency processing and pitch perception, *Journal of Neurophysiology* 87(1): 122–139.

Trehub, S. (2003). The developmental origins of musicality, *Nature Neuroscience* 6(7): 669–673.

Treisman, A. (1964). Selective attention in man, *British Medical Bulletin* 20(1): 12–16.

Tzur, G. & Berger, A. (2007). When things look wrong: An ERP study of perceived erroneous information, *Neuropsychologia* 45: 3122–3126.

Tzur, G. & Berger, A. (2009). Fast and slow brain rhythms in rule/expectation violation tasks: Focusing on evaluation processes by excluding motor action, *Behavioural Brain Research* 198(2): 420–428.

Uedo, N., Ishikawa, H., Morimoto, K., *et al.* (2004). Reduction in salivary cortisol level by music therapy during colonoscopic examination, *Hepatogastroenterology* 51(56): 451–453.

Van Berkum, J., van den Brink, D., Tesink, C., Kos, M. & Hagoort, P. (2008). The neural integration of speaker and message, *Journal of Cognitive Neuroscience* 20(4): 580–591.

Van Den Brink, D., Brown, C. & Hagoort, P. (2001). Electrophysiological evidence for early contextual influences during spoken-word recognition: N200 versus N400 effects, *Journal of Cognitive Neuroscience* 13(7): 967–985.

Van Herten, M., Kolk, H. & Chwilla, D. (2005). An erp study of p600 effects elicited by semantic anomalies, *Cognitive Brain Research* 22(2): 241–255.

Van Petten, C. & Kutas, M. (1990). Interactions between sentence context and word frequency in event-related brain potentials, *Memory and Cognition* 18(4): 380–393.

Van Petten, C. & Rheinfelder, H. (1995). Conceptual relationships between spoken words and environmental sounds: Event-related brain potential measures, *Neuropsychologia* 33(4): 485–508.

van Veen, V. & Carter, C. (2002). The timing of action-monitoring processes in the anterior cingulate cortex, *Journal of Cognitive Neuroscience* 14(4): 593–602.

Van Veen, V. & Carter, C. (2006). Error detection, correction, and prevention in the brain: A brief review of data and theories, *Clinical EEG and Neuroscience* 37(4): 330–335.

van Veen, V., Holroyd, C., Cohen, J., Stenger, V. & Carter, C. (2004). Errors without conflict: Implications for performance monitoring theories of anterior cingulate cortex, *Brain and Cognition* 56(2): 267–276.

VanderArk, S. & Ely, D. (1992). Biochemical and galvanic skin responses to music stimuli by college students in biology and music, *Perceptual and Motor Skills* 74(3c): 1079–1090.

Verleger, R. (1990). P3-evoking wrong notes: Unexpected, awaited, or arousing? *International Journal of Neuroscience* 55(2–4): 171–179.

Videbech, P. & Ravnkilde, B. (2004). Hippocampal volume and depression: A meta-analysis of MRI studies, *American Journal of Psychiatry* 161(11): 1957–1966.

Villarreal, E., Brattico, E., Leino, S., Østergaard, L. & Vuust, P. (2011). Distinct neural responses to chord violations: A multiple source analysis study, *Brain Research* 1389: 103–114.

Võ, M., Conrad, M., Kuchinke, L., *et al.* (2009). The Berlin Affective Word List Reloaded (BAWL-R), *Behavior Research Methods* 41(2): 534–538.

von Helmholtz, H. (1870). *Die Lehre von den Tonempfindungen als physiologische Grundlage für die Theorie der Musik*, F. Vieweg und sohn.

von Kriegstein, K., Eger, E., Kleinschmidt, A. & Giraud, A. (2003). Modulation of neural responses to speech by directing attention to voices or verbal content, *Cognitive Brain Research* 17(1): 48–55.

von Kriegstein, K., Smith, D., Patterson, R., Ives, D. & Griffiths, T. (2007). Neural representation of auditory size in the human voice and in sounds from other resonant sources, *Current Biology* 17(13): 1123–1128.

von Zuijen, T., Sussman, E., Winkler, I., Näätänen, R. & Tervaniemi, M. (2004). Grouping of sequential sounds-an event-related potential study comparing musicians and nonmusicians, *Journal of Cognitive Neuroscience* 16(2): 331–338.

von Zuijen, T., Sussman, E., Winkler, I., Näätänen, R. & Tervaniemi, M. (2005). Auditory organization of sound sequences by a temporal or numerical regularity–a mismatch negativity study comparing musicians and non-musicians, *Cognitive Brain Research* 23(2–3): 270–276.

Vrba, J. & Robinson, S. (2001). Signal processing in magnetoencephalography, *Methods* 25(2): 249–271.

Wambacq, I. & Jerger, J. (2004). Processing of affective prosody and lexical-semantics in spoken utterances as differentiated by event-related potentials, *Cognitive Brain Research* 20(3): 427–437.

Warner-Schmidt, J. & Duman, R. (2006). Hippocampal neurogenesis: Opposing effects of stress and antidepressant treatment, *Hippocampus* 16(3): 239–249.

Warren, J. D., Uppenkamp, S., Patterson, R. D. & Griffiths, T. D. (2003). Separating pitch chroma and pitch height in the human brain, *PNAS* 100(17): 10038–10042.

Warren, J., Sauter, D., Eisner, F., *et al.* (2006). Positive emotions preferentially engage an auditory-motor 'mirror' system, *Journal of Neuroscience* 26(50): 13067–13075.

Warren, J., Wise, R. & Warren, J. (2005). Sounds do-able: Auditory-motor transformations and the posterior temporal plane, *Trends in Neurosciences* 28(12): 636–643.

Warren, R., McLellarn, R. & Ponzoha, C. (1988). Rational-emotive therapy vs general cognitive-behavior therapy in the treatment of low self-esteem and related emotional disturbances, *Cognitive Therapy and Research* 12(1): 21–37.

Wassenaar, M. & Hagoort, P. (2007). Thematic role assignment in patients with broca's aphasia: Sentence-picture matching electrified, *Neuropsychologia* 45(4): 716–740.

Waszak, F. & Herwig, A. (2007). Effect anticipation modulates deviance processing in the brain, *Brain Research* 1183: 74–82.

Weber, G. (1817). *Versuch einer geordneten Theorie der Tonsetzkunst*, 2 vols. Mainz: B. Schott.

Wernicke, C. (1874). *Der aphasische Symptomencomplex*, Breslau: Cohn and Weigert.

West, J., Otte, C., Geher, K., Johnson, J. & Mohr, D. (2004). Effects of hatha yoga and african dance on perceived stress, affect, and salivary cortisol, *Annals of Behavioral Medicine* 28(2): 114–118.

Whitfield, I. (1980). Auditory cortex and the pitch of complex tones, *The Journal of the Acoustical Society of America* 67(2): 644–647.

Widmann, A., Kujala, T., Tervaniemi, M., Kujala, A. & Schröger, E. (2004). From symbols to sounds: Visual symbolic information activates sound representations, *Psychophysiology* 41(5): 709–715.

Willems, R., Özyürek, A. & Hagoort, P. (2008). Seeing and hearing meaning: ERP and fMRI evidence of word versus picture integration into a sentence context, *Journal of Cognitive Neuroscience* 20(7): 1235–1249.

Williamson, V., Baddeley, A. & Hitch, G. (2010a). Musicians' and nonmusicians' short-term memory for verbal and musical sequences: Comparing phonological similarity and pitch proximity, *Memory & Cognition* 38(2): 163–175.

Williamson, V., Mitchell, T., Hitch, G. & Baddeley, A. (2010b). Musicians' memory for verbal and tonal materials under conditions of irrelevant sound, *Psychology of Music* 38(3): 331–350.

Wiltermuth, S. & Heath, C. (2009). Synchrony and cooperation, *Psychological Science* 20(1): 1–5.

Winkler, I. (2007). Interpreting the mismatch negativity, *Journal of Psychophysiology* 21(3–4): 147–163.

Winkler, I., Denham, S. & Nelken, I. (2009a). Modeling the auditory scene: Predictive regularity representations and perceptual objects, *Trends in Cognitive Sciences* 13(12): 532–540.

Winkler, I., Háden, G., Ladinig, O., Sziller, I. & Honing, H. (2009b). Newborn infants detect the beat in music, *Proceedings of the National Academy of Sciences* 106(7): 2468–2471.

Winkler, I., Kujala, T., Tiitinen, H., *et al.* (1999). Brain responses reveal the learning of foreign language phonemes, *Psychophysiology* 36: 638–642.

Winkler, I., Kushnerenko, E., Horvath, J., *et al.* (2003). Newborn infants can organize the auditory world, *Proceedings of the National Academy of Sciences of the United States of America* 100(20): 11812–11815.

Wittfoth, M., Schröder, C., Schardt, D., *et al.* (2010). On emotional conflict: Interference resolution of happy and angry prosody reveals valence-specific effects, *Cerebral Cortex* 20(2): 383–392.

Wittgenstein, L. (1984). *Philosophische Untersuchungen*, Frankfurt: Suhrkamp.

Woldorff, M., Hansen, J. & Hillyard, S. (1987). Evidence for effects of selective attention in the mid-latency range of the human auditory event-related potential, *Electroencephalogr Clin Neurophysiol Suppl* 40: 146–154.

Wolpert, D. & Ghahramani, Z. (2000). Computational principles of movement neuroscience, *Nature Neuroscience* 3: 1212–1217.

Wolpert, D., Ghahramani, Z. & Jordan, M. (1995). An internal model for sensorimotor integration, *Science* 269(5232): 1880–1882.

Wolpert, D., Miall, R. & Kawato, M. (1998). Internal models in the cerebellum, *Trends in Cognitive Sciences* 2(9): 338–347.

Wong, P., Skoe, E., Russo, N., Dees, T. & Kraus, N. (2007). Musical experience shapes human brainstem encoding of linguistic pitch patterns, *Nature Neuroscience* 10(4): 420–422.

Woolhouse, M. & Cross, I. (2006). An interval cycle-based model of pitch attraction, *in Proceedings of the 9th International Conference on Music Perception and Cognition*, University of Bologna, pp. 763–771.

Yeung, N., Botvinick, M. & Cohen, J. (2004). The neural basis of error detection: Conflict monitoring and the error-related negativity, *Psychological Review* 111(4): 931–959.

Ylinen, S., Shestakova, A., Huotilainen, M., Alku, P. & Naatanen, R. (2006). Mismatch negativity (MMN) elicited by changes in phoneme length: A cross-linguistic study, *Brain Research* 1072(1): 175–185.

Zajonc, R. (2001). Mere exposure: A gateway to the subliminal, *Current Directions in Psychological Science* 10(6): 224–228.

Zatorre, R. (1988). Pitch perception of complex tones and human temporal-lobe function, *Journal of the Acoustic Society of America* 84: 566–572.

Zatorre, R. (2001). Neural specializations for tonal processing, *Annals of the New York Academy of Sciences* 930 (The Biological Foundations of Music): 193–210.

Zatorre, R., Belin, P. & Penhune, V. (2002). Structure and function of auditory cortex: Music and speech, *Trends in Cognitive Sciences* 6(1): 37–46.

Zatorre, R., Evans, A. & Meyer, E. (1994). Neural mechanisms underlying melodic perception and memory for pitch, *Journal of Neuroscience* 14(4): 1908–1919.

Zbikowski, L. (1998). Metaphor and music theory: Reflections from cognitive science, *Music Theory Online* 4(1): 1–8.

Zentner, M. & Eerola, T. (2010). Rhythmic engagement with music in infancy, *Proceedings of the National Academy of Sciences* 107(13): 5768–5773.

Zentner, M., Grandjean, D. & Scherer, K. (2008). Emotions evoked by the sound of music: Characterization, classification, and measurement, *Emotion* 8(4): 494–521.

Zwicker, E. (1961). Subdivision of the audible frequency range into critical bands (Frequenzgruppen), *Acoustical Society of America Journal* 33(2): 248–249.

Zwicker, E. & Terhardt, E. (1980). Analytical expressions for critical-band rate and critical bandwidth as a function of frequency, *The Journal of the Acoustical Society of America* 68: 1523–1525.

Index

γ-aminobutyric acid (GABA), 40

10-20 system, 43

a priori musical meaning, 180
Ablinger, P., 241
Action potentials, 42
Aesthetics, 216
Affektenlehre, 159
Aha moment, 216, 264
Alexithymia, 170
Alpha activity, 234
Alzheimer's disease, 238
Amplitude envelope, 12
Amygdala, 42, 85, 86, 217, 221–225, 227, 229, 230, 232, 239
  basolateral ~, 222
  centromedial ~, 222
  lateral ~, 6
Anterior cingulate cortex (ACC), 232, 233
Anxiety, 239
Arthritis, 239

Asymmetry
  in chord-pair ratings, 34
  in tone-pair ratings, 26
Attachment-related emotions, 230
Attention, 40, 52, 57, 59, 65, 71, 74, 137, 147–149, 155, 173, 175, 178, 236, 237, 250
Attention-deficit/hyperactivity disorder (ADHD), 237
Attentional bottleneck, 147
Auditory association cortex, 12, 15
Auditory cortex, 6, 12, 13, 90, 91, 95
Auditory grouping, 91
Auditory nerve, 5
Auditory nerve fibre, 5, 6
Auditory oddball paradigm, 54
Auditory parabelt, 12, 167
Auditory scene analysis, 15, 16, 91
Auditory sensory memory, 14, 29, 54, 57, 58, 90,

103–106, 116, 117, 135–137, 252
Auditory steady-state response (aSSR), 14
Auditory stream segregation, 91
Autistic Spectrum Disorder, 209
Autoimmune disease, 240
Autonomic nervous system, 217
Averaging, 47

B cells, 94
Basal ganglia, 93, 189, 194
Basic emotions, 160, 205
Basilar membrane, 3, 5, 6, 8, 9, 15
Beats, 9
Beethoven, L.v., 158
Behaviour-adaption hypothesis, 196
Bimanual coupling, 192
Binaural beats, 11
Biot-Savart law, 48
Bispectral Index, 132
Blood flow, 79

Blood-oxygen-level dependent contrast (BOLD), 79
Brainstem, 6, 11
Brainstem-responses, 42, 52
Broca's aphasia, 144, 145, 238
Broca's area, 92, 131, 133, 137, 145, 146
Brodmann
    area 6, 92, 137
    area 7, 94
    area 10, 217
    area 11, 217
    area 21, 166, 169
    area 22, 12, 91, 92, 96, 166
    area 37, 169
    area 40, 189
    area 41, 12, 90
    area 42, 12
    area 44, 91, 92, 131, 133, 138, 188
    area 45, 91
    area 47, 217
    area 52, 12

Capsaicin, 11
CARE system, 230
Caudal pontine reticular formation, 7
Cell assembly, 41
Central auditory pathway, 6
Cerebellum, 194
Cerebral cortex, 40
Characteristic dissonance
    of dominant, 21
    of subdominant, 21
Characteristic frequency, 6
Children, 62
Chill experience, 220
Chord
    profile, 30
Chord functions, 20
Chromatic scale, 17, 18, 27, 31
Chromatic supertonic, 22
Circle of fifths, 19, 27, 28, 74
Closure Positive Shift, 68, 77
Co-pathy, 209
Cochlea, 3

Cochlear nerve, 6
Cochlear nucleus, 8
Cognitive priming, 37
Cognitization, 237
Cohesion, 208, 209, 211, 213, 236, 250
Common average reference, 45
Common coding theory, 186, 201
Communication, 78, 162, 179, 180, 183, 210
Consonance
    harmonic ~, 10
    musical ~, 10
    sensory ~, 10
Contagion, 210
Context-free grammar, 107
Contextual Asymmetry, 34
Contextual Distance, 34
Contextual Identity, 35
Cooperation, 178, 179, 211, 236, 240, 250
Coordination, 160, 178, 179, 192, 193, 210, 211, 240
Coordination of actions, 210
Corollary discharge, 193
Correlation matrix, 27, 28
Cortex, 40
Cortisol, 94, 207
CPS, 67
Cranial nerve VIII, 6
Current source density, 50, 199

Degree, 20
Delta band oscillations, 129
Dentate gyrus, 229
Depression, 227, 239
Depth EEG, 42
Diatonic, 25–27, 34, 72
Diminished chord, 20, 30
Discourse, 62, 183
Dissonance
    sensory ~, 10
Dominant, 20, 31, 33
    ~ seventh chord, 21, 26
Dorsal cochlear nucleus, 9

Double dominant, 22, 115, 121, 122, 171
Dynamic Causal Modelling, 84

Ear, 3
Echoic memory, 113, 117, 118
Efference copy, 193, 194
Eigenvector centrality mapping (ECM), 84
Electroencephalogram (EEG), 42
Electrocorticography (ECoG), 42
Electrode, 43
Emotion, 53, 86, 160, 161, 179, 180, 184, 189, 203–207, 210, 212–216, 219, 221, 223, 225–228, 231–236, 238, 239, 250, 262, 263
Emotion principles, 212
Emotional contagion, 214
Emotional musicogenic meaning, 179
Empathy, 209
Equivalent current dipole (ECD), 48
equivalent rectangular bandwidth, 9
ERP-components
    CNV, 174
    CPS, 68, 77
    ELAN, 63–65
    endogenous ~, 51
    ERAN, 50, 71, 72, 74, 92, 109–111, 113, 114, 119–121, 123–127, 129–132, 134–138, 142–144, 146–155, 173, 174, 197–200
    ERN, 197–201
    ERN/Ne, 197
    Error positivity/Pe, 192
    exogenous ~, 51
    feedback ERN, 197, 198, 200
    LAN, 63–66, 140–143, 146, 149, 152–154, 173

LPC, 71, 72, 93
MMN, 14, 54–59, 75, 76, 90, 104, 111, 113, 114, 117, 130, 132–138, 142, 149, 155, 198–200
music-syntactic MMN, 130
N1, 52
N125, 128, 129
N180, 128, 129
N200, 197
N2b, 59, 72, 73, 121, 197, 199, 200
N400, 60–64, 70, 71, 76, 95, 96, 130, 139–141, 154, 163–170, 172–174, 181, 253
N5, 50, 74, 95, 96, 111, 119, 121, 124, 126, 129, 142–144, 152–154, 171–176, 181
Ne, 197
Novelty P3, 59
P1, 52
P2, 52, 53, 55, 130, 168, 171
P3a, 59, 74, 121, 192, 199
P3b, 59, 61, 67, 70, 71, 74, 121, 151, 198–200
P600, 66, 67, 74
RATN, 74, 92
RON, 174
syntactic MMN, 66
Error-awareness hypothesis, 196
Eureka moment, 216, 264
Event-related field (ERF), 48
Event-related potential (ERP), 47
Evoked potentials, 51
Excitatory postsynaptic potential (EPSP), 40
Expectancy, 36, 61, 71, 74, 104, 106, 111, 113, 114, 119, 136, 137, 176, 191, 199, 201, 202, 208, 211, 213, 215, 217, 218, 236–238

Exposition, 183
extended 10-20 system, 43
Extra-musical meaning, 163

F0 contour, 68
F0 frequency, 53
F5 area, 186
Ferneyhough, B., 246
Foetus, 57, 154, 242
Fibromyalgia, 239
fMRI, 30, 79–81, 84, 85, 107, 131, 146, 150, 151, 161, 162, 169, 187, 189, 214, 221–223, 228, 233, 235
Formal musical meaning, 171
Formal significance, 171
Frisson, 94, 220, 221, 225, 227, 228, 233
Fronto-midline theta activity, 234, 235
Functional connectivity analysis, 82
Functional magnetic resonance imaging (fMRI), 43

Gamma band oscillations, 129
Garden-path sentences, 93, 107, 143
Generative Syntax Model (GSM), 99, 109, 112, 116
Generative Theory of Tonal Music (GTTM), 99
Gestalt, 15, 58, 90, 91, 96, 97, 103, 104, 107, 135, 137, 176, 191, 208, 213, 236
Gestalt principles, 104
Gestures, 62
Glutamate, 5
Granger Causality Mapping, 84
Grouping, 58, 91, 103, 104
Gyrus
    angular ~, 13
    superior temporal ~, 12
    supramarginal ~, 13
    temporalis transversus (Heschl), 12

Hair cells, 4
Happiness, 205, 206, 208, 212, 225–227, 229–231, 234
Harmonic core, 33
Haydn, J., 158
Helicotrema, 3
Heschl's gyrus, 90
Hierarchical structure, 93, 98, 114
Hierarchy
    of stability, 31, 104, 105, 129, 155
    of tones, 25
Hippocampus, 44, 227–232
Hungarian minor gypsy scale, 18
Hypothalamus, 212, 226

i-dosing, 11
Iconic musical meaning, 158
Ideomotor principle, 201
Immune system, 208
Immunoglobulin A, 94
Incus, 3
Indexical musical meaning, 159
Inferior colliculus, 6, 7, 9
Inhibitory postsynaptic potential (IPSP), 41
Inion, 44
Insula, 69, 222, 225, 233
Intention, 162
Interval, 17–19, 21
Interval cycle-based model of pitch attraction, 106
Intonation (prosody), 245
Intonational phrase boundaries, 68
Intra-musical meaning, 170, 171
Intrakey Asymmetry, 34
Intrakey Distance, 33
Inverse problem, 49

Joint action, 178
Joy, 205, 208, 212, 214, 221, 227, 229–231

Key, 18–22, 24–35, 67, 74
  dominant ~, 19
  inter key distance, 27
  parallel ~, 28
  profile, 27
  regions, 29
  relative ~, 27, 28
  sense of ~, 30
  subdominant ~, 19
  membership, 32
Knowledge instinct, 216
Knowledge-free structuring, 103

Large-scale structuring, 108
Larynx, 189
Lateral amygdala, 6
Lemma, 107
Leukocyte, 94
Lexeme, 107, 146
Ligeti, G., 246
Logic, 133, 238
Love, 205, 212, 229, 230
Low pitch, 9
Lymphocyte, 94

Mafa people, 10, 102, 160
Malleus, 3
Markov model, 106
Mastoid, 44
Meaning
  designative ~, 158
  emotional musicogenic ~,
    179
  extra-musical ~, 95, 158,
    163, 165, 170, 173, 181,
    185, 253
  formal ~, 171
  formal significance, 171
  iconic musical ~, 171
  idiosyncratic symbolic musical
    ~, 162
  indexical musical ~, 171
  intra-musical ~, 25, 171,
    173–176, 181, 183–185,
    218
  motivational-structural ~, 159
  musicogenic ~, 177

personal musicogenic ~, 180
physical musicogenic ~, 177
socio-intentional musical ~,
  178
symbolic musical ~, 162
Meatus, 3
Medial forebrain bundle, 212
Medial geniculate body, 6
Mediant, 20, 31
MEG, 48–50, 76, 131, 146,
  150, 187
Melodic Intonation Therapy,
  238
Melody, 15, 23, 70–77, 91, 92,
  95, 126–129, 131, 132,
  134, 140, 150, 172, 188,
  189, 200, 225
Memory
  auditory sensory ~, 14
  echoic ~, 14
  episodic ~, 207, 213–215
  semantic ~, 213, 215
  Working ~, 107, 123, 152
Mere exposure effect, 215
Mesolimbic dopamine pathway,
  212
Mesolimbic reward circuit,
  226
Metre, 98, 102–104, 108, 155,
  189, 194, 253
Middle-latency responses, 52
Mimicry, 210
Major gypsy scale, 18
Mirror neurons, 187
Missing fundamental, 8
Modified Observer's Assessment
  of Alertness and Sedation
  Scale, 132
Modulation, 30
Monaural beats, 11
Morse code, 246
Motivational-structural musical
  meaning, 159
Motor theory of speech
  perception, 186, 243
Multidimensional scaling
  (MDS), 25, 27, 30, 32

Musculus zygomaticus major,
  210
Music Therapy, 203, 209, 210,
  212, 231, 236
musical consonance, 10
Musical expectancy, 71
Musical expectancy formation,
  106
Musical Frisson, 220
Musical training, 54, 58, 75,
  131, 149, 150, 152–154,
  188, 236–238, 264
Musicogenic meaning, 177
Mussorgsky, M., 158

n-gram model, 106
Nasion, 44
Neapolitan sixth chord, 21, 22,
  109–114, 119, 134, 148,
  151, 173, 217
Neuroplasticity, 58
Newborn infant, 151, 155
Noise, 46
Non-diatonic, 25, 27, 34,
  72
Nucleus, 6
  superior olivary ~, 6
  ventral cochlear ~, 6
Nucleus accumbens (NAc),
  218, 226, 227, 235
Nucleus ambiguus, 211

O-wave of the CNV, 174
Onomatopoetic, 159
Organ of Corti, 3
Oval window of the cochlea, 3
Overtone, 20
Oxytocin, 230

Pain, 239
Parahippocampal gyrus, 231,
  232
Parkinson's disease, 227, 239
Parra, H., 246
Pattern classification analysis,
  84
Periodicity pitch, 9

Personal musicogenic meaning, 208
PET, 80
Phoneme, 11, 15, 56, 58, 134, 151, 186, 241
Physical musicogenic meaning, 177
Pitch, 9
~ chroma, 24
~ double-helix, 24
~ height, 23
~ helix, 23
low ~, 9
virtual ~, 9
periodicity ~, 9
residue ~, 9
Pitch commonality, 117–119, 122
Place theory, 8
Planum polare, 15, 131, 188
Planum temporale, 12, 15, 16, 90, 91, 188
Plasticity, 58
Play, 212
Positron emission tomography (PET), 43
Post-error slowing, 192
Post-traumatic stress disorder, 229, 239
Pre-error slowing, 192
Prediction, 49, 111, 119, 135, 137, 201
Premature infants, 239
Premotor cortex (PMC), 137, 188, 189, 194
Primary motor cortex, 194
Production rules, 214
Propofol, 132
Proprioception, 237
Prosody, 68
Psycho-Physiological Interaction analysis, 82
Pyramidal cells, 40

Radiatio acustica, 6
Recapitulation, 183
Recursion, 107

Reference electrode, 43, 45
Regional cerebral blood flow (rCBF), 80
Residualton, 9
Residue pitch, 9
Reticular formation, 7
Reward circuit, 212
Rolandic operculum, 189
Root
position, 21
Rostral cingulate zone (RCZ), 199
roughness, 10

Sadness, 206, 219, 233
Saint-Saëns, C., 158
Scala media, 3
Scala tympani, 3
Scala vestibuli, 3
Scale, 17
Schönberg, A., 161
Secondary dominant, 22
SEEKING system, 226
Semantics, 59–62, 64, 70, 138, 142, 170, 172, 173, 181, 182, 185, 246, 247
Semiotics, 181
Semitone, 17
Sensory dissonance, 118
Sensory priming, 37
Sequential Check Theory of Emotion Differentiation, 214
Seven Cs, 208
Shared goal, 178
Shared intentionality, 178
Shared Syntactic Integration Resource Hypothesis, 146
Signal, 46
Signal-to-noise ratio (SNR), 47, 50
Similarity
matrix, 32
ratings, 32
Six-four chord, 21
Sixte ajoutée, 21

Sixth chord, 21
Skin conductance response, 217
Smetana, B., 158
Social cognition, 209
Social cohesion, 208, 211
Social contact, 208
Socio-intentional musical meaning, 178
Solresol language, 246
Sonata form, 183
Sound pressure level (SPL), 6
Specific language impairment, 151, 153
Spectrum envelope, 11
Spiral ganglion, 5
SQUID, 48
Stapes, 3
Stereocilia, 4
Stroke, 237
Subdominant, 20, 31, 33
Submediant, 20, 31
Superior olivary complex, 8
Superior temporal gyrus (STG), 12, 15, 45, 90–92, 131, 132, 166
Superior temporal sulcus (STS), 13, 162, 166, 169, 233
Supertonic, 20, 31, 104, 106, 115, 116, 119, 120, 129, 151
Supertonic chromatic, 22
Supplementary motor area (SMA), 189, 194
Supramarginal gyrus (SMG), 189
Surgery, 207
Symbolic musical meaning, 162
Symphonic poem, 183
Synchronization, 178
Syntactic Equivalence Hypothesis, 145
Syntactic integration, 108
Syntactic structure building, 107
Syntax, 36, 63, 64

Tactile perception, 237
Tactus, 103

Tectorial membrane, 5
Tempered intonation, 17
Temporal poles, 222, 223, 232
Temporal theory, 8
Tetrachord, 18
Thalamus, 6
Theory of Affections, 159
Theory of Mind, 161
Theta band oscillations, 129
Tinnitus, 237
Tonal context, 31
Tonal Pitch Space, 36
Tonal Pitch Space theory (TPS), 99
Tone painting, 158

Tonic
    chord, 20, 25, 29, 30, 33
    tone, 20, 21
Tonicization, 29
Tonotopy, 8, 13
Torus of keys, 26
Trapezoid body, 6
Travelling wave, 5
Tree-structure, 99–102, 107, 111, 116
Tympanic membrane, 3

Ursatz, 98

Ventral pallidum, 226

Ventral tegmental area, 212, 220, 226, 227, 230
Vertex, 44
Violence, 216
Virtual pitch, 9
Vitalization, 93, 108
Vivaldi, A., 158
Vocalization, 159, 167, 189, 232

Webern, A., 161
Wernicke's aphasia, 166
Wernicke's area, 12, 162, 166
Working Memory, 93, 107

Printed and bound by CPI Group (UK) Ltd, Croydon, CR0 4YY

16/04/2025

14658466-0001